Enabling Agri-entrepreneurship and Innovation

Empirical Evidence and Solutions for Conflict Regions and Transitioning Economies

Enabling Agri-entrepreneurship and Innovation

Empirical Evidence and Solutions for Conflict Regions and Transitioning Economies

---

Edited by

Catherine Chan

*Department of Natural Resources and Environmental Management,
College of Tropical Agriculture and Human Resources,
University of Hawai'i at Mānoa*

Brent Sipes

*Department of Plant and Environmental Protection Sciences,
College of Tropical Agriculture and Human Resources,
University of Hawai'i at Mānoa*

Tina S. Lee

*Department of Natural Resources and Environmental Management,
College of Tropical Agriculture and Human Resources,
University of Hawai'i at Mānoa*

CABI

CABI is a trading name of CAB International

CABI
Nosworthy Way
Wallingford
Oxfordshire OX10 8DE
UK

Tel: +44 (0)1491 832111
Fax: +44 (0)1491 833508
E-mail: info@cabi.org
Website: www.cabi.org

CABI
745 Atlantic Avenue
8th Floor
Boston, MA 02111
USA

Tel: +1 (617)682-9015
E-mail: cabi-nao@cabi.org

A catalogue record for this book is available from the British Library, London, UK.

Library of Congress Cataloging-in-Publication Data

Names: Chan, Catherine, 1949- editor. | Sipes, Brent S. (Brent Steven), 1960- editor. | Lee, Tina, 1985- editor.
Title: Enabling agri-entrepreneurship and innovation : empirical evidence and solutions for conflict regions and transitioning economies / [edited by] Dr. Catherine Chan, Department of Natural Resources and Environmental Management, College of Tropical Agriculture and Human Resources, University of Hawaii at Manoa; Dr. Brent Sipes, Department of Plant & Environmental Protection Sciences, College of Tropical Agriculture and Human Resources, University of Hawaii at Manoa; Tina Lee, Department of Natural Resources and Environmental Management, College of Tropical Agriculture and Human Resources, University of Hawaii at Manoa.
Description: Oxfordshire, UK ; Boston, MA : CABI, [2017] | Includes bibliographical references and index.
Identifiers: LCCN 2016035555 | ISBN 9781780647753 (alk. paper)
Subjects: LCSH: Agriculture--Economic aspects--Developing countries. | Entrepreneurship--Developing countries. | Agricultural industries--Developing countries. | Agricultural innovations--Developing countries.
Classification: LCC HD1417 .E556 2017 | DDC 338.109172/4--dc23 LC record available at https://lccn.loc.gov/2016035555

ISBN-13: 978 1 78064 775 3

Commissioning editor: Alex Hollingsworth
Editorial assistant: Emma McCann
Production editor: Lauren Povey

Typeset by SPi, Pondicherry, India.
Printed and bound in the UK by CPI Group (UK) Ltd, Croydon, CR0 4YY.

Contents

Contributors

Anera Alishani, University of Prizren, 1 Shkronjat, Prizren, 20000, Kosovo. E-mail: anera.alishani@gmail.com

Jovelyn Bantilan, Southern Christian College, Midsayap, North Cotabato, Philippines. E-mail: jovbantilan@gmail.com

Emilie Bayona, Southern Christian College, Midsayap, North Cotabato, Philippines. E-mail: bayonaemilief@gmail.com

Maurizio Canavari, Alma Mater Studiorum, University of Bologna, Department of Agricultural Sciences DIPSA, Viale Giuseppe Fanin, 50, 40127, Bologna, Italy. E-mail: maurizio.canavari@unibo.it

Catherine Chan, University of Hawai'i at Mānoa, Department of Natural Resources and Environmental Management, College of Tropical Agriculture and Human Resources, 1910 East West Road, Sherman 104, Honolulu, Hawaii, 96822, USA. E-mail: chanhalb@hawaii.edu

Domenico Dentoni, Management Studies Group, Wageningen University, 6708 PB, Wageningen, Netherlands. E-mail: domenico.dentoni@wur.nl

Ekrem Gjokaj, Department of Economic Analysis and Statistics, Ministry of Agriculture, Forestry and Rural Development (MAFRD), Rr. Nëna Terezë, 10000 Priština, Republic of Kosovo. E-mail: ekrem.gjokaj@rks-gov.net

Muje Gjonbalaj, University of Pristina, Faculty of Agriculture and Veterinary, Bill Clinton Str., 10.000 Priština, Kosovo. E-mail: muje.gjonbalaj@uni-pr.edu

Jacqueline Halbrendt, Farming Systems Ecology, Wageningen University, 6708 PB, Wageningen, Netherlands. E-mail: jhalbrendt@hawaii.edu

James R. Hollyer, College of Natural and Applied Sciences, University of Guam, UOG Station, Mangilao, Guam 96923. E-mail: hollyer@hawaii.edu

Drini Imami, Agricultural University of Tirana, Faculty of Economics and Agribusiness, Kodër Kamëz, Tirana, Albania. E-mail: dimami@ubt.edu.al

Cynthia Lai, University of Hawai'i at Mānoa, Department of Natural Resources and Environmental Management, College of Tropical Agriculture and Human Resources, 1910 East West Road, Sherman 104, Honolulu, Hawaii, 96822, USA. E-mail: laicynth@hawaii.edu

Tina S. Lee, University of Hawai'i at Mānoa, Department of Natural Resources and Environmental Management, College of Tropical Agriculture and Human Resources, 1910 East West Road, Sherman 104, Honolulu, Hawaii, 96822, USA. E-mail: tslee07@gmail.com

Kathleen Liang, North Carolina Agricultural and Technical State University, 120 Gibbs Hall, 1601 E. Market St, Greensboro, North Carolina, 27411. E-mail: chyi-lyi.liang@uvm.edu

Rusyan Jill Mamiit, University of Hawai'i at Mānoa, Department of Natural Resources and Environmental Management, College of Tropical Agriculture and Human Resources, 1910 East West Road, Sherman 104, Honolulu, Hawaii, 96822, USA. E-mail: rusyan@hawaii.edu

Lusille Mission, College of Business, Southern Christian College, Midsayap, North Cotabato, Philippines. E-mail: eaglebase_2005@yahoo.com

Elma Neyra, Office of the Vice President for Research and Extension, Southern Christian College, Midsayap, North Cotabato, Philippines. E-mail: elmamneyra@yahoo.com

Michelle Ragocos Ortez, College of Business, Southern Christian College, Midsayap, North Cotabato, Philippines. E-mail: mish_ragocos@yahoo.com

Bikash Paudel, Local Initiatives for Biodiversity, Research and Development, PO Box 324, Pokhara, Nepal. E-mail: bpaudel@libird.org

Mary M. Pleasant, Oregon State University, 1600 SW Western Boulevard, Corvallis, Oregon, 97333 USA. E-mail: mary.pleasant@oregonstate.edu

Klodjan Rama, European University of Tirana, CERGE-EI Teaching Fellow. Formerly Leibniz Institute of Agricultural Development in Transition Economies (IAMO) and Martin Luther University Halle-Wittenberg. E-mail: klodian.rama@gmail.com

Brent Sipes, University of Hawai'i at Mānoa, Department of Plant and Environmental Protection Sciences, 3190 Maile Way, Honolulu, HI 96822. E-mail: sipes@hawaii.edu

Pauline Sullivan, Tennessee State University, Department of Family and Consumer Sciences, 3500 John A Merritt Blvd, Nashville, Tennessee, 37209. E-mail: psullivan10605@gmail.com

Bir Bahadur Tamang, Local Initiatives for Biodiversity, Research and Development (LI-BIRD), Pokhara, Nepal. E-mail: btamang@libird.org

Katherine A. Wilson, University of Hawai'i at Mānoa, Department of Natural Resources and Environmental Management, Honolulu, Hawaii, USA. E-mail: kawils65@hawaii.edu

Edvin Zhllima, Agriculture University of Tirana, CERGE-EI Teaching Fellow, Tirana 1000, Albania. E-mail: ezhllima@ubt.edu.al

Preface

The concept behind this book is primarily derived from executing the USAID-funded project entitled "University Partnership Linking OSY (Out-of-School Youth) to Agri-Entrepreneurship and Development to Promote Job Opportunities and Business Scale-Up for Mindanao." The site of the project was in Mindanao, a perennial conflict region which poses many inherent challenges such as instability and lack of or slow economic growth. These situations often lead to much greater adverse impacts on marginalized groups such as women and youth. The charge of this project was to train youth agri-entrepreneurs; however, we realized early on that the US-based project team members would face great challenges as most had not worked in a conflict region. Through our experiences, coupled with the local team's wisdom, we managed a successful project recognized by the American Council on Education (ACE) Higher Education for Development (HED) and USAID. Thus, the editors and authors thought that publishing our journey from start to the completion of the project could offer practitioners working in conflict regions more knowledge and help to prepare them to meet their objectives. One thing is for certain: consumer preferences for agricultural food products in these regions must be discovered for agri-entrepreneurs to be successful.

This book provides information from various authors' work in conflict areas, transitional economies and poor countries. Some chapters, though not directly related to conflict regions or Mindanao, focus on areas that have transitional economies or are poor countries. The commonality among these places are that they have weak governments and poor physical infrastructure and they lack transparent policies for long-term planning. The book begins with a comparison of general entrepreneurs with agri-entrepreneurs, segmenting them by developing and developed countries. The next chapters discuss at some length how "skills training programs" require adjustments and modifications based on local knowledge, culture, the natural environment, and human-induced events due to unpredictable circumstances. In most examples the experiences have been with a youth audience who have many entrepreneurial experiences in dealing with life in conflict regions, but less experience with applying these skills to creative ideas for a business venture. Our project allowed the out-of-school youth of Mindanao to channel their unique life challenges and experiences into productive uses. Their successes and failures are documented in this book.

We could only have accomplished our goal with the generous support of USAID, our local team in Mindanao and collaborators from around the world. Dr Elma Neyra led the Mindanao team, who were the best collaborators. Their willingness to listen and their unyielding constructive suggestions made us all stronger and safer working in the conflict region of Mindanao. Overall, this book and the related project saw us grow closer, knowing we have each other's backs forever. It has been an incredible experience for us. We thank each of the chapter authors for their contributions and insights. We hope that by reading this book you will discover new knowledge for your own work.

Catherine Chan, Tina S. Lee, Brent Sipes and Katie A. Wilson
Honolulu, Hawaii
14 May 2016

1 Agri-entrepreneurs and Their Characteristics

Pauline Sullivan*
Tennessee State University, Nashville, Tennessee

1.1 Introduction

There are over three billion people working worldwide (World Bank, 2012). Of the three billion people with jobs, farming and household micro-businesses provide about 1.5 billion employment opportunities globally with 50% of employment in developing countries. Jobs are especially important for increasing numbers of underemployed youth in developing countries where there will not be enough jobs to employ the increasing number of unemployed. In these countries the largest group of those people is youth aged 15–24 (40%) (Kapsos, 2013). A disproportionate number of those people are young people, who will compete for ever fewer jobs in the future, as job creation lags population growth (Jones, 2015). Available jobs will require education and skills that poor people do not have. The few jobs accessible to people without an education or relevant skills do not provide livable wages. Entrepreneurship is a viable strategy for upward mobility, as a 1% increase in entrepreneurial activities decreases the poverty rate by 2% (Singh, 2014). Also, entrepreneurship is a means of identifying value-added activities that increase efficiency and employment opportunities. Entrepreneurs with successful businesses are self-employed. Thus, entrepreneurship decreases unemployment (Gorgievski *et al.*, 2011). Entrepreneurship is particularly important to agriculture for several reasons (Milestad *et al.*, 2011). Some challenges to agri-entrepreneurial success in rural areas include weak education systems, limited access to mail, and competition from large farms with technology that increases their productivity.

Stenholm and Hytti (2014) differentiate the entrepreneur-farmer from the producer-farmer. They suggest the entrepreneur-farmers establish their identity by addressing challenges in institutional norms and acting as change agents. Producer-farmers concentrate on following customary behavior. Since there are so many definitions of agri-entrepreneurs, in this chapter the agri-entrepreneur will be treated as an innovative change agent who finds opportunities to use land and its resources for specialized, value-added food and agriculture-based businesses. The individual assumes all risks and benefits derived from the business. Agri-entrepreneurs own and manage their business with the intent to make profit. Agri-entrepreneurship, overall, is a combination of agriculture and business (Bairwa *et al.*, 2014). This combination of agriculture and business fosters agri-entrepreneurs who innovate, identify markets, and satisfy needs by developing different ways of doing things. Agri-entrepreneurs

*psullivan10605@gmail.com

engage in a variety of activities in agriculture and related sectors.

This chapter discusses entrepreneurial activities related to agri-entrepreneurship since a majority of out-of-school youth (OSY) live in rural areas where agriculture, agricultural services, and farm-related product production are the primary source of jobs in developing countries. Most of the working-poor youth are employed in agriculture (Kapsos, 2013). Therefore, agri-entrepreneurship can increase employment opportunities, especially for out-of-work youth, and thus decrease poverty (Singh, 2014; Nagler and Naudé, 2015). The agricultural sector dominates economic activity in rural areas where a large number of OSY live and, as such, offers opportunities for entrepreneurship.

This chapter begins with an overview of the market conditions and challenges faced by agri-entrepreneurs (Sections 1.2 and 1.3). Sections 1.4 to 1.7 discuss generic entrepreneurship, entrepreneurial networks and innovation using Drucker's (1998) list of opportunities for entrepreneurships, supported by agriculture-related examples. Section 1.8 focuses on the differences between agri-entrepreneurship and new business creation in other industry sectors. Sections 1.9 to 1.11 discuss the challenges faced by agri-entrepreneurs, motivation for agribusiness start-ups and value-added agriculture, as well as characteristics of entrepreneurs engaged in situation-specific businesses (Miskin and Rose, 2015). Sections 1.12 and 1.13 provide examples of successful and unsuccessful agri-entrepreneurs, with a focus on their characteristics. Conclusions are discussed in Section 1.14.

1.2 Market Conditions and Challenges for Agri-entrepreneurs

Agribusiness is important to global food supplies and developing economies (Wilkinson, 2009). However, getting agricultural products from remote rural areas into larger markets has its challenges. These include a lack of intermediaries to develop relations between vendors, purchasers and contract systems (Parmigiani and Rivera-Santos, 2015). Other problems include inconsistent product quality and supply of products because of poor infrastructure. There is also competition in domestic markets from transnationals that can sell their products for less than domestic producers can.

In the 1970s, transnational agribusinesses entered markets in developing countries. The first foray of transnationals in emerging markets was to sell corporately branded products. When market share and profitability for transnational agribusinesses increased, these companies became involved in developing countries' agricultural supply-chain systems, which eventually provided the transnationals with opportunities to control supply-chain operations. This led to specialized food product exports, such as exotic or off-season vegetables, for consumers in affluent countries. None the less, diversion of agricultural resources increases poverty, displaces farmers, and marginalizes agrarian society and culture in developing countries (McMichael, 2005).

Increasingly, transnational food conglomerates are the dominant players in food production systems in developing countries (Wilkinson, 2009). Contract farming is one method that transnational businesses use to vertically integrate supply-chain production and ensure product supplies for supermarkets (Bijman, 2008). Suggested benefits of contract farming include access to a global market, risk reduction, technical assistance, reduced transaction costs, and access to credit. On the other hand, opportunistic behavior is a negative consequence from the use of contract farming by transnational agribusinesses.

Resource-poor farmers have challenges getting their products to market. Although transnational corporations could facilitate distribution of the poorest farmers' products in the global supply chain, contract farming often is unavailable to these farmers (Fraser, 2012). Transnational corporations limit contract farming in extremely poor countries because of the companies' concerns about geographic isolation, high transportation costs, communication difficulties, and lack of available technology. Moreover, large transnationals may perceive poor contract farmers as a risk in meeting their obligations. Transnational corporations seek more affluent contract farmers who are able to pay commitment fees and bear some of the crop-related risk.

While farmers can contract crops out to transnational corporations, their ability to be entrepreneurial decreases because of a need to focus on contract crops. Control of the global food supply chain by transnational corporations yields asymmetrical power relationships with contract farmers (Minot, 2007; Bijman, 2008). Asymmetrical power relationships between transnational corporations and contract farmers enable transnationals to use coercive power in contract negotiations (Maloni and Benton, 2000). For example, a transnational corporation applies coercive power when it dictates price reductions for contract farmers, which in turn challenges the profitability of independent farming. Farmers may have to sell products at prices below production costs, thus decreasing their earnings and increasing the probability of failure to break even on production costs (McMichael, 2005).

All payments to contract farmers are contingent on satisfaction of corporate quality standards, regardless of environmental conditions. Sometimes, contract farmers are not paid because their products do not meet corporate quality standards; for example, a tomato is too small or underweight (Global Policy Forum, 2015). Lost income can result in displacement of small farmers as they seek opportunities elsewhere. Arguably loss of farmers' income violates one of the fundamental premises of sustainability: social equity, which affects the well-being of individuals and families (WCED, 1987).

1.3 Micro, Small and Medium-size Enterprises and Entrepreneurship

Entrepreneurship is one way in which small-size farmers can confront issues mentioned above, such as OSY under-employment and need for new jobs. Opportunities exist for farmers to create value-added agricultural products that are sold in local markets. For example, consumer desire for transparency in the food supply chain contributes to localization of agri-food systems (LAS) (Requier-Desjardins *et al*., 2003). Rural entrepreneurs can compete against larger corporations through a business cluster that controls significant quantities of local farm products (Requier-Desjardins *et al*., 2003) and agribusinesses form Local Production Systems (LPS). Commercial relationships between LAS and LPS eventually create a supply chain, with added value at each level of activity. Clusters of micro-businesses making the same products establish a base of power in the marketplace through horizontal integration, if a strong vertical supply chain exists between its members wherein a manufacturer controls production, quality control, standardization, and packaging (Lamprinopoulou, *et al*., 2006; Lamprinopoulou and Teagear, 2011).

Agribusiness clusters increase efficiencies in production activities in micro/small/medium-size enterprises (MSMEs) and decrease transaction costs. Micro-businesses are defined as employing one to nine people, small-size businesses have 10–49 employees and the workforces of medium-size businesses range from 50 to 250 people (Andersén, 2012).

In addition, MSMEs establish cooperative–competitive relationships in these clusters as a result of workforce sharing (Trienekens, 2011). Workforce sharing is a reciprocal subcontracting system that enables MSMEs to be agile and adjust products or services in response to market demands. As additional new businesses with related products open, there is geographic clustering of farm products and processing operations. Clusters of businesses facilitate knowledge sharing and diffusion and innovation among businesses and countervail large monopolistic intermediaries.

Increased involvement of local businesses in LAS and LPS includes purchases and sales of goods or services between members of the local supply chain and establishes input–output relationships (Requier-Desjardins *et al*., 2003). The system works as follows. Inputs are costs associated with an MSME's purchases of raw materials and indirect costs from labor associated with product production as the business transforms raw materials into new or improved value-added products or services. An output is the value or sales revenue derived from the sum of goods produced by businesses in a cluster.

1.4 Generic Entrepreneurship

Entrepreneurship motivations are important in understanding how the entrepreneur operates their business or defines success. Traditionally, the definition of an entrepreneur is an individual who starts a new business, cognizant of associated risks and ambiguities, for profit and growth (Scarborough, 2012). The entrepreneur takes advantage of opportunities and gathers resources, usually money, and other necessary non-monetary resources to turn their ideas into reality. Motivation for entrepreneurship, its meaning and measures of success differ among people, as well as countries, and by income level.

In addition, agri-entrepreneurship is tied to personal identity (Stenholm and Hytti, 2014). Each entrepreneur seeks to legitimize their identity through business activities and thus their goals reflect the environment and social norms. For example, an entrepreneur in a developing country may view success as the ability to support oneself and one's family (Gindling and Newhouse, 2013). In contrast, another agri-entrepreneur may be competitive, seeking to be the largest and best in order to reach their goal to be a profitable self-supporting business, independent of others (Stenholm and Hytti, 2014). In the latter case, the agri-entrepreneur assumes all risks and develops structures that enable growth and profitability. The latter type of entrepreneur becomes a visible change agent in the community when others adopt their practices.

Other definitions consider entrepreneurs as active participants in their business, either as an employee or as a manager (Berglann *et al.*, 2011). One opinion is that entrepreneurs should control, either directly or indirectly, between 10% and 30% of their business. Access to wealth, albeit the individual's, spouse's, or parent's, influences the decision to become an entrepreneur. Individuals whose parents were entrepreneurs are likely to engage in entrepreneurship. An entrepreneur's experience in a particular industry is associated with new business start-ups. However, those who are unemployed are less likely to engage in entrepreneurship, while partially employed individuals are more apt to become an entrepreneur (Berglann *et al.*, 2011).

In some rural areas, for example in India, agri-entrepreneurship does not have enough available land to absorb the agricultural workforce, which contributes to unemployment and underemployment (Saxena, 2012). One solution to this problem is rural entrepreneurship from agricultural-based business. These businesses process and add value to local agricultural production as they move through the supply chain. They also contribute to local jobs. However, rural entrepreneurs rely on middlemen and middlewomen to expand their market area and get their product to market. Rural entrepreneurs have no control over their products when they are turned over to an intermediary for distribution.

Gartner (1990) identified eight dimensions of an entrepreneur. The first dimension, "The Entrepreneur," suggests that entrepreneurship attracts people with unique personalities and abilities. The second dimension, "Innovation," describes entrepreneurs' penchant for newness, whether an idea, product or service. The third dimension, "Organization Creation," describes behavior necessary for business start-ups. The fourth dimension, "Creating Value," seeks to generate economic wealth. The fifth dimension, "Profit or Non-profit," discusses issues pertaining to whether or not the entrepreneur's priority is profit or a social agenda. "Growth" is the sixth dimension and characteristic of entrepreneurial success. The seventh dimension, "Uniqueness," concerns the entrepreneur's ability to offer something distinctive that distinguishes them from the competition and allows them to best satisfy consumers' novelty-seeking behavior. The "Owner-Manager" is the eighth dimension and addresses the entrepreneur's dual tasks of business owner and manager. Gartner's study found that 79% of respondents considered entrepreneurship to be related to the characteristics of innovation, growth, and uniqueness. Another 21% of respondents were interested in outcomes of entrepreneurship, such as value creation and income gains.

Another type of entrepreneurship is social entrepreneurship, which seeks to improve social capital (Mair and Marti, 2006). Social entrepreneurship is the innovative use and mix of resources to take advantage of opportunities leading to social change or the ability to meet social needs. Relationships are important in social entrepreneurship, as its primary objective is to help others; hence profit is not the bottom line. One example of a type of social entrepreneur is an experienced village leader with outside networks who donates their time and expertise to help villagers set up new businesses and agribusiness cooperatives (Zhu *et al*., 2015).

Berglann *et al*. (2011) suggested that education is not a good predictor of whether or not an individual will become an entrepreneur. Rather, the key determinant of entrepreneurship is the type of education that the individual receives. Entrepreneurship programs are increasing, with about one-third of them taught outside business schools (Katz, 2008). One reason for this increase may be that the entrepreneurship programs and agricultural colleges are linked to extension outreach initiatives. Extension work with farmers provides opportunities to conduct research and learn about best practice in business/farm management which supports profitable farming (Dethier and Effenberger, 2012). Also, extension services and agents offer help to agri-entrepreneurs with the goal of avoiding business failures.

Entrepreneurship in developing countries is different from more mature economies (Lingelbach *et al*., 2005). Market inefficiencies in developing countries create more opportunities for entrepreneurial activities, when compared with developed economies. Situational influences, such as different degrees of market stability and consistency of products or services offered, provide additional prospects for entrepreneurship. Entrepreneurs in developing countries satisfy consumers' core needs better than in developed countries because of proximity to the market. Another difference between developing and more mature economies is that entrepreneurs in developing countries minimize risk by operating several different businesses, while those in mature economies tend to focus on one core business. Financial resources and access to loans are limited in developing countries, deterring new businesses. Entrepreneurial activities in developing countries are usually self-financed or funded by money from friends and family.

1.5 Entrepreneurship: Social Equity for Out-of-School Youth and Females

The proportions of unemployed young people in poorer countries are almost double those in affluent countries (GEM, 2015). Many youth will drop out of school in poor countries. Dropout reasons include poverty, parents wanting the youth to contribute to household income, and/or a lack of interest (Lalata *et al*., 2010). Limited education causes inadequate reading, writing, and arithmetic skills (UNESCO, 2014). Hence, OSY's ability to engage in entrepreneurial activities or find and keep jobs is low.

One study (Ahaibwe and Mbowa, 2014) found that about 24% of OSY were employed in wage-earning jobs. These employed youth usually had secondary education and worked in urban areas. The study also revealed that 66% of all jobs were in the agricultural sector, which employed a majority of OSY. However, only 5% of OSY do not have a full-time wage-earning job, therefore most live in poverty. In addition, OSY unemployment due to the gap between the number of OSY and the jobs available in the formal or public sector decreases the likelihood of entering the job market (GEM, 2015).

Consider the situation in Uganda, where young people represent 24% of those employed in the formal wage-paying job market (Ahaibwe and Mbowa, 2014). These young people are mostly urban workers with a secondary education. Rural OSY work in the agricultural sector. The agricultural sector provides 66% of all jobs, but only 5% of young workers earn wages. This indicates that farmer workers are not paid.

While entrepreneurship is a solution in poverty reduction and increasing income, teaching necessary skills to OSY is essential in preparing them for success (LaGraffe, 2012; Ahaibwe and Mbowa, 2014). Firstly, OSY should develop attitudes, knowledge, and awareness of available support services. Entrepreneurship training increases young people's prospects for employment, business start-ups, and maintaining successful businesses. Without this training, the probability of OSY finding wage-earning employment is low. OSY also need role models and access to capital in order to thrive as entrepreneurs (GEM, 2015). Unemployment and underemployment among OSY result in social problems, such as dependence on welfare, crime, and social unrest (LaGraffe, 2012; GEM, 2015).

For example, OSY aged 15–24 years in the Philippines account for 50.4% of all unemployment (Taborda, 2015). The overall employment rate for OSY is disproportionate with the Philippines' overall unemployment rate of 6.4% in July 2015 (Philippines Institute for Development Studies, 2015). In addition, there is high underemployment for OSY, and many OSY work in the agricultural sector.

A second group of people in developing countries at risk for unemployment and poverty is females. Females in developing countries have higher unemployment rates (5.8%) than males (5.3%) (ILO, 2012). There is no expectation that the employment situation for females will improve in the near future. When compared with males, slightly more females are in vulnerable employment situations, such as jobs without employment benefits or employment in unpaid work at home, while earning income in the labor force. The difference between males and females in vulnerable work situations can be as high as 24% in North Africa.

Just like OSY, females have limited employment opportunities (Bonkat, 2014; GEM, 2015). In addition, females face discrimination in the workplace. Employment opportunities can be gender specific, such as the expectation that Mexican females work in the service and retail sectors (GEM, 2015). Another problem is that women's wages often are less than the minimum wage, or in some cases nonexistent for their work. Females earn less than males, regardless of development levels (Verick, 2014).

Female entrepreneurship varies by country. The number of female entrepreneurs compared with males differs by country and generally there are fewer female entrepreneurs (48%) than males (52%) (Vossenberg, 2013). Female entrepreneurship participation rates around the world range from 1% in Pakistan to 40% in Zambia (GEM, 2015). Yet in Panama and Thailand, female entrepreneurs outnumber male entrepreneurs.

Most often the motivation for female entrepreneurship is survival, nonexistent job opportunities, or no other income source (Vossenberg, 2013). Female entrepreneurship may be motivated by any combination of reasons; for example, a hobby develops into a business, participation in an entrepreneurship training program, or as a means to cope with an economic shock in conflict zone (Bushell, 2008; Bonkat, 2014).

Primarily, female entrepreneurial activities are in the consumer sector (GEM, 2015). This includes food and related product services (Quinay, 2013). In rural conflict zones, female entrepreneurs sell vegetables and fruits to end-use consumers and engage in associated business activities (Bonkat, 2014).

Female entrepreneurs self-finance their businesses and are more risk averse than male entrepreneurs (Hampel-Milagrosa, 2014). Females face gender challenges in establishing business relationships (Bushell, 2008) and there are few successful role models for women. As is the case with OSY, women in developing countries have few opportunities for education or training related to entrepreneurship (GEM, 2015). Quinay (2013) suggested that empowering female entrepreneurs has three components: education, policy, and a social network. In order to succeed, female entrepreneurs need access to information communication technology (ITC) and assistance in identifying markets and finding distribution channels (Vossenberg, 2013). In addition, females face challenges in obtaining credit or financing for their businesses (Stupnytska *et al.*, 2014). Approximately 70% of female entrepreneurs in developing countries receive inadequate funding for their businesses.

1.6 Entrepreneurial Networks

Businesses that network outside their own business and share knowledge, expertise, and ideas create symbiotic relationships conducive to collaborative innovation (Ketchen *et al.*, 2007). Participation in these collaborative relationships helps entrepreneurs remain innovative. These networks offer a variety of content. Some networks have communication content, wherein information is shared. Another network offers exchange content conducive to the give and take of goods and services. Networks with normative content provide people with knowledge and support that meet the expectations of others who share an interest in special attributes (Aldrich and Zimmer, 1986). Thus networks are a valuable source of new ideas and information for entrepreneurs. For example, networks mentoring female entrepreneurs in developing countries contribute to their success and ability to obtain credit (Stupnytska *et al.*, 2014).

Entrepreneurs in developing countries also receive business information from personal and professional networks (Hampel-Milagrosa, 2014). Hampel-Milagrosa's study indicated about 59% of food-processing entrepreneurs who wanted to upgrade their business productivity, revenue, and size received market information from personal networks. Personal networks are an important contributor in the growth of food-processing businesses. For example, growth in food-processing businesses contributes to developing clusters of food-processing agri-entrepreneurs in Bukidnon on Mindanao Island in the Philippines (Philippines Institute for Development Studies, 2015). An entrepreneur who made linen and home-furnishings from natural fibers wanted to grow her business (Hampel-Milagrosa, 2014). She contacted a network of Mindanao tribes to help her design, source, and market her products. The tribal network's support influenced the entrepreneur's decision to partner with T'boli and Bagobo tribes in the creation of high-end personal handbags, as well as fashion and home-furnishing accessories with tribal designs. This collaboration was successful. The company began exporting their handcrafted products to Japan. Collaboration and mentoring of female entrepreneurs in developing countries helps improve their success rate and access to capital (Stupnytska *et al.*, 2014).

1.7 Entrepreneurship and Innovation

Schumpeter (1934) described entrepreneurs as innovative individuals who create new ideas or products, or a combination of both. These are introduced into the marketplace as innovations. Innovation disrupts market equilibrium, the status quo, but is essential as it moves the economy forward. Thus, entrepreneurship is the process through which innovation drives economic growth. Drucker (1998) argued that entrepreneurs systematically innovate or intentionally create focused activities that change the status quo. He cited opportunities for innovation as "unexpected occurrences, incongruities, process needs, industry and market changes, social and intellectual environment, demographic changes, changes in perception, and new knowledge" (Drucker, 1998).

Conflict negatively impacts the number of jobs and household income (World Bank, 2012). In areas where there are few jobs, entrepreneurship represents an alternative to poverty (Nagler and Naudé, 2015). Examples related to agriculture are provided for each innovation opportunity in Table 1.1.

1.8 Agri-entrepreneurs

The idea at the center of agri-entrepreneurship is the right to use land and its resources for agriculture, forestry, and related activities that generate income (Suárez, 1972). Also, agri-entrepreneurs actively engage in agriculture, use current technology to increase agricultural productivity, and adopt new systems of operations (Singh, 2014).

Farm-related entrepreneurship provides the economic foundation for rural economies (Hossain and Jaim, 2011). Agriculture employs 75% of people living in rural areas (World Bank, 2008). This includes poor small-size farmers and landless workers who earn a livelihood through agriculture. Research indicates that the agricultural sector is twice as effective in reducing poverty as any other sector. Increasing crop productivity reduces food prices for poor people and has the largest effect on decreasing poverty rates. Small-size farmers increase their ability to compete in the market with the aid of producer organizations and technology innovation.

Changes in agricultural markets, such as ICT, reduced subsidies for agriculture, shifts in consumer demand, new food production standards, and an emphasis on sustainability, require farmers to adapt to new circumstances in order to survive (Lans *et al.*, 2013). Agri-entrepreneurs need to be innovative, take risks, and be proactive in identifying opportunities for success. Agri-entrepreneurship is different from traditional entrepreneurship in that modernization of agriculture is narrowly focused on highly specialized areas, efficiency, and productivity in farm management. Also, agri-entrepreneurs' start-up costs, ranging up to US$600,000 for land and machinery, are significantly higher than non-farm businesses in developed countries (Richards and Bulkley, 2007). In contrast, start-up costs for street vendors in a conflict

Table 1.1. Innovation opportunities and examples.

Innovation opportunity	Example
Unexpected Occurrences	In 1873, a vineyard owner identified dried grapes that William Thompson marketed 2 years later as raisins (Kelly, 2013).
Incongruities	Innovations in agricultural biotechnology decrease crop variability for soybeans, cotton, corn and canola (Giddings and Chassy, 2009).
Process Needs	Robots will soon serve process needs by increasing efficiencies in fruit harvesting and livestock management (Harvey, 2014).
Industry and Market Changes	When blood cholesterol became associated with an increase in heart disease, consumers wanted beef with little fat and the beef industry raised cattle with less fat (Hoffman and Turnbull, 2007).
Social and Intellectual Environment	Farmers joined online communities to learn about market trends (Ganzell, 2009).
Demographic Changes	The Maryland Department of Agriculture began a programme that would identify local producers, sellers and buyers of specialty and ethnic food products. The most popular products sold were cilantro, specialty herbs, hot peppers and eggplants. The ethnicity of the specialty products and ethnic foods influenced the choice of products grown (Tubene *et al.*, 2015).
Changes in Perception	Most farmers in the Osun State of Nigeria perceive increased temperatures and deceased precipitation. The greatest number of farmers responded to climate change by planting crops earlier or later than previously done (Sofoluwe *et al.*, 2011).
New Knowledge	Alltech Crop Science developed mapping drone technology with real-time information transmission that identifies underperforming soil and crops. This technology enables the farmer to make an informed decision when employing management practices (Duboyne, 2015).

zone (Nepal) are about US$15.00 (Bushell, 2008). In contrast, a cottage industry, such as basket weaving, that can be operated from home, can cost around US$1100.

1.9 Challenges to Agri-entrepreneurship

Globally, agri-entrepreneurs need to respond to changes in consumption, products sought, distribution systems, new technologies, and industry structure (Boehlje *et al.*, 2011). In addition, there is a disconnect between the locations where agricultural products are produced, processed, and consumed. Going forward, agribusinesses face three significant issues: (i) increased risk and uncertainty in decision making; (ii) innovation, development, and adoption of new technologies that enable increased production efficiency and overall profitability; and (iii) agility in responding to change, competition, and evolving industry structures. Global climate change is another concern. Further challenges are the geographic location of the agribusiness, farm size, and land topography (McElwee and Smith, 2012). All three can have a negative effect on business success. Additional impediments to agri-entrepreneurship are volatile prices and product supplies, low and unpredictable earnings, and intensive farming with environmental costs (Pettinger, 2014). Although these issues are of universal concern, they disproportionately affect agri-entrepreneurs in developing countries.

Agri-entrepreneurs in developing countries face some challenges that are similar to those in more mature economies and other challenges that differ from more mature economies. The similarities include poverty, insecure livelihoods, lack of access to technology, socio-political instability, natural disasters, degradation of the environment, and gender discrimination (Faisal, 2010). Often it is hard for the agri-entrepreneur to meet with the experts or extension services that can provide guidance in developing their businesses. These agribusinesses have high start-up costs when compared with non-farm businesses (Richards and Bulkley, 2007). Rural agri-entrepreneurs have challenges obtaining credit for their start-up experiences and they are reticent in seeking help from others. Farmers who are not innovative use conventional farming methods that make them susceptible to price competition (Marsden and Smith, 2005).

A serious concern for agri-entrepreneurs in developing countries is transnationals seeking to purchase property in their country. Transnationals look for countries with weak governments and little regulation of direct foreign investment, because it makes it easier for them to enter a market and gain control of marketplace activities (Land Matrix, 2015). Property-seeking transnationals with deep pockets diminish opportunities for local agri-entrepreneurs who lack access to social media and money for traditional advertising. There are fewer female entrepreneurs in developing countries than developed countries (Stupnytska *et al.*, 2014).

Another issue affecting agri-entrepreneurs in transitional economies with vast numbers of rural communities, as well as poor countries, is undifferentiated agricultural products that account for 50% of total exports (UNIDO, 2015). In transitional economies, the export of undifferentiated agricultural products increases poverty, displaces farmers, and marginalizes agrarian society and culture in developing countries (McMichael, 2005). While transitional economies mostly export agricultural products, they also divert food products from their domestic markets (UNIDO, 2015). The result is a limited supply of food available to people living in these countries, especially those in rural communities. This affects local consumers' ability to buy food and feed their families.

1.10 Motivations for Agribusiness Start-ups

Given the challenges agri-entrepreneurship faces, what motivates aspiring agri-entrepreneurs to start their own business? Those considering a new agribusiness venture evaluate other job opportunities in order to determine which opportunity is in their best interest (Land Matrix, 2015). Potential agri-entrepreneurs also look at the business's impact on their family.

Agri-entrepreneurs can take advantage of opportunities to engage in new activities and increase household income, or expand farm operations so that additional family members can be employed (Agnete *et al.*, 2003). In this scenario, agri-entrepreneurship may be perceived as a means of providing a good living. This agri-entrepreneur perceives continuity of the farm and related businesses as an obligation or tradition. In some cases, the individual lacks other career options. A second reason to start an agribusiness is to take advantage of unique resources and use them for competitive advantage in business operations. In general, these businesses expand upon the entrepreneur's primarily agriculture-related business. Thirdly, the agri-entrepreneur may want to try out new business ideas.

Small farmers can remain economically viable when they diversify their activities and sell value-added products that best meet consumers' needs (Lowson, 2003; Haugen and Vik, 2008). Agri-entrepreneurs need to establish a competitive advantage by selling novel products not sold by agribusinesses. Product novelty will increase the likelihood that consumers will purchase their products at prices that exceed production costs, cover the cost of doing business, and provide sufficient profit.

1.11 Value-added Agribusiness

Consumer preferences for value-added and locally grown agricultural products create new opportunities for agri-entrepreneurs (DeLind, 2002). Agri-entrepreneurs can sell value-added products that exploit consumer demand for organic and local foods. Examples of value-added products include organic vegetables, BST-free cheese, unbleached flour, gluten-free food, and heritage apples. Value-added products are important because they generally have a high return on investment. Sales of these products generate additional income for agri-entrepreneurs and for the local and regional economy (Vogel, 2012). The different types of value-added agri-entrepreneurs include technical services and farm equipment rentals, agrotourism, forest products, direct-to-consumer sales, and distribution through community-supported agriculture. Farms with agribusinesses usually earn more income than those without these enterprises. For example, agri-entrepreneurs can earn up to 72% of their income selling products through community-supported agriculture, 50.7% from value-added farm produce products, and almost 25% from direct-to-consumer sales. Consumer demand for locally grown and organic agricultural products has increased over the past decade (Ikerd, 2011). Agribusinesses producing sustainable products need to label their products appropriately so that consumers will pay a premium price for the product. In addition, consumer relationships and ethical practices add value to sustainable agribusinesses.

Sustainable agricultural practices should meet needs for ecological integrity and economic gain (Ikerd, 2008). For example, agribusinesses using sustainable agricultural practices in Nicaragua, Guatemala and Honduras have better topsoil and moisture retention, with less erosion and economic loss, when compared with farmers not practicing sustainable agriculture (Altieri *et al.*, 2012). Small-farm operations in Chiapas growing coffee using sustainable agricultural practices find that their products have improved resilience to hurricanes as a result of the soil they are planted in and growing conditions. Agri-entrepreneurs in developing countries have an opportunity to benefit from consumers in developed countries who connect status with green/sustainable products and are willing to pay more for these products (Elliott, 2013).

Value-chain analysis is important to agri-entrepreneurs in emerging economies and developed countries (Trienekens, 2011). Agri-entrepreneurs need to identify constraints in developing value-added products. These constraints include access to market information and resources, in addition to infrastructure and institutional barriers. The next level of value-chain analysis requires the agri-entrepreneur to analyse how they can add value to a product and how to finance related costs. There are few loans available to agri-entrepreneurs in developing countries (Lingelbach *et al.*, 2005). Also, existing networks exist to ensure product distribution. In addition, the impact of governance on their business must be considered. In the last stage of the chain analysis, agri-entrepreneurs look for ways to upgrade their value-added activities. New products or services could be offered to consumers, the distribution network position could be improved, or partnerships can be established in order to grow their businesses.

Value-added products also establish a direct connection between the consumer, farmer, farm, and bioregion. This contributes to a sense of place and links consumer, producer, and place, which is important to one's social self (Feagan, 2007). An example of this would be childhood memories of pancakes with Canadian maple syrup. Place provides an additional way to differentiate products (Armington, 1969). This strategy is effective at the local level, as well as at the national level. A study of domestic computers of inferior quality versus imports of superior quality indicated that Australian consumers, who were not computer experts, preferred the low-quality computers made in Australia (Pecotich and Ward, 2007). This finding suggests that consumers pay more for products congruent with their perception of the product and how it supports the consumer's self-image. Country-of-origin branding has a different influence on product selection by computer novices and by experts. There is a dearth of available research about place and perceptions of product quality in developing countries, which suggests a need for marketing intelligence.

1.12 Characteristics of Successful Agri-entrepreneurs

Agri-entrepreneurs have personal characteristics that set them apart from others and influence the way they operate businesses. Bairwa *et al.* (2014) described agri-entrepreneurs as visionary, curious, proactive, determined, persistent, honest, hardworking individuals with integrity, as well as management and organization skills. They pointed out that agri-entrepreneurs often engage in a variety of activities, which suggests that they multi-task well. They recognized entrepreneurs as single-minded, driven, ambitious, creative problem solvers, yet practical and goal-oriented. In addition, they suggested that agri-entrepreneurs are able to recognize needs as opportunities. Lastly, agri-entrepreneurs willingly accept risks necessary to turn ideas into real and marketable products.

Sancho (2010) provided a comprehensive list of agri-entrepreneurial characteristics. These include initiative, sense of opportunity or intuitiveness, and independence. Agri-entrepreneurs are also viewed as dynamic, exhibiting leadership skills, a strong character, and goal driven. Other agri-entrepreneurial characteristics are responsibility, honesty, commited, energetic, hard work, and experience. Sancho (2010) also believed that agri-entrepreneurs are creative and imaginative problem solvers who can turn innovative ideas into reality and expect to succeed in their efforts. In addition, those involved in agri-entrepreneurship are knowledgeable and involved, and systematically evaluate their position. Moreover, agri-entrepreneurs value partnerships and teamwork. They readily take on tasks, but remain flexible and adaptable as they act as change agents.

Similarly Hossain and Jaim (2011) argued that, as owners of an organization, agri-entrepreneurs should take a personal stake in the enterprise, including all risks and uncertainties involved in running it. They contended that agri-entrepreneurial success results from firm performance. Anderson *et al.* (2006) suggested that entrepreneurial skills related to agribusiness are optimism, innovation, initiative, and risk-taking. They believed that agri-entrepreneurs willingly accept challenges, take responsibility and deal with the unknown and that they should be able to communicate effectively, have positive interpersonal skills, negotiate successfully, and lead their business operation. Anderson *et al.* (2006) shared the view of Hossain and Jaim (2011) that agri-entrepreneurs are opportunity seekers and change agents who can analyze, synthesize, evaluate, plan and execute ideas. These qualities ensure that the individual/actor/agent is relevant in their entrepreneurial role under any circumstance. For example, entrepreneurial soldiers traded cigarettes for products not easily obtained during World War II when consumer goods were rationed (albeit these entrepreneurial activities were short-term solutions to product shortages and ended with the end of the war).

A drive to succeed, a high degree of competitiveness, a high energy level, receptivity to change, and an appreciation of support networks are attributes of successful female entrepreneurs (Quinay, 2013). Other attributes are positive thinking, self-esteem, ability to see opportunities, willingness to work long hours, and recognition of the need for continuous entrepreneurship education. Flexibility is important to female entrepreneurs in conflict zones, as they need to be resourceful in developing their skills and business models so that they can earn an income if they are displaced.

Characteristics of successful entrepreneurship include a business that survives 3 years (Makhbul and Hasun, 2010), communicates with stakeholders (Islam *et al.*, 2011), establishes social networks, gains personal satisfaction (Gorgievski *et al.*, 2011), and balances work with life. An entrepreneur's success is judged by their ability to generate profit related to the business's break-even point (Gorgievski *et al.*, 2011; Miskin and Rose, 2015). Also, successful female entrepreneurs view business failures as a learning opportunity.

Agri-entrepreneurship education increases behavioral intentions towards opening a business (Mohamed *et al.*, 2012). Morgan *et al.* (2010) suggested that agri-entrepreneurial skill sets are situation specific. For example, conflict-zone entrepreneurs should sell products that are easily transported in order to earn income. Previous discussion (Section 1.6) of the partnership between an entrepreneur and T'boli and Bagobo tribes to create distinctive motifs for fashion handbags speaks to situation-specific skills (Hampel-Milagrosa, 2014). Agri-entrepreneurs need to feel confident in their abilities to develop a successful business (McElwee and Smith, 2012). They are internally driven to succeed. Characteristics most associated with agri-entrepreneurs (Fig. 1.1) include many of the terms associated with generic entrepreneurship.

Fig. 1.1. Word cloud of most common descriptors of agri-entrepreneurs.

1.13 Obstacles to Positive Agri-entrepreneurship Characteristics

While an entrepreneur's personal characteristics and experience contribute to agribusiness success, they also influence failed agribusiness. For example, risk aversion can contribute to unprofitable agribusinesses (Pannell *et al.*, 2000). Other characteristics of unsuccessful agri-entrepreneurs include poor financial management, inadequate cost analysis, and lack of attention to the product lifecycle, sales of unprofitable products, unprofitable unfavorable consumer perceptions, and inattentive customer service (Duft, 2015). Additionally, communities, organizations, infrastructure, and social capital that support entrepreneurship and job creation are damaged (Alexander, 2012). Ordinary crime, corruption, political and policy insecurity also hinder entrepreneurial success. Additionally, conflict disrupts markets and displaces people.

The failure of a Somalian slaughterhouse (Massimo *et al.*, 2013) shows that even when an opportunity is actualized as a business it can fail because of unanticipated threats. The slaughterhouse was needed to satisfy health standards for harvested animals and process local meat that could be used as a substitute for imported products or exported. It operated for a less than a year before closing. Access to a distribution system was a problem in terms of on-time meat delivery. The business did not have a contingency plan for increased competition in export markets. Moreover, there was a weather problem: an unanticipated drought was beyond the control of the business and negatively impacted harvest operations, as well as income.

1.14 Conclusion

Agriculture and household micro-businesses provide about 50% of employment in half of all jobs in developing countries (World Bank, 2012). However, most of these jobs do not generate enough income to lift people out of poverty. Entrepreneurial activities related to agriculture produce a solution for increasing household incomes and decreasing vulnerability in transitional economies and conflict zones (Singh, 2014). There is a problem, because poverty is inversely related to education, which limits opportunities for unemployed and underemployed individuals to succeed in entrepreneurship or finding a job (UNESCO, 2014). The type of entrepreneurship training received by interested entrepreneurs influences business success (Berglann *et al.*, 2011).

Aspiring entrepreneurs need skills to decrease risk and uncertainty in decision making and become agile in responding to change, competition, and evolving industry structures (Boehlje *et al.*, 2011). Agility and flexibility are important to conflict-zone entrepreneurs, in particular the most vulnerable—OSY and females—who may have to move business operations if they are displaced by violence (Quinay, 2013).

Successful entrepreneurs take advantage of shifts in consumer demand and satisfy consumers' needs with value-added products, such as organically grown coffee. Value-added products provide entrepreneurs with greater return-on-investment and profit margins. Entrepreneurs who succeed in business understand the importance of continuous education about markets, consumers, current practices, and the business environment. They also obtain information from personal and professional networks.

Key characteristics differentiate agri-entrepreneurs with a successful business from those who are unsuccessful. These characteristics include vision, curiosity, proactive behavior, determination, persistence in working towards goals, and management skills (Bairwa *et al.*, 2014). Successful entrepreneurs have integrity, are honest and persistent, and work towards their goals. Sometimes business failures are beyond an entrepreneur's control (Massimo *et al.*, 2013), but often business failures occur because entrepreneurs do not have a contingency plan. Other barriers to entrepreneurial success include an inability to manage money or company operations (Pannell, 2000). Lastly, the willingness and ability of an entrepreneur to learn from business failures explains why some entrepreneurs fail and others succeed.

References

Agnete Alsos, G., Ljunggren, E. and Pettersen, L. (2003) Farm-based entrepreneurs: what triggers the start-up of new business activities? *Journal of Small Business and Enterprise Development* 10, 435–443.

Ahaibwe, G. and Mbowa, S. (2014) Youth unemployment challenge in Uganda and the role of employment policies in jobs creation. *Africa in Focus* blog. Brookings Institution, Washington, DC. Available at: https://www.brookings.edu/blog/africa-in-focus/2014/08/26/youth-unemployment-challenge-in-uganda-and-the-role-of-employment-policies-in-jobs-creation/ (accessed 7 December 2016).

Aldrich, H. and Zimmer, C. (1986) Entrepreneurship through social networks. In: Sexton, D. and Smilor, R. (eds) *The Art and Science of Entrepreneurship*. Ballinger Publishing, Cambridge, Massachusetts, pp. 3–23.

Alexander, D.C. (2012) Facilitating economic activity in conflict zones. *Security Strategies* 8, 1–20.

Altieri, M.A., Funes-Monzote, F.R. and Petersen, P. (2012) Agroecologically efficient agricultural systems for smallholder farmers: contributions to food sovereignty. *Agronomy for Sustainable Development* 32, 1–13.

Andersén, J. (2012) A resource-based taxonomy of manufacturing MSMEs. *International Journal of Entrepreneurial Behavior and Research* 18, 98–122.

Anderson, A., Pyysiainen, J., Mcelwoo, G. and Vesala, K. (2006) Developing the entrepreneurial skills of farmers. *International Journal of Entrepreneurial Behavior and Research* 12, 24.

Armington, P.S. (1969) A theory of demand for products distinguished by place of production. *International Monetary Fund Staff Papers* XVI, 159–178.

Bairwa, S.L., Lakra, K., Kushwaha, S., Meena, L.K. and Kumar, P. (2014) Agripreneurship development as a tool to upliftment of agriculture. *International Journal of Scientific and Research Publications* 4, 1–4.

Berglann, H., Moen, E.R., Røed, K. and Skogstrøm, J.F. (2011) Entrepreneurship: origins and returns. *Labour Economics* 18, 180–193.

Bijman, J. (2008) *Contract farming in developing countries. An overview of the literature.* Working paper. Department of Business Administration, Wageningen University, Wageningen, Netherlands. Available at: http://www.wageningenur.nl/upload_mm/5/c/b/79333121-6f4b-4f86-9e8e-0a1782e784d6_ReviewContractFarming.pdf (accessed 24 July 2015).

Boehlje, M., Roucan-Kane, M. and Bröring, S. (2011) Future agribusiness challenges: strategic uncertainty, innovation and structural change. *International Food and Agribusiness Management Review* 14, 53–82.

Bonkat, L. (2014) Survival strategies of market women and violent conflicts in Jos, Nigeria. *Journal of Asia Pacific Studies* 3, 281–299.

Bushell, B. (2008) Women entrepreneurs in Nepal: what prevents them from leading the sector? *Gender and Development* 16, 549–564.

DeLind, L.B. (2002) Place, work, and civic agriculture: common fields for cultivation. *Agriculture and Human Values* 19, 217–224.

Dethier, J.J. and Effenberger, A. (2012) Agriculture and development: a brief review of the literature. *Economic Systems* 36, 175–205.

Drucker, P.F. (1998) The discipline of innovation. *Harvard Business Review* 76, 149–157.

Duboyne, I. (2015) Live drone demonstration. *Wisconsin State Farmer News*. Available at: http://www.wisfarmer.com/business/live-drone-demonstration-b99520052z1-307411751.html (accessed 16 June 2015).

Duft, K. (2015) *Managing for success and failure in a small agri-business firm*. Agri-business Management, School of Economic Sciences, Washington State University, Pullman, Washington. Available at: http://www.agribusiness-mgmt.wsu.edu/ExtensionNewsletters/price-cost/managesuccess.pdf (accessed 20 July 2015).

Elliott, R. (2013) The taste for green: the possibilities and dynamics of status differentiation through 'green' consumption. *Poetics* 41, 294–322.

Faisal, I. (2010) Institutionalization of agricultural knowledge: management system for digital marginalized rural farming community. In: Coudel, E., Devautour, H., Soulard, C. and Hubert, B. (eds) *Proceedings of a symposium on Innovation and Sustainable Development in Agriculture and Food, Montpellier, France, 28 June to 1st July 2010*, pp. 10–22. Available at: https://hal.archives-ouvertes.fr/hal-00516458/document (accessed 8 December 2016).

Feagan, R. (2007) The place of food: mapping out the 'local' in local food systems. *Progress in Human Geography* 31, 23–42.

Fraser, A. (2012) *Four ways to make contract farming work for the very poor.* Global Envision: Value Chains blog feed. Mercy Corps, Portland, Oregon. Available at: www.globalenvision.org/2012/12/12/four-ways-make-contract-farming-work-very-poor (accessed 21 July 2015).

Ganzell, B. (2009) *Farming and rural life in the 1970s to today*. Wessels Living History Farm, York, Nebraska. Available at: http://www.livinghistoryfarm.org/farminginthe70s/farminginthe1970s.html (accessed 15 June 2015).

Gartner, W.B. (1990) What are we talking about when we talk about entrepreneurship? *Journal of Business Venturing* 5, 15–28.

GEM (2015) *Future potential: a GEM perspective on youth entrepreneurship*. Global Entrepreneurship Monitor, Global Entrepreneurship Research Association, London. Available at: http://www.innovacion.gob.cl/wp-content/uploads/2015/08/gem-2015-youth-report-1436523546.pdf (accessed 8 December 2016).

Giddings, L.V. and Chassy, B. (2009) Igniting agricultural innovation: biotechnology policy prescriptions for a new administration. *Science Progress*. Available at: http://scienceprogress.org/2009/07/igniting-agricultural-innovation (accessed 15 June 2015).

Gindling, T.H. and Newhouse, D. (2013) Self-employment in the developing world. Background paper for the *World Development Report 2013*. World Bank, Washington, DC. Available at: http://siteresources.worldbank.org/EXTNWDR2013/Resources/8258024-1320950747192/8260293-1320956712276/8261091-1348683883703/WDR2013_bp_Self_Employment_in_the_Developing_World.pdf (accessed 23 July 2015).

Global Policy Forum (2015) *Agribusiness companies*. Global Policy Forum, New York. Available at: https://www.globalpolicy.org/social-and-economic-policy/world-hunger/agribusiness-companies.html (accessed 12 June 2015).

Gorgievski, M.J., Ascalon, M.E. and Stephan, U. (2011) Small business owners' success criteria, a values approach to personal differences. *Journal of Small Business Management* 49, 207–232.

Hampel-Milagrosa, A. (2014) *Micro and small enterprise upgrading in the Philippines: the role of the entrepreneur, enterprise, networks and business environment*. Deutsches Institut für Entwicklungspolitik, Bonn, Germany. Available at: http://www.die-gdi.de/uploads/media/Studies_86_neu.pdf (accessed 24 July 2015).

Harvey, F. (2014) Robot farmers are the future of agriculture, says government. *The Guardian*, 9 January 2014. Available at: http://www.theguardian.com/environment/2014/jan/09/robots-farm-future (accessed 23 July 2015).

Haugen, M.S. and Vik, J. (2008) Farmers as entrepreneurs: the case of farm-based tourism. *International Journal of Entrepreneurship and Small Business* 6, 321–336.

Hoffman, T. and Turnbull, J. (2007) *Farming in the 1950s and 1960s: changing consumer desires*. Wessels Living History Farm, York, Nebraska. Available at: http://www.livinghistoryfarm.org/farminginthe50s/crops_03.html (accessed 15 June 2015).

Hossain, M. and Jaim, W.M.H. (2011) *Empowering Women to Become Farmer Entrepreneurs. Case study of a NGO supported program in Bangladesh*. Paper presented at the IFAD Conference on New Directions for Smallholder Agriculture, 24–25 January 2011. International Fund for Agricultural Development, Rome.

Ikerd, J. (2008) Sustainable capitalism: a matter of ethics and morality. *Problems of Sustainable Development* 3, 13–22.

Ikerd, J.E. (2011) Local food: revolution and reality. *Journal of Agricultural and Food Information* 12, 49–57.

ILO (2012) *Global Employment Trends for Women 2012*. International Labour Organization, Geneva. Available at: http://www.ilo.org/global/about-the-ilo/newsroom/news/WCMS_195445/lang--en/index.htm (accessed 16 July 2015).

Islam, M.A., Khan, M.A., Obaidullah, A.Z.M. and Alam, M.S. (2011) Effect of entrepreneur and firm characteristics on the business success of small and medium enterprises (SMEs) in Bangladesh. *International Journal of Business and Management* 6, 289–299.

Jones, S. (2015) Youth unemployment 'a timebomb' in developing countries, UK MPs say. *The Guardian*, 24 March 2015. Available at: http://www.theguardian.com/global-development/2015/mar/24/youth-unemployment-timebomb-developing-countries-committee (accessed 20 July 2015).

Kapsos, S. (2013) *Global Employment Trends for Youth*. International Labour Organization (ILO), Geneva, Switzerland.

Katz, J.A. (2008) Fully mature but not fully legitimate: a different perspective on the state of entrepreneurship education. *Journal of Small Business Management* 46, 550–566.

Kelly, L. (2013) 8 famous foods discovered by mistake. *Fox News*, 12 November 2013. Available at: http://www.foxnews.com/leisure/2013/11/12/8-famous-foods-discovered-by-mistake/ (accessed 11 June 2015).

Ketchen, D.J., Ireland, R.D. and Snow, C.C. (2007) Strategic entrepreneurship, collaborative innovation, and wealth creation. *Strategic Entrepreneurship Journal* 1, 371–385.

LaGraffe, D. (2012) The youth bulge in Egypt: an intersection of demographics, security, and the Arab Spring. *Journal of Strategic Security* 5, 65–80.

Lalata, D.M., Despi, D.C. and Carlos, P.R. (2010) *Democratic Deficits in the Philippines: What is to be Done?* Konrad-Adenauer-Stiftung (KAS) (Konrad Adenauer Foundation), Makati City, Philippines. Available at: http://www.kas.de/wf/doc/kas_21327-1522-1-30.pdf?110126063409 (accessed 19 June 2015).

Lamprinopoulou, C. and Tregear, A. (2011) Inter-firm relations in SME clusters and the link to marketing performance. *Journal of Business and Industrial Marketing* 26, 421–429.

Lamprinopoulou, C., Tregear, A. and Ness, M. (2006) Agrifood SMEs in Greece: the role of collective action. *British Food Journal* 108, 663–676.

Land Matrix (2015) *Agricultural drivers*. Land Matrix Global Observatory. Available at: http://www.landmatrix.org/en/get-the-idea/agricultural-drivers/ (accessed 15 June 2015).

Lans, T., Seuneke, P. and Klerkx, L. (2013) Agricultural entrepreneurship. In: Carayannis, E.G. (ed.) *Encyclopedia of Creativity, Invention, Innovation and Entrepreneurship*. Springer, New York, pp. 44–49.

Lingelbach, D.C., De La Vina, L. and Asel, P. (2005) *What's distinctive about growth-oriented entrepreneurship in developing countries?* Working paper no. 1, Center for Global Entrepreneurship, UTSA College of Business. University of Texas, San Antonio, Texas. Available at: http://papers.ssrn.com/sol3/papers.cfm?abstract_id=742605 (accessed 15 June 2015).

Lowson, R.H. (2003) *Strategic Operations Management: the New Competitive Advantage*. Routledge, London.

Mair, J. and Marti, I. (2006) Social entrepreneurship research: a source of explanation, prediction, and delight. *Journal of World Business* 41, 36–44.

Makhbul, Z.M. and Hasun, F.M. (2010) Entrepreneurial success: an exploratory study among entrepreneurs. *International Journal of Business and Management* 6, 116.

Maloni, M. and Benton, W.C. (2000) Power influences in the supply chain. *Journal of Business Logistics* 21, 49–74.

Marsden, T. and Smith, E. (2005) Ecological entrepreneurship: sustainable development in local communities through quality food production and local branding. *Geoforum* 36, 440–451.

Massimo, C., Kinyanjui, W., Njue, S., Joseph, G., Nyorangi, O.J. and William, O. (2013) Opportunities and challenges of setting up HACCP system in slaughterhouses in countries emerging from conflict: an appraisal of North-Western Somalia. *Food and Nutritional Sciences* 4, 126–130.

McElwee, G. and Smith, R. (2012) Classifying the strategic capability of farmers: a segmentation framework. *International Journal of Entrepreneurial Venturing* 4, 111–131.

McMichael, P. (2005) Global development and the corporate food regime. *Research in Rural Sociology and Development* 11, 269–303.

Milestad, R., Ahnström, J. and Björklund, J. (2011). Essential multiple functions of farms in rural communities and landscapes. *Renewable Agriculture and Food Systems* 26, 137–148.

Minot, N. (2007) *Contract farming in developing countries: patterns, impact, and policy implications*. Case Study, 6-3, Food Policy for Developing Countries program (The Role of Government in the Global Food System). Division of Nutritional Sciences, Cornell University, Ithaca, New York.

Miskin, V. and Rose, J. (2015) New venture initiation: factors influencing success. *Journal of Small Business Strategy* 1, 1–9.

Mohamed, Z., Rezai, G., Nasir Shamsudin, M. and Mu'az Mahmud, M. (2012) Enhancing young graduates' intention towards entrepreneurship development in Malaysia. *Education + Training* 54, 605–618.

Morgan, S.L., Marsden, T., Miele, M. and Morley, A. (2010) Agricultural multifunctionality and farmers' entrepreneurial skills: a study of Tuscan and Welsh farmers. *Journal of Rural Studies* 26, 116–129.

Nagler, P. and Naudé, W. (2015) *Counting Africa's rural entrepreneurs*. African Can End Poverty (a blog about the economic challenges and opportunities facing Africa). The World Bank, Washington, DC. Available at: http://blogs.worldbank.org/africacan/counting-africas-rural-entrepreneurs (accessed 20 July 2015).

Pannell, D.J., Malcolm, B. and Kingwell, R.S. (2000) Are we risking too much? Perspectives on risk in farm modeling. *Agricultural Economics* 23, 69–78.

Parmigiani, A. and Rivera-Santos, M. (2015) Sourcing for the base of the pyramid: constructing supply chains to address voids in subsistence markets. *Journal of Operations Management* 33, 60–70.

Pecotich, A. and Ward, S. (2007) Global branding, country of origin and expertise: an experimental evaluation. *International Marketing Review* 24, 271–296.

Pettinger, T. (2014) *Problems of agriculture – market failure*. EconomicsHelp.org. Available at: http://www.economicshelp.org/blog/4977/economics/problems-of-agriculture-market-failure/ (accessed 20 June 2015).

Philippines Institute for Development Studies (2015) *Unemployment rate by region*. Economic and Social Database. Available at: http://econdb.pids.gov.ph/tablelists/table/627 (accessed 19 July 2015).

Quinay, E.B. (2013) Current status and attributes of Filipino women entrepreneurs: basis for an entrepreneurial model. In: *Proceedings of 4th Asia Pacific Business Research Conference, September 30-October 1, Bay View Hotel, Singapore*. Association for Small Business and Entrepreneurship, USF St Petersburg, Florida. Available at: http://www.wbiworldconpro.com/uploads/singapore-conference-2013/management/1380703423_428-Erma.pdf (accessed 25 July 2015).

Requier-Desjardins, D., Boucher, F. and Cerdan, C. (2003) Globalization, competitive advantages and the evolution of production systems: rural food processing and localized agri-food systems in Latin-American countries. *Entrepreneurship and Regional Development* 15, 49–67.

Richards, S.T. and Bulkley, S. (2007) *Agricultural entrepreneurs: the first and the forgotten?* Research Paper 07-01, Center for Employment Policy Entrepreneur Series. Hudson Institute, Washington, DC.

Sancho, F. (2010) Agricultural and rural entrepreneurship: concepts for modeling development. *Comuniica Magazine* 6, 1–10.

Saxena, S. (2012) Problems faced by rural entrepreneurs and remedies to solve it. *Journal of Business and Management* 3, 23–29.

Scarborough, N.M. (2012) *Effective Small Business*, 10th edn. Prentice Hall, Upper Saddle River, New Jersey.

Schumpeter, J.A. (1934) *The Theory of Economic Development*. Harvard University Press, Cambridge, Massachusetts.

Singh, M.P. (2014) Entrepreneur and economic development: a study of role of various forms of entrepreneurs in economic development. *Global Journal of Multidisciplinary Studies*, 3(5). GJMS, Jabalpur MP, India. Available at: http://www.gjms.co.in/index.php/gjms/article/view/169 (accessed 16 July 2015).

Sofoluwe, N.A., Tijani, A.A. and Baruwa, O.I. (2011) Farmers' perception and adaptation to climate change in Osun State, Nigeria. *African Journal of Agricultural Research* 6, 4789–4794.

Stenholm, P. and Hytti, U. (2014) In search of legitimacy under institutional pressures: a case study of producer and entrepreneur farmer identities. *Journal of Rural Studies* 35, 133–142.

Stupnytska, A., Koch, K., MacBeath, A., Lawson, S. and Matsui, K. (2014) *Giving credit where it is due: how closing the credit gap for women-owned SMEs can drive global growth*. Goldman Sachs Global Market Institute, New York. Available at: http://www.goldmansachs.com/our-thinking/public-policy/gmi-folder/gmi-report-pdf.pdf (accessed 8 December 2016).

Suárez, M. (1972) Campesino communitarian enterprises in Latin America. In: Araujo, J.E.G. (ed.) *The Community Enterprise: a Model for the Reform Process in Latin America*. IICA Institutional Development, no. 4. Inter-America Institute for Cooperation on Agriculture, Washington, DC, pp. 297–335.

Taborda, J. (2015) *Philippines unemployment rate 1994–2015*. Trading Economics, New York. Available at: http://www.tradingeconomics.com/philippines/unemployment-rate (accessed 20 July 2015).

Trienekens, J.H. (2011) Agricultural value chains in developing countries: a framework for analysis. *International Food and Agribusiness Management Review* 14, 51–83. Available at: https://www.ifama.org/resources/Documents/v14i2/Trienekens.pdf (accessed 8 December 2016).

Tubene, S.L., Myers, R.D., McClurg, C. and Afantchao, Y. (2015) Buyers and sellers' responses to ethnic and specialty produce in the Baltimore–Washington area. MarylandEthnicVegetable.com project of the University of Maryland Extension and Small Farm Institute, Upper Marlboro, Maryland. Available at: http://www.marylandethnicvegetable.com/production/ (accessed 15 June 2015).

UNESCO (2014) *Teaching and Learning: Achieving Quality for All*. EFA Global Monitoring Report 2013/4. Education for All, United Nations Educational, Scientific, and Cultural Organization, Paris. Available at: http://unesdoc.unesco.org/images/0022/002256/225660e.pdf (accessed 25 July 2015).

UNIDO (2015) Agribusiness and rural entrepreneurship development. United Nations Industrial Development Organization, New York. Available at: http://www.unido.org/agro.html (accessed 16 July 2015).

Verick, S. (2014) Female labor force participation in developing countries. IZA (Institute for the Study of Labor) World of Labor, Bonn, Germany. Available at: http://wol.iza.org/articles/female-labor-force-participation-in-developing-countries.pdf (accessed 18 July 2015).

Vogel, S.J. (2012) *Multi-enterprising farm households: the importance of their alternative business ventures in the rural economy*. USDA-ERS Economic Information Bulletin 101. USDA Economic Research Service, Washington, DC. Available at: http://www.uvm.edu/~kmmorin/cs008/assign4/Multi-EnterprisingFarmHouseholds-The%20ImportanceofTheirAlternativeBuisnessVenturesintheRuralEconomy.pdf (accessed 20 July 2015).

Vossenberg, S. (2013) *Women entrepreneurship promotion in developing countries: what explains the gender gap in entrepreneurship and how to close it*. Working paper no. 2013/08. MSM Working Paper Series. Maastricht School of Management, Maastricht, Netherlands. Available at: ftp://ftp.repec.org/opt/ReDIF/RePEc/msm/wpaper/MSM-WP2013-08.pdf (accessed 7 August 2016).

WCED (1987) *Our Common Future (Brundtland Report)*. UN World Commission on Environment and Development. Oxford University Press, Oxford, UK, pp. 437–495.

Wilkinson, J. (2009) The globalization of agribusiness and developing world food systems. *Monthly Review* 61, 38–49.

World Bank (2008) *Agriculture and Poverty Reduction*. Policy brief extracted from *World Development Report 2008: Agriculture for Development*. World Bank, Washington, DC. Available at: http://go.worldbank.org/OPEDJWW8W0 (accessed 8 December 2016).

World Bank (2012) *World Development Report 2013: Jobs*. World Bank, Washington, DC. doi: 10.1596/978-0-8213-9575-2. License: Creative Commons Attribution CC BY 3.0. Available at: http://econ.worldbank.org/WBSITE/EXTERNAL/EXTDEC/EXTRESEARCH/EXTWDRS/EXTNWDR2013/0,,contentMDK:23044836~pagePK:8258258~piPK:8258412~theSitePK:8258025,00.html (accessed 8 December 2016).

Zhu, Y., Lan, H., Ness, D.A., Xing, K., Schneider, K., Lee, S.H. and Ge, J. (2015) Rural community development and the role of social entrepreneurs. *Transforming Rural Communities in China and Beyond*. Springer International Publishing, New York, pp. 17–34.

2 Comparing Agri-entrepreneurs in Non-conflict Regions versus Conflict and Transitional Economies

Pauline Sullivan*
Tennessee State University, Nashville, Tennessee

2.1 Introduction

This chapter discusses agri-entrepreneurship in non-conflict zones, conflict zones, and transitional economies. The introduction provides an overview of global agriculture and entrepreneurship, followed by separate sections for non-conflict zones, conflict zones, and transitional economies. Each section describes the region in general, then examines the agricultural sector in each group, and lastly discusses agri-entrepreneurship. The final section compares similarities and dissimilarities among agri-entrepreneurs in the three groups.

Agri-entrepreneurs around the world have much in common because of growing demand for food due to predictions that the world's population will reach nine billion people by 2050 (FAO, 2009). Population growth will be the greatest in Africa and Asia, exceeding the ability to produce food in these countries (UN, 2015). The global geographical shift in population requires food security. However, the need for food security is an opportunity for agri-entrepreneurs to use transparent production systems and gain market share.

Other changes related to population growth and geography affect agri-entrepreneurs and the type of products they produce. For example, the ethnic minority population is growing and young, with comparatively large families (Euromonitor International, 2015). This group of consumers are an attractive market for agri-entrepreneurs with niche products that are culturally relevant to these customers. Similarly, there is an increasing number of Muslims around the world, some of whom observe dietary traditions. The demand for value-added Halal foods will grow.

Global warming will affect agri-entrepreneurs in non-conflict zones, conflict zones, and transitional economies (World Bank, 2011). Weather patterns are increasingly volatile and affect growing seasons. There is a need for agri-entrepreneurs to find innovative solutions that enable crops to be viable, given climate change. Another concern, shared by all agri-entrepreneurs, will be a diminished water supply needed for crop production and scarcity of agricultural land due to development.

Agri-entrepreneurs face changes in product characteristics and consumer preferences, geographic location of crop growth and food processing, and global distribution systems (Boehlje *et al.*, 2011). There will also be structural changes in the agricultural sector, including small and medium-size agribusinesses. In addition, new technologies that increase efficiencies in agricultural production will be developed.

*psullivan10605@gmail.com

Agri-entrepreneurs navigate through many obstacles in order to achieve their goals and remain strong while facing uncertainty (García and Calvo, 2013). Their experiences, ability to adapt to change, and confidence in achieving success make agri-entrepreneurs more resilient. Their resiliency is psychological capital, which helps when dealing with uncertainties encountered by businesses (De Hoe and Janssen, 2014). This psychological capital is conducive to optimism and is advantageous when facing situations that require quick decisions. Psychological capital is associated with optimism, an asset that helps agri-entrepreneurs innovate, which is necessary to remain competitive and successfully compete in the global marketplace.

Agri-entrepreneurs will need to rely on their knowledge and experience as they continuously evaluate market change. These changes include methods of business operation, financial management practices, and agricultural policies and regulations (Boehlje *et al.*, 2011). Collaborative partnerships among agri-entrepreneurs become important because they often lead to new opportunities.

One example of a partnership that has evolved over time is the cut-flower market controlled by the Netherlands (Benson-Rea and Stringer, 2015). The cut-flower market was a key player in the global value chain. The market position of the Netherlands is a result of the country's natural competitive advantages, such as its reputation as a pioneer in horticulture and its proximity to the airport in Amsterdam. The cut-flower market in Aalsmeer is about 30 minutes from Schiphol Airport, a transportation hub. Aalsmeer is an ideal location for a sophisticated international distribution center for cut flowers. Originally flowers from the Netherlands were primarily sold at this distribution center. At the time the distribution center was built, the Netherlands captured 60% of Europe's cut-flower sales.

The Aalsmeer distribution center and the dominant role of the Netherlands in the cut-flower market provided producers from emerging product markets with an opportunity to sell their products internationally (Benson-Rea and Stringer, 2015). Small-size market value-added flower producers in developing countries had little experience with international markets, so sending their products to the Dutch distribution center decreased their risk and costs of market entry. This arrangement is mutually beneficial as it allows year-round sales of cut flowers supplied from producers in developing countries and distribution using the Dutch competitive advantage in the supply-chain distribution. This partnership between the Dutch flower distribution center and agri-entrepreneurs in developing countries helped the Aalsmeer distribution center save money on its two largest inputs, labor and energy, which were increasing.

This innovative partnership between small and medium-size value-added flower producers in emerging markets and a distribution center that dominates the supply chain is an example of a creative win–win situation. It also suggests that agri-entrepreneurs in non-conflict zones, conflict regions, and transitional economies can profitably collaborate when a means of transporting products exists.

2.2 Non-conflict Zones

This section describes non-conflict zones as "positive peace" countries. The Institute for Economics and Peace (IEP, 2015) described positive peace as "the attitudes, institutions and structures which create and sustain peaceful societies. These same factors also lead to many other positive outcomes that support the optimum environment for human potential to flourish." To measure positive peace, the IEP developed a Positive Peace Index (PPI). The PPI has 24 peace indicators that can be grouped into eight domains. The eight domains are: well-functioning government; sound business environment; equitable distribution of resources; acceptance of the rights of others; good relations with neighbors; free flow of information; high levels of human capital; and low levels of corruption. Only 25 countries, out of 162, in the world have the highest scores on the PPI. Positive peace countries include Iceland, Denmark, Austria, New Zealand, Switzerland, Finland and Canada. Iceland received the best PPI. Generally, the positive peace countries are also developed countries. The term developed countries refers to all industrialized countries, as well as those in transition (FAO, 2015). These include countries in Europe, Canada, the USA, Australia, and Japan.

2.2.1 Agriculture and agribusiness in developed countries

The number of individuals in developed countries who work in agriculture represents less than 10% of all workers (World Bank, 2015a). Value-added agricultural products add less than 10% to GDP in developed countries. Table 2.1 shows the annual growth rate of value-added agriculture for the ten most peaceful countries, as identified in the PPI (IEP, 2015). This table indicates that the sector for agricultural and value-added products has a low growth rate in non-conflict countries.

One example of agriculture in developed countries is discussed within the context of Canada's economy. Only 1% of Canadians work in the agricultural sector as farmworkers (Statistics Canada, 2015). However, Canadian farmers' receipts increased by 11.2% from 2010 to 2011, yielding earnings of Can$46 billion. Also, operating expenses decreased by 2.9%. Nevertheless, farmers whose cash receipts were between Can$10,040 and Can$49,000 did not break even and lost Can$0.0628 per dollar spent on farming. In comparison, large farms earning between Can$250,000 and Can$499,999 make Can$0.214 on the dollar.

In order to increase their return on investment, small Canadian farmers focus on niche value-added products (Government of Canada, 2013). Their two concentrations are: (i) functional foods; and (ii) organics. Organic value-added products created by Canadian agri-entrepreneurs range from fruits and fruit juices to cereals. Given the growth rate and profitability of the organic product market, Canada exports these products to the USA, the European Union, and Japan. Canada's Organic Products Regulations ensure that food products are produced through sustainable, healthy, and productive ecosystems. Also, Canada promotes "buy Canadian" products and official "made in Canada" labels for food products.

Table 2.1. Annual agriculture growth rate in world's ten most peaceful countries (from IEP, 2015).

	Rank PPI			
Country	2010	2011	2012	2013
Iceland	–8.4	3.9	3.8	N/A
Denmark	30.3	–10.3	14.7	–19.7
Austria	–4.7	15.8	–7.9	–3.5
New Zealand	–6.7	12.4	–2.7	3
Switzerland	–3.3	7.7	–4.6	–2
Finland	0.5	4.4	–6.2	0.2
Canada	N/A	N/A	N/A	N/A
Japan	–1	2	0.3	N/A
Australia	–0.9	3.5	1.4	–0.6
Czech Republic	–16.3	1.8	2.6	–9

2.2.2 Agri-entrepreneurs in developed countries

Developed countries support innovation and commercialization of agriculture (Sandu and Ryzhenkova, 2015). Examples of how governments in these countries support small and medium-sized businesses include tax reductions, reduced customs taxes, and tax free loans for small businesses purchasing technology.

In Europe, farms have been in the family for years (FAO, 2014). Farm owners, particularly young farmers, consider changing the types of products produced, finding new buyers for their products, and starting farm-related direct selling. Challenges for agri-entrepreneurs in Europe include lack of access to land, capital, and credit, as well as increasing farming costs; however, some loans are available for young farmers.

Another example of agri-entrepreneurship in non-conflict zones is female agri-entrepreneurs in Japan. About 3% of Japanese women are agri-entrepreneurs and approximately another 10% are involved with food entrepreneurship (Welsh *et al*., 2013). The Japanese government provides support and assistance for female entrepreneurs. Lack of financial capital, management skills, knowledge of business operations, time management skills, and life/work balance skills are challenges for Japanese female entrepreneurs. They are interested in learning about marketing and management, and would join an entrepreneurship support group, but they prefer it be an online organization.

Producing agritourism goods and services can be a lucrative venture for micro, small and medium-size enterprises in developed countries. Micro and small Tennessee agribusinesses accounted for 51.6% of respondents in a study about agritourism (Jensen *et al*., 2013).

Approximately 80% of these agri-entrepreneurs explicitly sold products and services to generate income. Pick-your-own berries and pumpkins were the top crops grown. Average income from agritourists per farm visit was about $35 from product sales. Souvenirs accounted for only 4.2% of agritourist expenditure. The state of Tennessee provided agri-entrepreneurs with promotional support and a "Pick Tennessee" program. Tennessee's agritourism sector employed 521 people and directly generated US$16,453,265 from value-added sales. Agri-entrepreneurs earned about 40% of their total revenues from agritourism. Overall, this study demonstrates the importance of value-added agritourism products and services to augment farm earnings.

In summary, governments in non-conflict zones support agri-entrepreneurship through incentives and loans. Agri-entrepreneurs are willing to diversify product offers, although access to land and start-up costs are fundamental barriers to agri-entrepreneurship.

2.3 Conflict Zones

Global conflict costs about $100 billion annually (Alexzander, 2012). Drahoňovská (2010) described places where violence occurs as conflict zones. These areas have organized and intense armed confrontations, between government and other groups (Alexzander, 2012). Indicators of a conflict zone include ongoing domestic and international conflict, concerns about safety and security, and militarization (IEP, 2015). In conflict zones, people are continually aware of the potential for violent demonstrations, political instability, terrorism, violent crime, homicide, and struggles with neighboring countries. About 33% of the world's population lives in conflict zones (IDS, 2015). Countries with the worst conflict zones represent 6.7% of the world's 162 countries, while about another 28% have some level of regularly occurring conflict. The annual rate of change in the contribution of conflict-zone countries to world output was −1.1% between 2014 and 2015, and −2% for 2013 to 2014 (UN, 2014a). In addition, exports by countries in conflict zones add less to gross domestic product (GDP) when compared with non-conflict countries (Speakman and Rysova, 2014). Entrepreneurs in conflict zones concentrate on products for domestic markets.

People living in conflict zones have low incomes and are among the poorest in the world (UN, 2014b). Most conflict-zone residents live in rural areas of developing countries. Agriculture and agri-entrepreneurship are important to the livelihoods of people living in these areas. Being in a conflict zone, with associated lost income, makes people vulnerable and impacts their quality of life (UN, 2015). In rural conflict zones, there is low employment and there are few opportunities for education (Drost *et al.*, 2013). None the less, people living in rural conflict zones want to learn about agribusiness as it potentially offers a means of earning income (Alexzander, 2012).

2.3.1 Agriculture and agribusiness in conflict zones

Conflict-zone violence interferes with market transactions and influences changes in household behavior (Arias *et al.*, 2013). Sometimes households are pressured into growing illegal crops, or intimidated into paying taxes to non-governmental actors controlling the region.

Crops are conflict-zone farmers' investments (Arias *et al.*, 2013). Conflict and violence influence farmers' crop-planting decisions. For example, when there is no governmental control of an area, farmers plant seasonal crops, use land for pasturing, or do nothing with the land. They often plant crops or livestock that can be sold quickly under conditions of uncertainty. On the other hand, when farmers have less concern about conflict and violence they plant perennial crops. Farmers in conflict zones are resilient and adapt agricultural activity to reflect the conflict situation.

Initially farmers' investments in agriculture decrease due to armed non-governmental actors controlling the region. However, when these actors remain in control of an area for more than 7 years, farmers incrementally increase their investment in agriculture.

Table 2.2 shows the annual growth rate in agriculture for the ten least peaceful countries, as identified in the PPI (IEP, 2015). Data indicate

Table 2.2. Annual agricultural growth rate: world's ten least peaceful countries. Growth rates from World Bank (2015b).

	Rank PPI			
Country	2010	2011	2012	2013
Syria	N/A	N/A	N/A	N/A
Iraq	1.1	16.6	4.3	2.8
Afghanistan	–6.4	–7.8	18.2	–0.2
South Sudan	N/A	N/A	N/A	N/A
Central African Republic	N/A	N/A	N/A	N/A
Somalia	N/A	N/A	N/A	N/A
Sudan	–1.2	–0.9	48.7	N/A
Democratic Republic of the Congo	3.6	3.6	3.5	4.2
Pakistan	0.2	2	3.6	2.9
North Korea	N/A	N/A	N/A	N/A
Russia	–11.3	13.9	–3.5	–2

that agricultural and value-added products have a low growth rate, or none at all. Available data show the volatility of the agricultural sector in conflict zones.

2.3.2 Agri-entrepreneurs in conflict zones

Entrepreneurs in conflict zones face limited choices and/or mobility (Speakman and Rysova, 2014). Their businesses tend to be informal, ranging in size from a small family business to medium-size enterprises. Agricultural entrepreneurship efforts comprise about 99% of businesses in conflict zones. Agri-entrepreneurs with large firms primarily engage in manufacturing activities and need to be in areas with a high concentration of potential factory workers.

In Afghanistan, businesses are resilient and they do not stop operating during outbreaks of active conflict (Cusack and Malmstrom, 2010). Agri-entrepreneurs' concerns include uncertainty about the business environment, inability to receive necessary items for production, access to finance, and the power supply. A lot of money is invested in Afghanistan for business development, so entrepreneurs seek opportunities with the highest margins. Often entrepreneurs start a business outside the agricultural sector and then transition into agri-entrepreneurship. Because of the war, money was invested in sophisticated telecommunications and media sectors. Telecommunications and media provide a way for Afghani agri-entrepreneurs to reach markets outside their country. Thus the quality of telecommunications and media in Afghanistan provides agri-entrepreneurs with a competitive advantage in reaching target markets.

Only 13% of Afghanistan has arable land, yet agriculture provides the largest income source for the domestic economy and creates jobs for 75% of the population (Wadsam, 2014). Afghanistan's agri-entrepreneurs focus on niche products such as pomegranates, raisins, apples, almonds, and apricots. They specialize in producing high-quality dried fruits; they have a reputation for quality dried fruits from Afghanistan and are exploring international marketing opportunities for these products.

Issues related to the production of agricultural products that concern agri-entrepreneurs in Afghanistan include limited available storage, inefficient transportation systems, lack of safety certifications for their products, and entrepreneurship training.

South Sudan provides another example of agriculture-related entrepreneurial activities in a conflict zone (Chokerah and Horvath, 2012). It is estimated that about 90% of South Sudan's 640,000 km^2 of land can be farmed, of which 50% is prime agricultural land. More than 80% of people living in South Sudan earn a subsistence living from growing crops and livestock. Climate conditions are perfect for growing a broad variety of crops, but over half the crops produced are for personal consumption. About 32% of agricultural products are marketed outside the farm. Agriculture represents one-third of South Sudan's GDP and employs about 67% of the population.

Long-term conflict decreases agricultural production in South Sudan, which contributes to a 51% poverty rate in the country (Chokerah and Horvath, 2012). Challenges to the agricultural sector include creation of new value-added products, funding available for agri-entrepreneurs, limited access to credit, and an inadequate infrastructure. Agri-entrepreneurs do not have extension services or available research that could help farmers increase crop resistance. In addition to the threat of conflict, natural disasters can affect farm profitability. Above all, inadequate or a lack of appropriate transportation systems makes it difficult for products to reach intended markets.

South Sudan's agri-entrepreneurs seek new opportunities for value-added products and services (Chokerah and Horvath, 2012). One option that South Sudanese agri-entrepreneurs consider is to join an agricultural cooperative where farmers may share materials and information, as well as work together to obtain market recognition and control costs. These cooperatives provide a place where agri-entrepreneurs can develop skills necessary to operate their businesses and market their products.

In conclusion, agri-entrepreneurs in conflict zones have a multitude of problems to solve in order to increase profitability. These include violence, transportation systems, credit or funding for their business, proper training for running a business, and increasing the profitability of agriculture and related activities.

2.4 Transitional Economies

Transitional economies are formerly communist nations moving towards market economies (IMF, 2000). These include countries in south-eastern Europe, as well as the Commonwealth of Independent States (CIS, including the Russian Federation) and Georgia. Overall, transitional economies contribute to increasing the world's agricultural output. Between 2014 and 2015 the agricultural output in transitional countries grew 2% (UN, 2014b). Contribution by geographic area for the same time period is as follows: south-eastern Europe 3.1%; CIS and Georgia 2.1% (Russian Federation 1.2%). The slow rate in transitional economies is attributed to high unemployment, low consumer demand, fiscal gaps, geopolitical tensions, and high levels of uncertainties. Transitional economies also contend with natural disasters that affect agricultural productivity (UN, 2014b).

Although agricultural output increased after 2001, it remains below levels in the Soviet era (Cvijanović *et al.*, 2013). Also, agriculture's contribution to GDP, gross national income and gross value-added decreased over time (1995–2012) (Vasile and Ungureanu, 2014).

2.4.1 Agriculture and agribusiness in transitional economies

Table 2.3 provides information about the annual growth rate of value-added agriculture in the ten most peaceful countries in transitional economies (IEP, 2015). The table indicates the volatility in transitional economies' agricultural profitability.

2.4.2 Agri-food sector development and importance

Agricultural enterprises contribute to economic development in transitional economies (Cvijanović *et al.*, 2013). In transitional economies, agri-entrepreneurs need to increase their knowledge of market economy and best business practices. They require market research in order to identify products that satisfy consumers' needs. In addition, they also need financial information to remain cognizant of cost-saving mechanisms that contribute to business profitability. Market research also provides necessary information in making business decisions, such as type of products sold. New marketing techniques can efficiently inform consumers about an agri-entrepreneur's products, which can create demand. In particular, agri-entrepreneurs should use labels to communicate product safety to

Table 2.3. Annual agricultural growth rate: world's ten most peaceful transitional countries (from World Bank, 2015b).

	Rank PPI			
Country	2010	2011	2012	2013
Czech Republic	−16.3	1.8	2	−9
Slovenia	2.3	7.7	−8	−4.7
Poland	−6.7	6.4	−4.7	4.5
Slovakia	−14.8	18.7	2.8	19.5
Romania	−5.3	13.3	−22.4	N/A
Croatia	−6.6	−3.8	−14.7	−1.6
Bulgaria	−6.4	−1.5	−7.3	3.3
Lithuania	−3.3	N/A	N/A	N/A
Estonia	10.2	16.4	16.6	−3.9
Kazakhstan	−4.6	26.5	−17.4	11.2
Tajikistan	6.8	7.9	10.4	18.8

consumers to decrease risks associated with purchases and establish a brand identity.

Challenges faced by agri-entrepreneurs in transitional economies include limited financial resources, understanding of market economy business practices, and market research for new product development (Cvijanović *et al.*, 2013). That is also the case in Albania, a transitional economy that ended communism and emerged from a strictly planned economy in early 1990s. Entry into a free market economy required Albania to undergo drastic changes.

Agriculture in Albania is one of the most important sectors from a socio-economic view point, as it remains the largest employer in Albania. Almost half of the employed population is engaged in the agricultural sector, with more women employed than men. Agriculture contributes about 20% of Albania's GDP, thus being an important source of income for the rural population (almost 50% of the total population live in rural areas, with most rural households engaged in agriculture) (MARDWA, 2014).

In the beginning of the Albanian transition to a market economy, agricultural land was distributed to local residents through the land reform of 1991. This was somewhat problematic as farmers (who until that year had been working as employees of state-run farms or cooperatives) did not have the capital or technical and market know-how to operate in the context of the emerging market economy. In the early phase, most agricultural holdings were operating as subsistence farms, producing food for household needs.

Agricultural productivity has improved in recent years, due to the capital accumulation and growing production and market experience of Albanian farmers. However, a substantial part of the agricultural production remains oriented to self-subsistence and takes place just to provide food and limited income support for rural families. Also, the production that is destined for market is largely sold on the spot market, often disregarding market demand, product quality, and consistent standards.

Important constraints to competitiveness must be addressed in Albania. These constraints include agricultural productivity, increased regional competition, and harmonization with EU legislation and standards. In order to meet these challenges, consolidation and modernization of the agri-food sector are necessary (World Bank, 2007). Lack of vertical and horizontal coordination in the agri-food value chain, combined with limited financial and technical support from the government, has hampered the further development of the sector. The effectiveness of the public agricultural extension system has been characterized by gaps in terms of infrastructure on the one hand and quantitative and qualitative levels of available human resources on the other. These constraints have resulted in the limited impact of the public extension services on performance and innovation in the agricultural sector (MARDWA, 2014).

2.4.3 Agri-entrepreneurship and gender in Albania

The transition in Albania mostly affected vulnerable social groups, including women and children. Rural regions tend to be challenging environments for the promotion of entrepreneurship, especially for women. Lack of economic diversification and high dependence on farming is creating a harsh environment for entrepreneurs. Attitudes towards women are said to be less egalitarian than in urban areas and traditional notions about gender roles may act as barriers to female business owners.

The figures derived from the Business Register issued by Albania's Institute of Statistics (INSTAT) for 2014 provide further evidence for Albania (Table 2.4). Only 22.4 % of local units are concentrated in rural areas. Overall women constitute a large underrepresented group in Albania in terms of business start-up and ownership. Women-run businesses make up 28% of the total business and are located mostly (87%) in urban areas and concentrated in Tirana (47.5%). This reflects the scarce economic diversification in rural areas as well as the unfertile environment for entrepreneurship activities in off-farm business.

The female-run small and medium-size enterprises (SMEs) appear mostly in highly fragmented units of production operating at lower value added and highly informal sectors such as retail trade, hotels and restaurants, as well as other service sectors. In rural areas, most sectors are underrepresented (less than 25%) (Table 2.5).

Table 2.4. Women-led businesses to total businesses in Albania, by region (from INSTAT, 2014).

Prefectures	Total	Women-led	Location	
			Municipality	Commune
Total	112,537	32,023	27,778	4,245
Berat	4,302	1,229	989	240
Dibër	2,164	404	344	60
Durrës	9,578	2,640	2,362	278
Elbasan	7,859	2,016	1,698	318
Fier	9,693	2,576	1,911	665
Gjirokastër	2,681	644	545	99
Korçë	7,311	1,965	1,679	286
Kukës	1,068	167	140	27
Lezhë	3,388	970	766	204
Shkodër	5,446	1,508	1,251	257
Tiranë	49,467	15,197	13,645	1,552
Vlorë	9,580	2,707	2,448	259

Table 2.5. Number of companies operated by women in Albania, by economic activity and location (from INSTAT, 2014).

Sector	Operated by women	%	Municipality	Commune
Agriculture, forestry, fishing	167	7.39	32.3	67.7
Industry	2,205	21.72	83.9	16.1
Construction	387	8.46	82.9	17.1
Trade	15,366	34.17	87.2	12.8
Transport and storage	294	3.89	77.6	22.4
Accommodation and food service activities	5,119	28.34	80.3	19.7
Information and communication	477	18.42	86.4	13.6
Other services	8,008	35.8	92.4	7.6
Total number	32,023			

Approximately 95% of the enterprises managed by women have fewer than five workers. This share is much larger compared with enterprises headed by men (Table 2.6).

The business environment in Albania is not very encouraging or supportive for women entrepreneurs. Moreover, rural business environments provide less space for women (Table 2.7). The majority of these highly fragmented units are located in urban areas. It means that the majority of these entrepreneurs are microbusinesseses supported by family labor and using modest quantity of inputs.

Data show that services represent the most successful sub-sector for women entrepreneurs in Albania. Government studies show that small-scale activities in tourism, trade and artifacts provide fewer barriers to women (UNDP, 2013). Also there are strong cultural barriers for women entrepreneurs, who are often excluded from business activities such as sales. However, women can be successful when coached effectively by organizations or mentors. There are policies that support removing obstacles to entrepreneurial activity by women and provide services that help business start-up. Services that can contribute to the success of female entrepreneurs are available credit, advisory services, networking, and participation in professional associations and local governance. The advisory services rarely include women. Fewer than half of participants in mobile training during the year 2015 were women (Fig. 2.1). Main trends show that there is a distinction between women from Albania's northern remote areas and women from southern plains areas. In the

Table 2.6. Women-managed businesses in Albania, by number of employees (from INSTAT, 2014).

Gender of business owner	Number of employees 1–4	5–9	10–49	50+	Total
Females (no.)	30,154	908	723	238	32,023
Share of total (%)	94.2	2.8	2.3	0.7	100
Males (no.)	70,871	4,479	3,924	1,240	80,514
Share of total (%)	88.0	5.6	4.9	1.5	100
Total	101,025	5,387	4,647	1,478	112,537

Table 2.7. Women-led businesses in Albania, by location (municipality, commune) (from INSTAT, 2014).

Dimension (number of workers)	Location: Municipality	Commune
1–4	26,209	3,945
5–9	789	119
10–49	563	160
50+	217	21

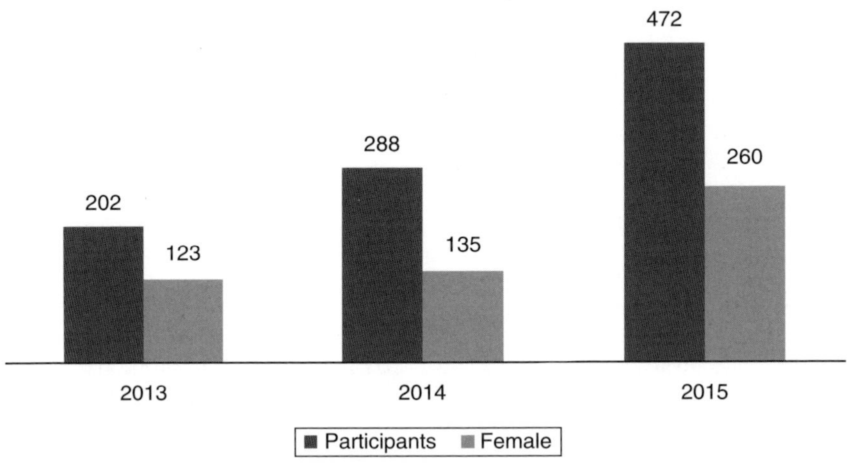

Fig. 2.1. Enrollment in mobile training center by gender during the period 2013–2015 (from MTC *Yearly Analysis Report*).

northern areas, a patriarchal culture limits women's ability to participate in public business transactions. Thus, in northern areas, female entrepreneurs are limited to business activities in their homes, such as tailoring and embroidery. In contrast, women in southern Albania can engage in service-based activities, such as hotel and restaurant services. Rural women are a vulnerable group as they need training in order to become profitable entrepreneurs. Additionally, Albanian agriculture is still not very sophisticated.

Often a lower participation by women in training workshops is witnessed. There is a disparity of unequal access of women versus men farmers with regards to the agricultural extension service. UN Women (2013) shows that females in rural areas, especially those born in the past three decades and without access to agricultural vocational education, were particularly more vulnerable to lack of access to agricultural advisory services.

Women's access to any services is further constrained by improper venues for training, the timing for the meetings, and lack of equal attention paid to men and women in regard to their concerns at the workshops or training. The perception that the head of the household should attend communal meetings on behalf of the household often limits women's attendance. Such beliefs are reinforced when meetings of extension services or other advisory services are held at bars or cafes—spaces often regarded as less than appropriate for women.

Similar results are confirmed by a comparative survey when asking respondents how often they participated in training. Males statistically have higher attendance compared with females regarding trainings or group meetings. This is similar to other surveys and studies (USAID/Albania, 2008; FAO, 2015) showing a very low percentage of women taking part in training.

Another important issue with regard to vocational training is the quality and the topics of the training. The types of program offered to women are often in the low-wage categories (MoYSW, 2014). According to Development Solutions Associates (DSA, 2016), training courses that cater for women to participate are tailoring and handicraft making. Furthermore, training without coaching is not very fruitful in most cases. A suggestion is to provide training and coaching to groups, rather than individuals, in areas of high wage categories such as management and entrepreneurship.

2.5 Comparing Agri-entrepreneurs: Non-conflict Zones, Conflict Zones, and Transitional Economies

Agri-entrepreneurs in non-conflict zones, conflict zones, and transitional economies share similarities and dissimilarities in the way that they operate businesses, as well as challenges that they face. Similarities and dissimilarities amongst the three groups are due to economic circumstances, educational opportunities, geography, conflict and political environments, and other situational factors. Agri-entrepreneurs in all three groups are vulnerable to natural disasters; however, the amount of government aid required to reopen a business, or make up for crop loss, differs among the three groups.

Agri-entrepreneurs in non-conflict zones receive incentives, such as lower taxes and loans to buy technology, to start a business. For those living in conflict zones there are few economic incentives, other than income. It is interesting to note that agri-entrepreneurs in Afghanistan have technology that can be used for business purposes. Transitional countries receive some incentives, but agri-entrepreneurship might not be the highest priority.

In addition, agri-entrepreneurs in all three groups consider access to credit and start-up costs as barriers to market entry. Those in non-conflict zones have limited access to land, which is expensive. Those in other groups may not have usable land nearby. Furthermore, women in developing transitional countries not only have limited access to land, but also face barriers obtaining titles to their land.

Available entrepreneurship education is another difference between the three groups. In non-conflict zones there are many educational opportunities and mentoring programs for entrepreneurs. In conflict zones or transitional economies, few entrepreneurship education or mentoring programs are available. In cases where there are training programs and workshops, the focus is more on production and less on business and entrepreneurship training. Furthermore, trainings are often given by staff who have not been well trained. Women are often left out of the training due to perception of who gets trained and to lack of sensitivity in terms of when and where the training is held. None the less, agri-entrepreneurs in these areas have a need and want to learn about best business practices.

A big difference between agri-entrepreneurs in the three groups is the quality of their supply chain and access to markets. Agri-entrepreneurs in non-conflict zones have sophisticated distribution systems available in addition to market intelligence information on consumer demand. Supply-chain distribution systems are problematic for agri-entrepreneurs in conflict zones and transitional economies. The lack of consumer preference information further hampers the value-chain actors on meeting the desired demand. Lack of standardization and safety certification on food products are a major roadblock, creating a demand for imports. Governments in transitional economies have to invest in institutions to ensure standards and safety in order to compete with imports.

A willingness to innovate is shared among agri-entrepreneurs in non-conflict zones, conflict zones and transitional economies. Agri-entrepreneurs are willing to change the types of products sold and services offered. They recognize the need to specialize in niche value-added products and services that increase business profitability. Above all, agri-entrepreneurs in all groups are resilient, as they consistently have to address changes in their environments.

References

Alexzander, D. (2012) Facilitating economic activity in conflict zones. *Security Strategies Journal* 8(16), 1–20.

Arias, M.A., Ibáñez, A.M. and Zambrano, A. (2013) *Agricultural production amidst conflict: the effects of shocks, uncertainty and governance of non-state armed actors*. Serie Documento CEDE 2014-8. CEDE, Universidad de los Andes, Bogotá.

Benson-Rea, M. and Stringer, C. (2015) Small firm specialization in global value chains: evidence from the cut flower industry. *International Journal of Business and Economics* 14(1), 43–62.

Boehlje, M., Roucan-Kane, M. and Bröring, S. (2011) Future agri-businesses challenges: the strategic uncertainty, innovation and structural change. *International Food and Agribusiness Management Review* 14(5), 53–82.

Chokerah, J. and Horvath, S. (2012) *Investing in agriculture for food security and economic transformation*. UNDP South Sudan, Juba. Available at: http://www.ss.undp.org/content/dam/southsudan/library/Reports/southsudanotherdocuments/Investing%20in%20Agriculture%20for%20Food%20Security%20and%20Economic%20Transformation%20-%20November%202012.pdf (accessed 8 December 2016).

Cusack, J. and Malmstrom, E. (2010) *Afghanistan's willing entrepreneurs. Supporting private-sector growth in the Afghan economy*. Policy brief for Center for New American Security, Washington, DC.

Cvijanović, D., Trandafilović, S. and Imamović, N. (2013) Marketing concept in terms of agricultural enterprises development in transitional countries. *Economics of Agriculture (Ekonomika Poljoprivrede)* 60(1), 113–122.

De Hoe, R. and Janssen, F. (2014) Developing psychological capital to learn and bounce back from entrepreneurial failure. In: *Fourth International Workshop Entrepreneurship, Culture, Finance and Economic Development Proceedings*. Alpen-Adria University, Klagenfurt, Austria.

Drahoňovská, S. (2010) *Women Rights in Conflict Zones: a Focus on India*. Human Rights Law Network, New Delhi. Available at: http://www.hrln.org/hrln/images/stories/pdf/Report%20-%20Women%20Rights%20in%20Conflict%20Zones%20-%20A%20Focus%20on%20India.pdf (accessed 8 December 2016).

Drost, S., de Boer, D. and van Wijk, J. (2013) *Including conflict affected producers in agri-food chains: honey business in Northern Uganda*. IS Academy on Human Security in Fragile States, Wageningen, Netherlands. Available at: C:\Users\PoveyL\AppData\Local\Temp\Occasional_Paper_05_Honey_business_in_northern_Uganda-3.pdf (accessed 8 December 2016).

DSA (2016) *Study on situation of women and economic diversification in rural areas, prepared for UN Women*. Development Solutions Associates, Tirana, Albania.

Euromonitor International (2015) *Cultural diversity and its impact on global consumer markets*. Available at: http://www.euromonitor.com/cultural-diversity-and-its-impact-on-global-consumer-markets/report (accessed 24 March 2016).

FAO (2009) *Global agriculture towards 2050. High-level expert forum*. Food and Agriculture Organization of the United Nations, Rome. Available at: http://www.fao.org/fileadmin/templates/wsfs/docs/Issues_papers/HLEF2050_Global_Agriculture.pdf

FAO (2014) *Deep Roots*. Food and Agriculture Organization of the United Nations, Rome. Available at: www.fao.org/3/a-i3976e.pdf (accessed 24 June 2015).

FAO (2015) Annex 2: Definitions. In: *State of the World's Forests*. Food and Agriculture Organization of the United Nations. Available at: http://www.fao.org/docrep/w4345e/w4345e0l.htm (accessed 24 June 2015).

Fuglie, K. and Sun Ling Wang, S.L. (2012) New evidence points to robust but uneven productivity growth in global agriculture. *Amber Waves* 10(3), 1. Available at: http://baobab.uc3m.es/monet/monnet/IMG/pdf/globalag.pdf (accessed 9 December 2016).

García, G.M. and Calvo, J.C.A. (2013) Psychometric properties of Connor-Davidson Resilience Scale in a Spanish sample of entrepreneurs. *Psicothema* 25(2), 245–251.

Gellynck, X., Cárdenas, J., Pieniak, Z. and Verbeke, W. (2015) Association between innovative entrepreneurial orientation, Absorptive capacity, and farm business performance. *Agribusiness* 31(1), 91–106.

Government of Canada (2013) *Canadian Food: eatCanadian*. Agriculture and Agri-food Canada, Ottawa. Available at: http://www.agr.gc.ca/eng/industry-markets-and-trade/canadian-food-eatcanadian/?id=1392216500650 (accessed 25 June 2015).

IDS (2015) *Conflict and violence*. Conflict and Violence Research Cluster, Institute of Development Studies, Brighton, UK. Available at: http://www.ids.ac.uk/team/conflict-and-violence (accessed 25 June 2015).

IEP (2015) *Global Peace Index*. Institute for Economics and Peace, Sydney, Australia. Available at: http://www.visionofhumanity.org/#/page/our-gpi-findings (accessed 24 June 2015).

IMF (2000) *IMF transition economies: an IMF perspective on progress and prospects*. International Monetary Fund, Washington, DC. Available at: http://www.imf.org/external/np/exr/ib/2000/110300.htm (accessed 25 June 2015).

INSTAT (2014) Business register statistics, available on request. Institute of Statistics, Tirana, Albania.

Jensen, K., Bruch, M., Menard, J. and English, B. (2013) *A snapshot of Tennessee agritourism: 2013 update*. Center for Profitable Agriculture, University of Tennessee Institute of Agriculture, Spring Hill, Tennessee. Available at: https://ag.tennessee.edu/cpa/CPA%20Publications/2013%20Agritourism%20Study%20Final%20Report%206%2020%2013.pdf (accessed 25 March 2015).

Kahan, D. (2012) *Entrepreneurship in Farming*. Food and Agriculture Organization of the United Nations, Rome. Available at: http://www.fao.org/uploads/media/5-EntrepreneurshipInternLores.pdf (accessed 8 December 2016).

MARDWA (2014) *Inter-sectoral Strategy for Agriculture and Rural Development for the period 2014–2020*. Albanian Ministry of Agriculture, Rural Development and Water Administration, Tirana, Albania.

MoYSW (2014) *Strategy for employment and skills 2014–2020*. Ministry of Youth and Social Welfare, Tirana, Albania. Available at: http://www.seecel.hr/UserDocsImages/Documents/EMP-SKILLS-STRATEGY_Albania.pdf (accessed 8 December 2016).

MTC (2015) *Yearly Analysis Report*. Available on request from Mobile Training Center, Tirana, Albania.

Prishchepov, A.V., Müller, D., Dubinin, M., Baumann, M. and Radeloff, V.C. (2013) Determinants of agricultural land abandonment in post-Soviet European Russia. *Land Use Policy* 30(1), 873–884.

Sandu, I.S. and Ryzhenkova, N.E. (2015) Review of market mechanism features of innovation support in agriculture industry in more economically developed countries: strategies for Russian Federation. *Współczesna Gospodarka* 6(1), 13–18.

Speakman, J. and Rysova, A. (2014) *The small entrepreneur in fragile and conflict-affected situations*. Directions in Development, Private Sector Development, World Bank, Washington, DC. Available at: http://dx.doi.org/10.1596/978-1-4648-0018-4 (accessed 25 June 2015).

Statistics Canada (2015) *Agriculture*. Statistics Canada, Ottawa. Available at: http://www5.statcan.gc.ca/subject-sujet/theme-theme.action?pid=920andlang=engandmore=0andMM (accessed 24 June 2015).

UN (2005) *Population challenges and development goals*. United Nations, New York. Available at: http://www.un.org/esa/population/publications/pop_challenges/Population_Challenges.pdf (accessed 24 March 2016).

UN (2014a) *World Investment Report 2014: Investing in the SDGs: An Action Plan*. UNCTAD (UN Conference on Trade and Development), Geneva.

UN (2014b) *World economic situation and prospects, Global Outlook*. Department of Economic and Social Affairs, United Nations, New York. Available at: http://www.un.org/en/development/desa/policy/wesp/wesp_current/2016wesp_full_en.pdf (accessed 8 December 2016).

UN (2015) *Global Issues: Agriculture*. United Nations, New York. Available at: http://www.un.org/en/globalissues/agriculture/ (accessed 25 June 2015).

UN Women (2013) *Action Plan for women entrepreneurship in Albania*. UN Equity for Gender Equality and the Empowerment of Women (UN Women), New York. Available at: http://www.ekonomia.gov.al/files/userfiles/Plani_Veprimit_Ang.pdf (accessed March 2016).

UNDP (2013) *Comparative analysis of best cases for women entrepreneurs in Albania*. Unpublished report prepared for UNDP by Edvin Zhllima.

USAID/Albania (2008) *Albanian Women in Agriculture: Case Studies*. Produced for USAID by Kara Nichols Barrett, Development & Training Services, Inc., under the Greater Access to Trade Expansion (GATE) Project, WID Office IQC contract number GEW-I-00-02-00018-00, Task Order No. 02. USAID, Washington, DC.

Vasile, A. and Ungureanu, A. (2014) The importance of economic structure evolution in achieving performance – from agrarian economy to competitiveness in Romanian economy. *Economics of Agriculture* 61(4), 945–957.

Wadsam (2014) *Agriculture as a Growth Industry in Afghanistan: Challenges and Possibilities*. Wadsam Afghan Business News portal. Available at: http://wadsam.com/afghan-business-news/agriculture-as-a-growth-industry-in-afghanistan-challenges-and-possibilities-232/ (accessed 24 June 2015).

Welsh, D.H.B., Memili, E. and Ochi, M. (2013) Japanese women entrepreneurs: Prospects for growth. In: *Proceedings at USASBE Conference*, San Francisco, California. US Association for Small Business and Entrepreneurship, University of Wisconsin, Whitewater, Wisconsin. Available at: https://libres.uncg.edu/ir/uncg/f/E_Memili_Japanese_2013.pdf (accessed 25 June 2015).

World Bank (2007) *Strategic Policies for a More Competitive Agriculture Sector in Albania*, Report No. AAA18-AL. Tirana, Albania.

World Bank (2011) *Growth and productivity in agriculture and agribusiness: evaluative lessons from World Bank Group experience*. World Bank, Washington, DC. Available at: http://siteresources.worldbank.org/EXTGPAA/Resources/Agribusiness_eval.pdf (accessed 9 December 2016).

World Bank (2015a) *Enabling the Business of Agriculture 2015: Progress Report*. World Bank, Washington, DC. Available at: http://eba.worldbank.org/~/media/WBG/AgriBusiness/Documents/Reports/2015/Enabling-the-Business-of-Agriculture-2015.pdf (accessed 9 December 2016).

World Bank (2015b) *Value-added (annual % growth) Agriculture*. World Bank, Washington, DC. Available at: http://data.worldbank.org/indicator/NV.AGR.TOTL.KD.ZG (accessed 24 June 2015).

Yaseen, A. and Somogyi, S. (2015) *Promoting entrepreneurship in the Pakistan's dairy industry: an empirical testing of two predictive models of entrepreneurial intentions*. Paper presented at 25th International Food and Agribusiness Management Association Congress, Minneapolis, June 10–13, 2015. Available at: www.ifama.org/files/conf/2015%20Conference/1226_paper_Yaseen.pdf (accessed 24 June 2015).

Zhllima, E. (2013) Analysis of successful business cases for women in rural Albania. Internal document prepared by for UNDP.

3 Agri-entrepreneurship Enabling Program Design in Conflict Regions for Youth Development: Best Practices and Lessons Learned

Kathleen Liang[1]* and Tina S. Lee[2]
[1]*North Carolina Agricultural and Technical State University, Greensboro, North Carolina, USA;* [2]*University of Hawai'i at Mānoa, Honolulu, Hawaii*

3.1 Introduction

Recent happenings have revealed significantly growing concerns of economic and social instabilities around the world. Refugees, many children among them, take risks to travel across land and sea to seek new economic opportunities. Most of the information shared by press releases or social media only tells a fraction of stories about issues in conflict regions. The *World Development Report 2011* discussed several issues with respect to unstable state-of-conflict regions (World Bank, 2011). For example, approximately 1.5 billion people live in conflict regions where countries experience repeated cycles of political and criminal violence. Civilians who live in the conflict regions often experience famine and brutal attacks of political crossfire. Youth are particularly vulnerable due to lack of support from a steady system and safe environment to obtain education and training to achieve economic mobility.

Conflicts and instability in an area not only eliminate resources and opportunities for their own people, but also threaten peace and economic growth for the rest of the world. The ripple effects of wars, crimes, and poverty often generate more fragile states for other countries and people, and further muddle judgment and decisions to move forward in seeking peace and economic growth. "To break cycles of insecurity and reduce the risk of their recurrence, national reformers and their international partners need to build the legitimate institutions that can provide a sustained level of citizen security, justice, and jobs" (World Bank, 2011). The same report (World Bank, 2011) also pointed out that "children living in conflict regions are twice as likely to be undernourished, and three times as likely to be out of school." Key sources of societal stress and organized violence have interrupted the learning process of the youth. Tensions among ethnic and religious groups and trafficking networks have lured young people to participate in criminal activities, and misled youth to the wrong path of development.

Healthy and well educated young people are our future for prosperity and growth. Unfortunately many countries have a high unemployment rate among youth. In 2012, nearly 75 million young people between the ages of 15 and 24 were unemployed worldwide

*chyi-lyi.Liang@uvm.edu

 Enabling Agri-entrepreneurship and Innovation: Empirical Evidence and Solutions for Conflict Regions and Transitioning Economies (eds C. Chan *et al.*)

(Council on Hemispheric Affairs, 2015). Globally the youth unemployment rate has been about 12% since 2011, and has been projected to remain at a high level through 2016 (ILO, 2012). It is alarming to researchers, educators and policy makers when young people are discouraged by poor job market performance due to economic recession and societal tensions. It is even more worrisome from the economic development perspective to see so many young people drop out of the school system during economic crisis. Education is supposed to offer knowledge, skills, and experiences for young people to explore opportunities as they develop life and career competency. Sufficient and effective knowledge, skills, and experiences often lead to employment, sustainability, and self-confidence. Unfortunately, given the resource scarcity and lack of stable support from public institutions, children living in conflict regions are not able to access knowledge and training to develop skills. Whether it is due to being discouraged by the system or lack of resources, the more young people drop out schools, the less likely a country is to achieve economic growth and social mobility.

Many have suspected that climate change has some impacts on social, economic, and political instability. Growing seasons and environmental factors (rainfall, temperature, heat index, etc.) have shifted significantly over time. Many countries have experienced drought, flood, and other unprecedented natural disasters that have severely interfered with food production and distribution. Rural villages and communities traditionally relying on agriculture have gradually lost jobs and income sources. When hunger strikes, societal and political order can no longer be maintained.

One way to stop the downward-spiral scenario as well as to inspire and motivate youth in learning (in school or out of school) is through entrepreneurship. Entrepreneurship plays a key role in promoting economic development and societal prosperity globally. Many young people in developed countries aged 18–25 have been offered opportunities or attempted to engage in entrepreneurial activities that would lead to their dream career paths. There are many successful youth entrepreneurship programs, organizations and curricula in the USA, for example. The Network for Teaching Entrepreneurship (NFTE) was established to "provide programs that inspire young people from low-income communities to stay in school, to recognize business opportunities and to plan for successful futures" (https://www.nfte.com). Youth About Business (YAB) is another great example: "Our mission is to expose youth to the business world through our experiential learning model, fostering the development of business literacy and leadership skills necessary to be more successful in school and in life. We teach students about the world of business one merger and acquisition at a time." (http://www.youthaboutbusiness.org/about.php). Finally a Junior Achievement (JA) program supported by the SCORE, a non-profit youth program that fosters entrepreneurship, has been inspiring students for nearly 100 years. "During the most recent school year, JA programs influenced over 4 million students in nearly 200,000 classrooms across the U.S. Its volunteer driven programs reach kids from K-12, fostering not just entrepreneurship, but work readiness and financial literacy skills as well." (https://www.score.org/resources/update-youth-entrepreneurship-program). These youth entrepreneurship programs aim to encourage and guide young people to acquire knowledge, skills, and experiences through training, courses, and internship positions that motivate them to create career opportunities, whether it is working for others or starting their own businesses. This chapter introduces some background and literature of positive youth development, linking to a best practice of utilizing youth entrepreneurship education programs to change lives for young people in conflict regions. The chapter offers an overview of the issues of out-of-school youth in the conflict region where a youth entrepreneurship program has been established in Mindanao, Philippines. A description of the youth entrepreneurship program, UPLOAD JOBS, will be provided to introduce people and program contents. Outcomes and evaluations of the UPLOAD JOBS will be summarized. Finally discussions of the impacts and future development will conclude this chapter.

3.2 Theories Supporting Positive Youth Development and Entrepreneurship Education

Many scholars have presented a systematic framework to investigate community change. Social capital has been identified as one of the most critical resources that have the potential to reverse the downward spiral of loss in jobs, income, and opportunities and hope (Emery and Flora, 2006). Youth groups in conflict regions often experience such downward spirals due to unstable and uncertain social, economic, and political environments. The unemployment rate among youth has been high and growing in many conflict regions around the world. The uncertain and insecure nature of the environment has exposed youth to chronic poverty, famine, and wars that impose high risks, for young people respond with rage, distrust, hopelessness, and violence (Greene, 1993). Many countries have introduced entrepreneurial skills development and educational programs to motivate young people to establish small businesses that will improve their income and welfare (Nafukho, 1998). Other factors to support youth development involve support from family and friends, access to resources and opportunities, and organized and guided learning environments.

There are two schools of theories supporting positive youth development developed by scholars. Earlier theories established in the 1950s focused on interventions and preventions targeting issues or behavior to support families and children by responding to existing crises such as social conflicts, family distraction, education failure, economic immobility, cultural segregation, or opportunity inequality (Catalano *et al.*, 2004). As knowledge and information have grown in scientific studies from analyzing longitudinal data, the second theory of supporting youth development has evolved to incorporate multiple factors, co-occurrence of problems, and environmental predictors related to social, economic, and community interactions in addition to prevention and intervention. Scholars and practitioners have begun to explore and examine successful approaches leading to a new era of positive youth development that promotes bonding, resilience, social and emotional competence, moral and behavioral competence, self-determination, and self-efficacy. This new movement toward positive youth development is predicted to generate positive outcomes in social and economic mobility for families and society (Catalano *et al.*, 2004).

A parallel theory of entrepreneurship education has been developed to focus on positive youth development by improving entrepreneurial mindset, entrepreneurial skills/experiences, and entrepreneurial opportunities (Liang, 2015). Figure 3.1 shows how entrepreneurship education concepts and strategies directly link to positive youth development and community prosperity. Young generations need to be encouraged and guided to dream—dream up ideas that might be new products and services, or new ways to improve existing products and services. Once the ideas are formed and presented, the next steps involve the following.

- **Design** concepts, conduct experiments to test the concepts, and generate research and practical-based information to validate the new concepts—this stage relates to identifying community interests, needs, and opportunities that will drive the success of job creation, as well as creating joint benefits and wealth for individuals, families, and the community. This stage is often led by a group of community leaders, educators, and policy makers who are involved in community development planning and services. Most of the time this group evolves into an advisory board for potential collaborations.
- Prepare a **plan** that leads to feasibility studies and enterprise development—this stage involves intensive actions in identification of resources (people, capital, funding, etc.), market opportunities, management strategies, operational procedures, and financial assessment. The advisory board members will collaborate with entrepreneurs to prepare for the plan of overall community development goals. The entrepreneurs will develop individual

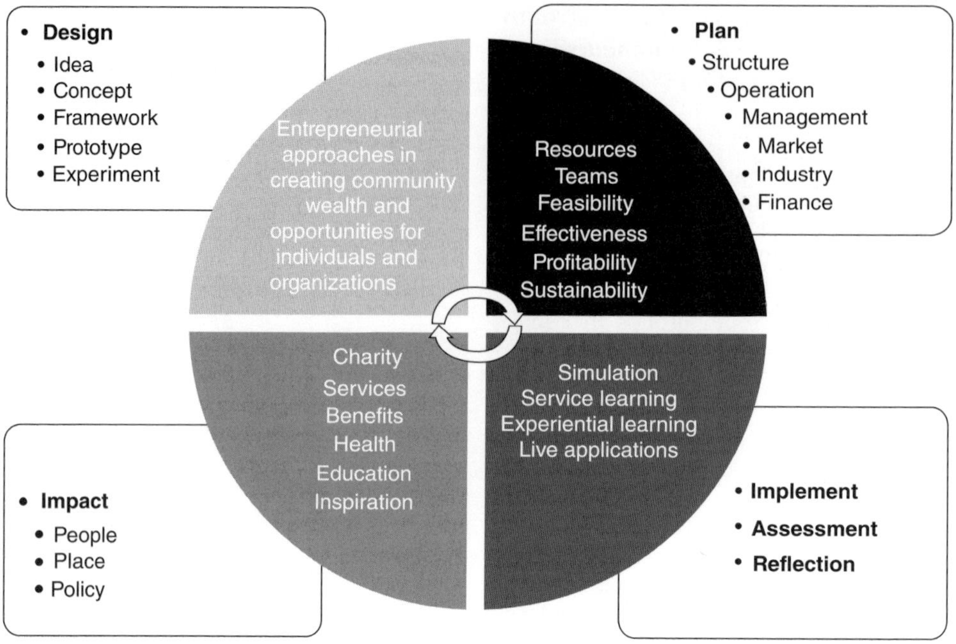

Fig. 3.1. Theory framework linking positive youth development, entrepreneurship, and community development (Liang, 2015).

business plans for enterprise development that fit into the mission of community development in the long term.

- **Implement** direct contacts to apply skills and concepts—it is critical for young people to connect directly with varieties of organizations and partners to get hands-on experiences and to apply knowledge and skills to help others. Simulations, service-learning, and experiential learning are a few popular models designed by the training programs at educational institutions or offered by real enterprises at communities to assist young people to develop self-competency, confidence, and critical thinking when they are solving real problems. **Assessment and reflections** on learning, challenges, and interactions are expected to enhance young people's ability to become proficient in decision making.
- Generate positive **impacts** of learning for youth—an interactive and well balanced curriculum will serve the greater interests and need for communities and countries through a healthy, knowledgeable, and competitive labor force.

This framework has been successfully implemented to design, create, and sustain one of the best entrepreneurship programs in the USA: Dollar Enterprise (*Inc. Magazine*, 2014). Introduced to the Community Entrepreneurship program at the University of Vermont in 2005, Dollar Enterprise has offered more than 3000 young learners the opportunity to experience entrepreneurial mindset, critical thinking, and new venture creation process with $1 seed money for each individual. The idea of Dollar Enterprise is to motivate learners to create products following opportunity and market orientation, and actually make prototypes that will be tested with market experiments ("Design" stage). At the "Design" stage, individuals sharing similar ideas or interests work in teams to reach consensus of goals, objectives, and mission of their business.

After understanding what would be the most popular products for the market, young learners work together as teams to prepare a business plan that will describe their business structure, operation style, leadership positions, management of workers and inventory, market positions, advertising and promotion strategies, competition and responses, pricing and cost structures, and risk and contingency planning ("Planning" stage). In the "Planning" stage, team members need to work with limited resources such as using recycled materials or other available materials/ingredients to create their products, evaluate feasibility and effectiveness of their production and management, and identify the most profitable and sustainable practices to run their business. The ultimate purpose of Dollar Enterprise is to reduce waste and up-cycle waste materials to become new value-added products.

Once teams go through the "Design" and "Planning" stages, they move on to real enterprise operations for 4 weeks during one semester, and work with local community partners to deliver benefits ("Implement, Assessment, and Reflection" stage). Team members hold weekly meetings to conduct peer reviews, team achievement assessment, goals and objectives review, and setting reasonable expectations and strategies to improve. For all Dollar Enterprise teams, they need to donate all proceeds to local charity organizations at the end of 4-week operations. Each team also needs to contribute service-learning hours for charity organizations by participating in activities, events, festivals, workshops, trainings, or other types of services that benefit more clients. Dollar Enterprise activities aim to generate positive social impacts in addition to creating new value-added products from the economic point of view. Dollar Enterprise offers young people an opportunity to be trained by a new entrepreneurship paradigm with a general understanding of joint benefit and value added between individuals and charity missions. Successful entrepreneurs are often the most generous contributors in our society. The Dollar Enterprise program has donated more than $60,000 and over 30,000 service-learning hours to support more than 400 charity organizations since 2005 ("Impact" stage). Some of the most successful Dollar Enterprise examples include:

- using recycled bike parts to create jewelry and decorative accessories;
- growing own vegetables in gardens, and make fresh pastries, pies, or pickled products;
- using recycled wrapping papers, magazines, and newspaper to create containers, jewelry, vase, or shopping bags;
- re-designing and sewing old clothes to become new outfits, handbags, or book covers; and
- creating gluten-free or vegan products by transforming conventional recipes and using non-conventional ingredients (e.g. using banana or other fresh fruits to replace sugar, eggs, and milk).

3.3 Methods for Designing and Implementing an Entrepreneurship Program

Creating and implementing a successful entrepreneurship program is a complex process. It is particularly challenging to establish a system to introduce, promote, and sustain entrepreneurial activities in conflict regions where people and organizations constantly encounter violence and wars. Scholars agree that it is essential to facilitate individual learning in an entrepreneurship program, and to offer opportunities for learners to be proactively engaged in entrepreneurial thinking and activities. Entrepreneurial knowledge and skills include multidisciplinary training plus field work. In the USA, only one entrepreneurship course was offered in 1955 (Vesper, 1982). Currently over 1200 courses and programs are offered in American universities (Solomon, 2007). Given a growing number of courses offered in entrepreneurship, scholars have developed a rich body of research on effective methods of instruction. A UK-based International Entrepreneurship Educator's Programme (IEEP) was one of the most recognized systems in creating entrepreneurship education, and it recommends leading individuals through a series of problem-solving training through different stages of finding and evaluating opportunities, designing an implementation plan, negotiating for resources, creating products and services, and evaluation of success (Box 3.1) (Gibb, 2011).

Box 3.1. Learning goals of entrepreneurship education programs (Gibb, 2011).

- To glean information personally from any and everywhere, and to weigh it,
- To learn while and through doing,
- To learn by solving problems and learning from failure,
- To develop "gut feel" decision making with limited information,
- To develop the most appropriate solution under pressure,
- To recognize the widely varied goals of others,
- To evaluate by judgment of people and events through direct feedback.

Many young people in conflict regions probably have never been exposed to any entrepreneurial concepts at all. Being a successful entrepreneur is beyond the completion of courses and training sessions. There are certain stages of learning that need to occur for young people to be inspired and guided to become entrepreneurs. The learning stages are described below and the methods are described in Table 3.1.

1. Help them recognize opportunities.
2. Help them acquire the necessary knowledge and skills such as business processes, management, operations, production, marketing, networking, and financial analysis.
3. Help them understand the risks and consequences (both positive and negative outcomes), and how they can deal with negative outcomes and generate more positive learning experiences.
4. Support them to access infrastructure and resources.
5. Establish mentor–protégé system for long-term development.
6. Empower youth to be accountable and responsible for their own decisions, to make their own decisions, to take control and to become independent financially.

3.4 Example of Positive Youth Development and Agri-entrepreneurship Education: UPLOAD JOBS for Mindanao

The theory framework described in Fig. 3.1 can be directly applied to promote and motivate positive youth learning in a conflict region. Given the resource limitations and opportunity constraints, young people in conflict regions would not be exposed to innovative, flexible and unconventional learning environments. However, it is very possible to design and deliver effective programs that will encourage and stimulate entrepreneurial thinking and activities beyond traditional classroom learning. Using the theory framework described in Fig. 3.1, educators and trainers in a conflict region need to be creative and willing to modify pedagogy. This section introduces an example, UPLOAD JOBS for Mindanao, Philippines, to demonstrate how entrepreneurship concepts can encourage and stimulate positive youth development in a conflict region (Fig. 3.2).

Table 3.1. Methods of entrepreneurship education (from Garavan and O'Cinneide, 1994; Solomon, 2007).

Formal instruction	Hands-on experience
Readings	Experiments
Experienced guest speaker lectures	Workshops
Analysis papers	Field projects
Computer simulations	Student businesses start-ups
Discussions	Field trips
Role plays	Management
Case studies	Coaching
Learning diaries	Consultations with entrepreneurs
Venture planning	–

3.4.1 Background for study region

Mindanao, Philippines, the study region (Fig. 3.3), is the second largest island in the Philippines and is highly dependent on the agricultural industry. Davao City is the largest city in Mindanao. According to the 2010 census, the Mindanao island group has a population of approximately 22 million. Offering major crops such as tropical

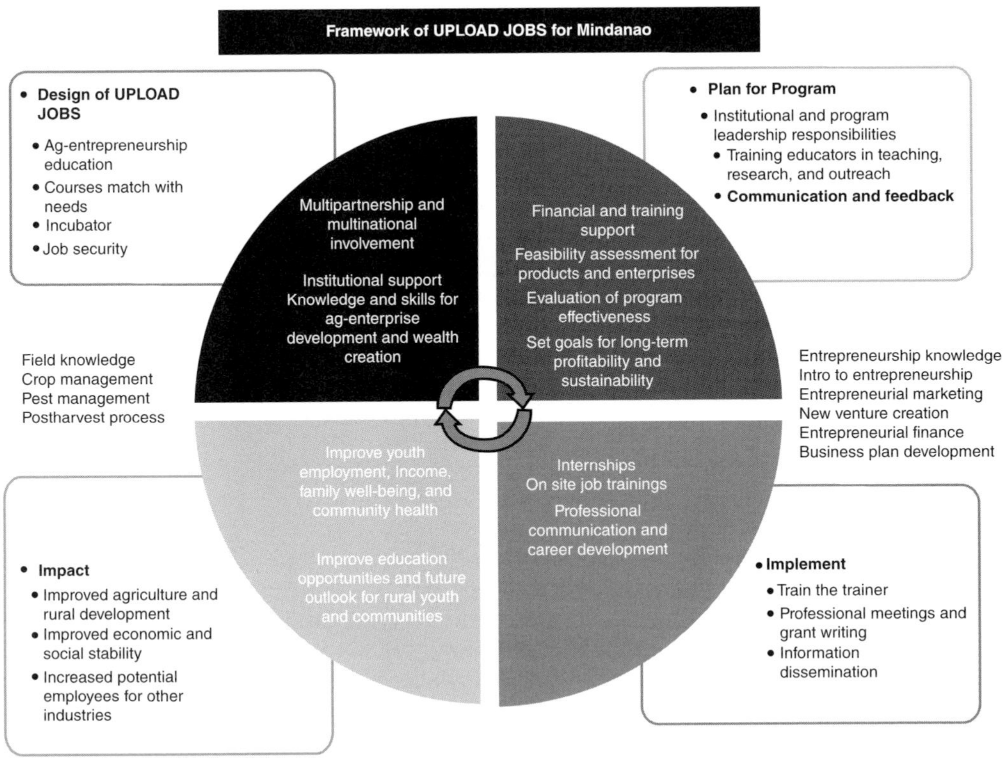

Fig. 3.2. A conceptual framework describing the UPLOAD JOBS program.

fruits, vegetables, sugarcane, coconut, coffee, freshwater fish, and livestock, Mindanao has been well known as the Food Basket supporting the country's farming landscape. Central Mindanao is a rural area where producers have limited access to capital, information, education, training, and other resources. On top of limitations in accessing resources and opportunities, Central Mindanao also faces a constant state of conflict, which is deeply rooted in the history of conflicts between early settlers of Christian and Catholic migration to predominantly Muslim sultanates of Maguindanao and Sulu (Chan and Neyra, 2015). More violence and crimes have been further triggered by modern-day clans that have smuggled illegal items such as weapons to threaten peace and livelihood of local communities. Young people living in Central Mindanao become vulnerable and fragile after being exposed to economic and social distractions. Many young people have been forced or involuntarily pulled away from normal education systems, because they must share the responsibilities to earn income for their families.

According to the US Agency for International Development (USAID, 2007), the youth unemployment rate was 42% in the Autonomous Region of Muslim Mindanao, compared with the national average of 6%. "In Central Mindanao, only 52 percent of its young adults age 15–24 register to secondary schools, and the average Filipino farmer only reaches the 5th grade" (Chan and Neyra, 2015). The Global Entrepreneurship Monitor (GEM) 2014 global report (Singer *et al.*, 2015) has collected some evidence on employment and new venture creation activities in the Philippines, identified by the GEM as a country in transition to efficiency-driven economies. Many Filipinos who participated in the GEM surveys believed that entrepreneurship was a good career choice, and individuals who were entrepreneurial earned high status in the society. Survey

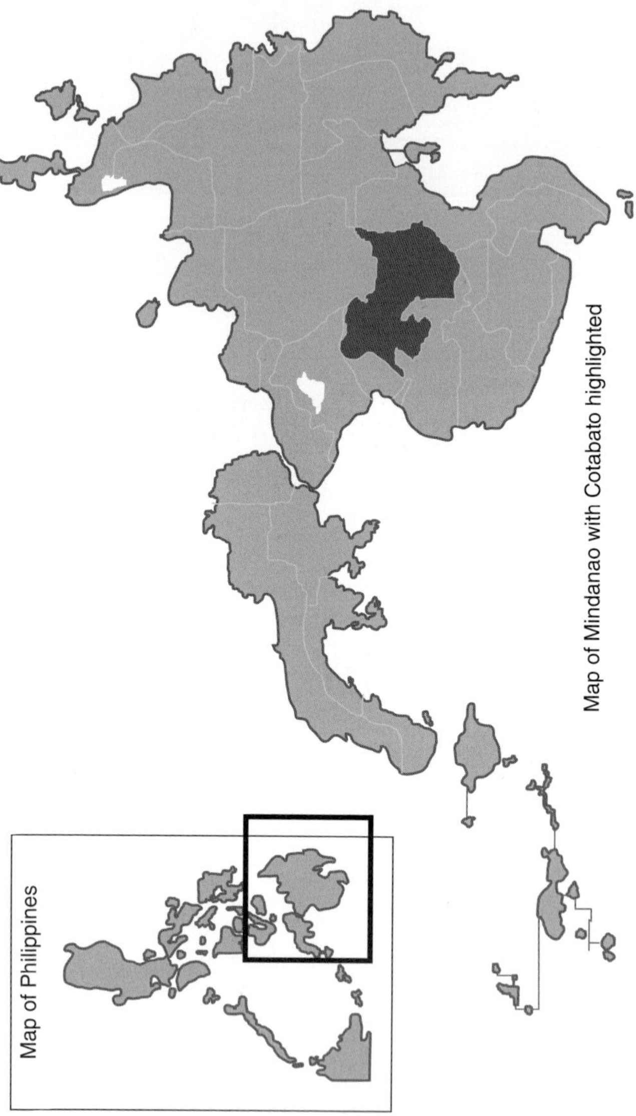

Fig. 3.3. Map of Mindanao, Philippines with the Province of North Cotabato highlighted.

results also showed a high percentage of Filipinos had individual attributes toward entrepreneurship, and many were willing to create, or were in the early stage of creating, their own ventures. Furthermore, many survey participants in early-stage entrepreneurial activities were driven by opportunities and seeking improvement for their future.

Some institutions established in the Philippines have attempted to address the issues of youth education and unemployment. For example, the "General Education Curriculum: Holistic Understandings, Intellectual and Civic Competencies" policy was introduced and implemented in 2013 to improve competency in social and personal skills beyond traditional math, science, English, and history. Entrepreneurship education has been established in higher education system in the Philippines since the early 1980s. Several institutions have offered bachelor's degrees and master degrees in entrepreneurship. Unfortunately people, particularly the youth, in rural and excluded regions have very limited access to entrepreneurial training. There has been a growing need and interest for more institutions to provide technical training programs and certificates to out-of-school youth to improve their skills in job search. Both public and private organizations understand the importance of nurturing and supporting young people in acquiring sufficient knowledge and skills that will alleviate future poverty and unemployment concerns. Many urge the establishment of programs specifically focused on agri-entrepreneurial training for out-of-school youth, hoping to stimulate learning and practices for young people who are seeking business opportunities in the agricultural industry.

3.5 Utilizing a Conceptual Framework to Describe the UPLOAD JOBS for Mindanao Project

3.5.1 Design of UPLOAD JOBS for Mindanao

UPLOAD JOBS is the acronym for *U*niversity *P*artnership *L*inking *O*ut-of-school youth to *A*gri-entrepreneurship and *D*evelopment to promote *J*ob *O*pportunities and *B*usiness *S*cale-up for Mindanao. It is a partnership between the University of Hawai'i at Mānoa (UHM) in the USA and Southern Christian College (SCC) in the Philippines. The project was funded by the United States Agency for International Development – Philippines Mission (USAID Philippines), and administered by Higher Education for Development (HED) of the American Council for Education. The project leaders include faculty and staff members from UHM College of Tropical Agriculture and Human Resources/Natural Resources and Environmental Management Department, and faculty and staff members from the SCC Community Education, Extension, and Research Administration.

Project design was driven in large part by indicators within USAID Philippines Mission's Development Indicator #2 – Peace and Stability in Mindanao. The project was initiated in 2012 to respond to the need for (i) increased informal and extension education opportunities in agriculture and entrepreneurship targeted towards farmers and youth in conflict regions, and (ii) institutional and human capacity building to provide agri-entrepreneurship training programs in conflict regions. The goals of this project include: (i) to improve quality of life and employment for rural youth population in conflict regions; (ii) to provide out-of-school youth with workforce skills that would enhance their capacity to become more productive in the labor force; and (iii) to raise agricultural productivity and employment opportunities for youth population in conflict regions.

The original project design was submitted as a proposal for funding to Higher Education for Development; upon selection of the proposal, HED, UHM faculty, and SCC faculty developed indicators of success. A list of indicators was identified and agreed upon by all project leaders and stakeholders, such as number of faculty and staff trained for SCC, a number of out-of-school youth trained in Mindanao, design and creation of a Center for Agriculture and Farming Entrepreneurs (CAFE), and degrees of satisfaction with the training programs among all project participants.

3.5.2 Plan for program

UPLOAD JOBS training programs followed a stepwise agri-entrepreneurship training design that would: (i) develop a dynamic and modern program by considering regional and local contexts; (ii) monitor and evaluate the progress of the training program as well as examining effectiveness for all participants; and (iii) adapt training programs as needed without compromising the overall desired impacts. UPLOAD JOBS was designed to be a series of short-term extension courses that educate, assist, and support out-of-school youth to become micro agricultural-based entrepreneurs.

Understand regional training needs

To understand the needs and existing challenges in Mindanao region, project leaders conducted a baseline assessment with out-of-school youth, faculty, and businesses. Among 30 out-of-school youth interviewed, they revealed interests in acquiring knowledge and skills with respect to crop management, pest management, post-harvest processes, and supply-value chain. Forty-two faculty members were interviewed in SCC, and they responded positively to receiving training related to entrepreneurship concepts and practices, food processing, market research, and operations and strategies. Sixteen of the local businesses interviewed were seeking positive support for out-of-school youth, skilled workers for all purposes, and recommended certifications upon completing the training courses.

The project leaders reviewed 32 entrepreneurship certificate programs in the USA, and discovered that: (i) average number of courses offered was five; (ii) average credit hours offered were 15; and (iii) total number of course hours on average was 217. Some of the most commonly offered courses through these entrepreneurship certificate programs were Introduction to Entrepreneurship, Entrepreneurial Marketing, Entrepreneurial Finance, Business Plan and Model Development, and Entrepreneurial Venture Creation.

Course design

Using information gathered from focus groups and program reviews, the project leaders determined a general approach to include specific courses in the UPLOAD JOBS program that would align baseline assessments and curriculum contents for local youth (Fig. 3.4):

- for Entrepreneurial Concepts and Training: Introduction to Entrepreneurship, Entrepreneurial Marketing, New Venture Creation (how to start a new business), Entrepreneurial Finance, and Business Plan Development; and
- for Agricultural Concepts and Training: Crop Management, Post-Harvest Process, and Pest Management.

Each individual (youth) who successfully participates and completes all courses will receive a certificate.

Recruit trainers

The content of the training program was created at first by US-based faculty experts in the fields of agriculture, agricultural economics, marketing, and entrepreneurship, and a team of local faculty from SCC helped make the program relevant to the local context. UPLOAD JOBS used a teach–coach–mentor framework to deliver the training and for capacity building:

- **Stage 1:** UHM faculty offered training courses and introduced training materials to out-of-school youth. SCC faculty worked alongside the University of Hawai'i faculty to assist in translation and technical support. While UHM faculty delivered training courses, SCC faculty members observed and obtained skills in assisting training. This offered an opportunity for SCC faculty to participate, to be familiar with training courses and materials, and to be trained as instructors.
- **Stage 2:** After participating in training sessions, SCC faculty served as primary trainers to work with the out-of-school youth directly. UHM faculty observed and supported the training programs as secondary trainers. This stage also included new SCC faculty members and selected out-of-school youth who had successfully completed the training programs in stage 1 to serve as secondary trainers in this stage. This would build continuous effort and

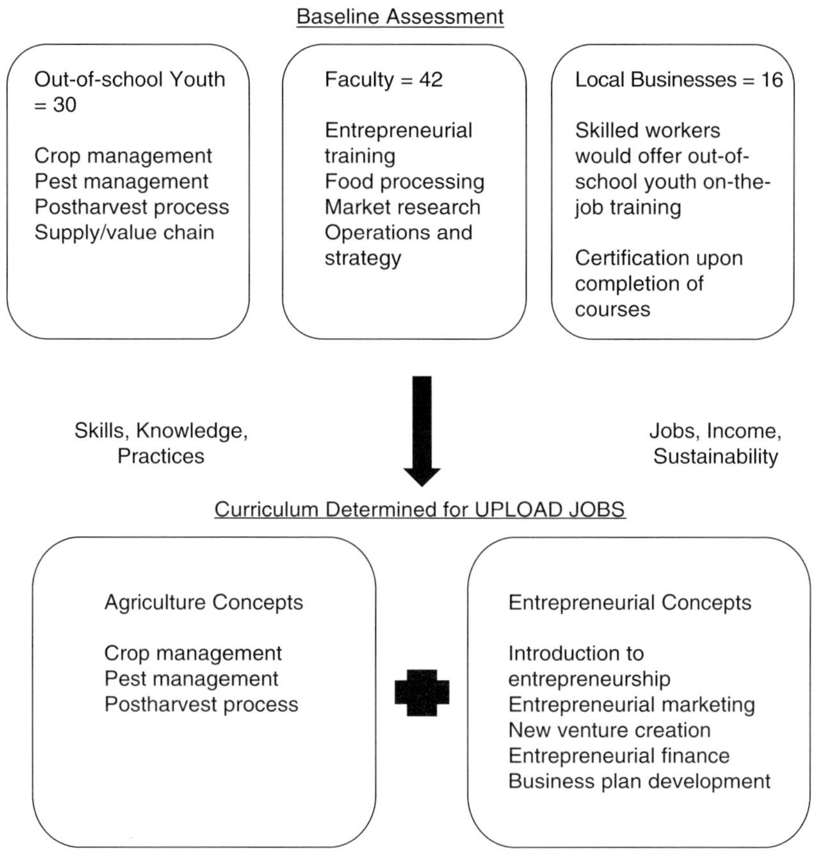

Fig. 3.4. Aligning an on-the-ground baseline assessment with existing entrepreneurship curriculum to design an Agri-entrepreneurship Training Program for a conflict region.

capacity for local participants to grow the program in the future.

- **Stage 3:** Once local trainers were established among SCC faculty and selected out-of-school youth leaders, the UHM faculty became mentors of the UPLOAD JOBS programs by offering consultation services and contributing new ideas and new training concepts. UHM faculty also tested and monitored the training materials after translation to ensure the quality and consistency of the training concepts across translations.

In the long term, it would be ideal for SCC faculty members working with youth leaders to offer continuous training curriculum in their own communities to build social capital and capacity.

Staffing

The project was supported by a small staff team on both sides of the leading partnership institutions. At SCC, there was a staff team of four full-time employees and one to two student helpers. SCC employed a Project Coordinator, Communications Manager, Center for Agriculture and Farming Entrepreneurship Director, and Project Director. The UHM project team had a similar composition with a full-time Project Coordinator, a part-time Communications Manager, a Project Director, and one to two graduate assistants. The UHM Project Director was the Principal Investigator for the project.

Community outreach and marketing

The project partners designed and organized annual colloquium and annual trade shows between 2014 and 2015 for UPLOAD JOBS program participants. These events promoted program accomplishment, introduced new products created by agri-entrepreneurship program attendees, recruited more young people to attend training courses, and recruited more organizations to support young people to participate in future training. Young people used these events to sell their products, and to receive feedback on funding, business development, and product design.

Continue program evaluation and make necessary modifications

The UHM and SCC teams met regularly to have open discussions about challenges and reflections from the youth, their communities, and within each partnership team's institutions. In addition, the staff designed and conducted surveys to gather information from out-of-school youth before and after attending training courses to evaluate whether the training program was fitting their needs. At the end of the project, a professional consulting firm, Evaluation Research Services, LLC, was contracted to evaluate the project outcomes for UPLOAD JOBS between July 2012 and July 2015. Qualitative information and quantitative data were collected through personal interviews, case studies, records of training participants, and reviewing documentation and reports through project participants.

3.5.3 Implement

Throughout the implementation of the project, the project leads faced a wide range of circumstances that necessitated both UHM and SCC to continuously adapt program design to fit project goals. Because the project was implemented in a conflict region and project responsibilities were shared between diverse partners, actual results from the design framework were unpredictable. Maintaining flexibility when starting and implementing the training program was important, and throughout the 3-year implementation period of the project, the program continued to adapt based on evaluation, assessment, and reflection (Table 3.2).

3.5.4 Impact

The primary objectives of UPLOAD JOBS were: (i) to improve institutional and human capacity at Southern Christian College in the area of agri-entrepreneurship; and (ii) to improve the livelihoods of out-of-school youth and others in conflict regions through improved entrepreneurship farmland practices.

Several key achievements were summarized as follows.

1. Created a Center for Agriculture and Farmland Entrepreneurship (CAFE) to support education and training programs in Mindanao

Prior to establishing this partnership project, there had been very limited entrepreneurship education and training in Mindanao. Community stakeholders suggested the SCC serve as a center for advancing entrepreneurship and providing resources and new information to communities. The project team began working on designing a framework for CAFE as soon as the project was funded, establishing a charter and a board of directors to identify responsibilities and duties of individuals and organizations involved in CAFE. An advisory board was formed that included a Bangsamoro representative, out-of-school youth representative, NGO representative, banking sector representative, higher education representative, and government representative. Advisory board members assisted CAFE to develop by creating networking capacity, providing advice and guidance, and representing CAFE publicly. Many advisory board members also contributed to planning and organizing events, programs, and other activities.

2. Trained new environmental educators and created a training program in a conflict region

UHM faculty/experts and SCC partners collaboratively designed and developed these materials for advancing knowledge and skills in agricultural entrepreneurship. A total of 17 short-term courses, modules, and workshops were developed and delivered between 2013 and 2015. These courses were introduced to out-of-school youth. Additional courses were designed, based on community partners' feedback, and these courses included sales training, leadership training, natural farming, value-added production, and pest management.

Table 3.2. Adaptations to the UPLOAD JOBS for Mindanao project in response to unpredictable challenges faced during the implementation of the project.

Assessment and reflection	Strategy implemented
Out-of-school youth were encouraged to create an enterprise that filled a niche or demand in their communities, but the actual market and demand for products were not well understood by both the youth and trainers.	University of Hawai'i faculty and staff offered research and professional development opportunities for Southern Christian College faculty beyond agri-entrepreneurship training courses. UHM faculty prepared and designed research skill training for SCC faculty to conduct market research for products developed by the out-of-school youth groups through the entrepreneurship education program. The results of nine market research projects had been presented in national and international conferences. This outcome was not described in the original project objectives. However, it was important to develop and enhance research, outreach, and teaching capacity for SCC faculty and staff, if they were going to participate in future grant writing and program design opportunities.
Community stakeholders identified a need for a sustained long-term entrepreneurship support for out-of-school youth and other community members.	The project teams conceptualized the Center for Agriculture and Farmland Entrepreneurship (CAFE) to serve as a hub for entrepreneurship training and resources in Central Mindanao. One key component of CAFE was to host a Community Kitchen for out-of-school youth and community members to utilize shared equipment and resources specifically for food production.
Conceptual entrepreneurship training was difficult for out-of-school youth to apply to their day-to-day lives.	The project teams implemented a Business Plan Competition for each cohort. At the end of each training session, youth participants needed to complete a business plan, and they would receive funding to actually start a business. CAFE supported each student group with networking opportunities, guidance, and technical/enterprise development. The goal for CAFE was to further improve the business plan developed in the training sessions to build a more realistic, robust enterprise in the real world. Twenty-five business plans were completed and improved between 2013 and 2015.
Even after learning entrepreneurship skills, trainees lacked the confidence and self-awareness to lead their own enterprises.	Although not identified in the original 2012 literature review of entrepreneurship training programs, the project team added Sales Training and Leadership Training for youth entrepreneurs. Both trainings focused on role-playing sales and team management situations to build confidence and enhance their ability to communicate.
Length of training (9 days) was disruptive to out-of-school youth family obligations, and many youth went home half way through.	The project team divided the training into two parts: one 5-day segment, and one 4-day segment. Between each segment, the youth had 1 month to complete their business plans and start working on their businesses.
SCC Project Team and UHM Project Teams had different work styles and capacity that resulted in misunderstandings and inefficiency. Due to existing hierarchical relationships at universities, SCC faculty and staff did not feel they could speak openly to project supervisors. Conflicts started brewing on payment/salary-related issues.	UHM and SCC Project Staff implemented a communications protocol that included scheduled monthly Skype calls and communication through e-mail. Project teams developed close friendly relationships, which assisted with communication challenges. At the end of the project, UHM invited a leadership trainer to conduct a DiSC analysis (a professional management style analysis) to help break down barriers of communication among the project team and SCC university leadership.

Eleven faculty members and extension staff from SCC were trained to be trainers. These faculty members were primarily teaching faculty members in SCC, and the administrators had not required or expected their faculty members to attend training sessions or become trainers for out-of-school youth beyond their teaching responsibilities. Based on the project structure and operation agreements between UHM and SCC, faculty and staff in SCC voluntarily participated in training to become trainers. At the beginning of the project, SCC faculty and staff participated in training sessions delivered by UHM faculty to assist in translation materials while observing the training process (Cohort 1). In the next step, SCC faculty and staff delivered training courses to out-of-school youth, while UHM faculty coached in the training process (Cohort 2). Some out-of-school youth who completed the full set of training courses successfully were recruited to become trainers. Eventually SCC faculty and staff worked with selected youth leaders to deliver training sessions to others, and UHM faculty served as mentors in the program (Cohort 3).

In total, 202 out-of-school youth were trained during the project time period between 2012 and 2015. Approximately 52% of these young people received an agri-entrepreneurship certificate. For non-out-of-school youth, a total of 783 individuals participated in a variety of partnership activities between 2012 and 2015 and seven out-of-school youth became trainers and leaders within the UPLOAD JOBS program.

3. Increased the confidence and competency for out-of-school youth

New partnerships with community organizations such as banks, businesses, government agencies, and other universities offered youth who had completed agri-entrepreneurship training programs more opportunities to use their knowledge and skills. Some local organizations provided on-the-job training programs to these young people, while others provided employment opportunities directly.

3.6 Best Practices and Lessons Learned

3.6.1 Best practices

Evaluation and assessment were conducted using two procedures: (i) the project team periodically gathered learning experiences and feedback from out-of-school youth who participated in agri-entrepreneurship training; and (ii) external evaluation was conducted by consultants at the end of the project in 2015. Information gathered from both procedures revealed mostly positive feedback and experiences with project partners, workshop participants, and out-of-school youth, as follows.

1. This project directly offered opportunities for faculty and staff in institutions in different countries to collaborate, recognizing that both institutions were building capacity through the partnership. As an institution focused on teaching students, faculty and staff in SCC had little experience in or programs to teach agri-entrepreneurship to out-of-school youth. Experts and the faculty from UHM had experience in project management, entrepreneurship education, and international development, but lacked local knowledge of culture and socio-economic conditions on-the-ground that impacted entrepreneurship education. Through the partnership, both partners were able to learn from each other to significantly improve and expand education programs and professional development for faculty, staff, and youth in local communities.

Strong, respectful working relationships were helpful for two diverse partner institutions to implement a project together. For faculty and staff representing UHM, they accomplished a great deal through strong leadership and outcome-driven objectives. They maintained a positive attitude in coordinating with partners in Mindanao regardless of security issues, weather concerns, and other challenges.

Implementing a new training program helped SCC build and strengthen alliances with local businesses, regional corporations, other educational institutions, and non-profit organizations. SCC was also already a partner that had strong existing alliances in the region. The project introduced a creative public–private partnership with many local organizations such as Technical Skills and Development Authority, Agricultural Training Institute, Department of Trade and Industry, and other municipal offices. Partnering with institutions that worked more with out-of-school youth and community members helped the project leverage existing resources and programs for out-of-school youth instead of inventing a new, stand-alone training program.

2. The guided train-the-trainer framework evolved into a mentor–protégé system that will continue to support local people and community organizations to expand agri-entrepreneurship education across region.

3.6.2 Challenges and barriers

It has not been an easy task to establish and implement a new program in a conflict region. UPLOAD JOBS targeted out-of-school youth in an area that historically endured poverty, economic depression, cultural conflict, and political turmoil. It would be expected for project team members to encounter rejection, doubt, and distrust among local people and organizations. Some of the challenges and barriers revealed by project participants included the following.

1. It was difficult to recruit, retain and require out-of-school youth to participate and complete training sessions. Many young people, aged 18–25 living in rural communities, were already married and needed to attend to family issues. Many youth participated in training, started businesses, and quit after a few months. There was lack of support and understanding from their family members. There was also insufficient infrastructure and resources to support youth to continue and expand their business operations.

2. Training time period could be too short for youth to learn. A training course was often delivered in 2 days, while many concepts like marketing and financial analysis took one semester for youth to learn. There were serious concerns with respect to learning efficiency, knowledge retention, and ability to practice in the real world.

3. It was challenging to sustain consistent and appropriate communication among project partners. It is common for international collaborators to experience communication gaps due to time differences, language barriers, cultural dissimilarity and technology incompatibility. The project partners often relied on Skype and email to communicate. Field visits and face-to-face meetings were arranged to deliver training materials, offer support, and exchange information. While most of the communication methods focused on sharing and exchange of information, personality and communication styles are as important while building international relationships. The UHM team utilized a personality assessment tool to understand team members' preferences and styles to improve teamwork and communication. For SCC faculty and staff, most of the communication between team members occurred at local level. SCC faculty serving as UPLOAD JOBS trainers only communicated with local members, and did not directly communicate with UHM project leaders. There was confusion and misunderstanding regarding tasks and responsibilities, decision making, and budget management.

4. The administrators of SCC needed to continuously and consistently offer support for faculty and staff to engage in project activities. The training activities, research, and professional development through participation in national and international conferences were not fully included in roles and responsibilities for SCC faculty and staff in their daily work.

5. There were few trainers involved with training activities. The number of faculty and staff attending training sessions declined over the project time period due to safety and security concerns. Regional and local volatile environments disrupted learning and training for many project participants. Not all youth and adults were able or willing to travel to training sites to continue and complete the training.

6. There was a lack of balance in gender and educational background among youth participants. The project team encountered challenges to recruit a balanced diversified out-of-school youth to participate in UPLOAD JOBS programs due to cultural factors such as religion and ethnicity. Girls and women were not encouraged to participate in the training, and they had very limited access to education in general. The project encouraged more female participants via a quota system of at least 40% female to male ratio. During training periods, many youth participants did not have the same or similar education background prior to attending training sessions but most have some high school and beyond education level. Some participants completed elementary education (8%). Some completed high school diploma (71%). A few completed some vocational school or college education (21%). The final participants were selected based on pre-training tests on their entrepreneurial potential.

7. It was difficult to document and trace real economic and social impacts of UPLOAD JOBS among youth in the long term. Entrepreneurship education is very different from traditional disciplines such as math, science, literature, or business. Entrepreneurship is an integrated field combining the development of the individual's mindset, ability, knowledge, skill, and application. The project partners agreed that they had achieved significant milestones compared with the project outcome-driven objectives. However, it is still early to conclude that UPLOAD JOBS participants are indeed able to generate higher income or obtain better employment opportunities in the long term, though some data has been collected to indicate participants' average income has improved and more participants are active in the workplace or have returned to school. Moreover, the quality of training materials and delivery methods also need to be monitored and modified over time to fulfill true needs of communities. Existing trainers were given additional training through attending workshops and conferences to enhance their ability to motivate themselves to improve the quality of their course delivery.

Limitations of the continuous assessment and evaluation included: short time frame for data collection, using secondary information to approximate reality, and limited on-site interviews with all participants and collaborators; information gathered only reflecting the few individuals who were willing and able to attend evaluation sessions; language barriers; and potentially inconsistent and biased statements offered by a few individuals. These are the realities of working in conflict regions where people are mobile because of situations and necessities and it is a challenge to conduct assessments and evaluations comprehensively.

3.7 Conclusions and Implications

There should be three levels of entrepreneurship training for any institution to create and implement a successful entrepreneurship program.

1. Create entrepreneurial educators

Educators receive degrees in traditional disciplines such as math, history, music, or economics. We rarely explore or are encouraged to explore new opportunities to be innovative or creative in developing curricula. Since entrepreneurship requires interdisciplinary and multidisciplinary thinking and design, most educators have never been exposed to situations where teaching becomes a risky event when teachers and learners need to deal with new opportunities and uncertainty. In reality, entrepreneurship exists in all fields and disciplines if only we have opportunities, resources and support to discover new paradigms and processes to develop new knowledge and experiences for all learners through collaboration.

2. Create entrepreneurial curriculum

A successful entrepreneurship curriculum needs to accomplish three goals: (i) to deliver knowledge and skills; (ii) to nurture mindset, characteristics, heart, and spirit; and (iii) to engage in actions, decision making, and reflections on success and failure.

3. Create entrepreneurial environment

It is essential to create and support an entrepreneurial environment where educators, learners, and community organizations working together form a strong alliance in pursuing entrepreneurial endeavors. The best strategy is to keep this system informative, interactive, inexpensive, and integrated into various development policies to achieve positive outcomes and benefits.

The UPLOAD JOBS program offers a good example to demonstrate the procedures and experiences in creating an agri-entrepreneurship program in conflict regions. Other institutions and organizations can benefit from this example while establishing similar entrepreneurship programs to motivate and empower young people. In conclusion, several key elements must be in place.

1. To begin with building a transparent, collaborative, consistent and trusting relationship between partners. All partners must share institutional rules and policies prior to collaboration: management, operations, financial compensation, payment plan, workload, expected outcomes (measurable, reasonable, time commitment with or without other obligations), leadership, decision-making structure, and information exchange protocols.

2. To create an effective communication venue that is culturally appropriate and professionally respectful with full clarity. All partners must design and agree on a series of acceptable communication methods and with a professional manner. All partners need to be honest and respectful to cultural differences, using language and terms that are acceptable to all participants.

3. To establish effective strategies in decision making to encourage all participants and program leaders to support each other. There are different levels of decision making among collaborators. Project or program outcomes need to be discussed and explained clearly when recruiting project collaborators. Board of Directors and Advisory Board members must generate a meaningful and mutually agreed system to support program leaders. Program leaders need to share information openly with all project participants, and reach agreeable working relationships to clarify who would make what decisions based on a common ground. Management, operation, and financial decisions may be shared across project leaders. However, there has to be a transparent procedure to share information.

4. To design outcome-driven objectives which are timely, measurable, reasonable, realistic, and achievable. Cultural factors could significantly influence participation and outcomes. When defining objectives, the project leaders need to accommodate local culture and environment. Many objectives in the short-term training sessions do not match with long-term planning in community and enterprise development. Clearly articulated objectives by stages of development will be more appropriate.

5. To design and implement curricula by considering local culture and needs, participants' demographics, stakeholders' preferences, and learners' availability and learning capacity. Include stakeholders and targeted audience in designing and delivering materials. Cultural sensitivity is critical to success.

6. To support and develop confidence and competency for youth participants and to hold them more accountable during and after training. Young people must have the necessary knowledge and skills needed to succeed and be confident in the use of that knowledge and skills. They should be encouraged to continue and complete all training sessions. Family members of youth should be involved in training sessions, or at least be informed about training purposes and expectations.

7. To maintain flexibility and interdependency in creating new programs, curricula, and collaborative efforts. When barriers or problems arise, all partners should participate in recognizing and resolving them.

Acknowledgments

Information on project location, development, outcomes, and assessment has been summarized from the report by Fellner and Zelezny-Green (2015). The authors would also like to acknowledge Dr Catherine Chan and Dr Elma Neyra, the Principal Investigator and Project Directors for UPLOAD JOBS for Mindanao, Higher Education for Development, and USAID Philippines.

References

Catalano, R.F., Berglund, M.L., Ryan, J.A.M., Lonczak, H.S. and Hawkins, J.D. (2004) Positive youth development in the United States: research findings on evaluations of positive youth development programs. *Annals of the American Academy of Political and Social Science* 591, 98–124. Available at: http://impact.sp2.upenn.edu/ostrc/doclibrary/documents/PositiveYouthDevelopmentResearch.pdf (accessed 9 December 2016).

Chan, C. and Neyra, E. (2015) UPLOAD JOBS for Mindanao: Final Partnership Report. Appendix D in: Job Opportunities for Business Scale-up (JOBS) for Mindanao. Final Associate Award Report, September 30, 2011 – September 29, 2015. *USAID and HED*, Washington, DC.

Council on Hemispheric Affairs (2015) *Youth Unemployment and Entrepreneurship in the Caribbean*. COHA, Washington, DC. Available at: http://www.coha.org/youth-unemployment-and-entrepreneurship-in-the-caribbean/ (accessed 8 December 2016).

Emery, M. and Flora, C. (2006) Spiraling-up: mapping community transformation with community capitals framework. *Community Development* 37(1), 19–35.

Fellner, T. and Zelezny-Green, R. (2015) UPLOAD JOBS for Mindanao External Evaluation. Technical Report submitted to Higher Education for Development. Appendix B in: Job Opportunities for Business Scale-up (JOBS) for Mindanao. Final Associate Award Report, September 30, 2011 – September 29, 2015. *USAID and HED*, Washington, DC.

Garavan, T.N. and O'Cinneide, B. (1994) Entrepreneurship Education and Training Programs: a review and evaluation. *Journal of European Industrial Training* 18(8), 3–12.

Gibb, A. (2011) Concepts into practice: meeting the challenge of development of entrepreneruship educators around an innovative paradigm: the case of the International Entrepreneurship Educators' Program (IEEP). *International Journal of Entrepreneurial Behavior and Research* 17(2), 146–165.

Greene, M. (1993) Chronic exposure to violence and poverty: interventions that work for youth. *Crime & Delinquency* 39(1), 106–124.

ILO (2012) *Global Employment Trends for Youth*. International Labour Organization, Geneva, Switzerland. Available at: http://www.rcsa.com.au/documents/other_docos/ILO%20Report%20Youth%20Employment%20May%2012.pdf (accessed 8 December 2016).

Inc. Magazine (2014) *Best Entrepreneurship Courses in America*. Available at: http://www.inc.com/ss/best-entrepreneurship-courses-america#3 (accessed 8 December 2016).

Liang, C. (2015) *Dollar Enterprise – From Theory to Reality*. Kendall Hunt Publishing, Dubuque, Iowa.

Liang, C. and Dunn, P. (2012) Entrepreneurial learning and business assessment – what new decisions would be made after learning? *Journal of Business and Entrepreneurship* 24(1), 105–122.

Nafukho, F.M. (1998) Entrepreneurial skills development programs for unemployed youth in Africa: a second look. *Journal of Small Business Management* 36(1), 100–103.

Singer, S., Amorós, J.E. and Moska, D. (2015) *Global Entrepreneurship Monitor 2014 Global Report*. Global Entrepreneurship Research Association, London Business School, London.

Solomon, G. (2007) An examination of entrepreneurship education in the United States. *Journal of Small Business and Enterprise Development* 14(2), 168–182.

USAID (2007) *Jobs for the 21st Century: Philippines/Autonomous Region of Muslim Mindanao (ARMM) Assessment. United States Agency for International Development*, Washington, DC. Available at: http://pdf.usaid.gov/pdf_docs/Pnadj499.pdf (accessed 9 December 2016).

Vesper, K.H. (1982) *Research of Education for Entrepreneurship*. Prentice Hall, Englewood Cliffs, New Jersey.

World Bank (2011) *World Development Report 2011: Conflict, Security, and Development*. World Bank, Washington, DC.

4 A Capabilities Approach to Designing Agri-entrepreneurship Training Programs for Conflict-affected Regions: The Case of Central Mindanao, Philippines

Mary M. Pleasant[1]* **and Rusyan Jill Mamiit**[2]
[1]*Oregon State University, Corvallis, Oregon, USA;*
[2]*University of Hawai'i at Mānoa, Honolulu, Hawaii*

4.1 Introduction

Around the world, people are exposed to armed conflict leading to adverse impact on the socio-economic, environmental, and institutional systems in which they live. This is particularly true among countries in the Global South, which includes most of Asia, Africa, South and Central America, and the Middle East, and refers to those countries that are more politically unstable, less technologically innovative, and poorer than those in the Global North (Odeh, 2010). Armed conflict is a broad-scale and far-reaching problem—approximately one-third of the world's population is exposed to it, and half of the world's poorest countries endured war or civil conflict at some point in the last three decades (Goodhand, 2001). Among the general population, conflict dismantles various forms of stability, and conversely engenders livelihood insecurities including threats to personal safety, income, social networks, and infrastructure, as well as increasing the possibility of displacement. Goodhand (2001) provided evidence that insecure livelihoods increase chronic poverty, and that chronic poverty can, in turn, predispose societies to further conflict. Entrepreneurial training development efforts should break the destructive cycle linking instability, chronic poverty, and armed conflict in regions where war and violence persist.

This chapter focuses on the creation of training programs for people living in rural agrarian-based economies of conflict regions, as small farms and agribusinesses can be a source of resilience at the community level, providing a means to regenerate stable incomes, social networks, and food security (UNEP, 2013). Countries in the Global South are typically agrarian-based with 58% of those in the labor force engaged in agriculture (Odeh, 2010). In developing countries, smallholder farmers make up the majority of the farming population (UNEP, 2013). With the promise of a continued and expanded need to provide food and fiber to the exponentially growing population, the agri-food sector has potential to be a market hotspot in developing countries, under government and societal contexts that promote and protect smallholder farms. However, the current socio-political contexts of many developing countries make it difficult for smallholder farms and agribusinesses to thrive. Large corporations are assuming and concentrating the

*mary.pleasant@oregonstate.edu

lands and production of many smallholder farms. Farmers are encouraged to monocrop as contractors for larger exporting industries. Imports are subsidized for developed countries, making their products cheap and difficult for small farms competing in the market (UNEP, 2013). In the era of global markets and agricultural industrialization, government policies continue to marginalize those in the rural-agrarian sector. Chemical inputs may have increased the yield of many of the world's major crops; however, the distribution of income and food has been increasingly unequal, perpetuating further conflict (UNEP, 2013). While the agri-food sector is a potential source of community-level stabilization, economic regeneration, and food supply in conflict and post-conflict regions, there are many barriers preventing new entry into agribusiness markets. Developing strategies to improve livelihoods of those in rural areas requires overcoming these social and economic constraints.

Agricultural entrepreneurship programs can offer a means to overcome some social and material constraints caused by conflict in developing countries through opportunities to engage in entrepreneurial livelihood activities that generate long-term social stability and economic growth. Following the various interpretations in the literature, agri-entrepreneurship is an individual's effort to pursue an agriculture-oriented economic opportunity by being creative and innovative, and taking risks (Sancho, 2010; Alsos *et al.*, 2011). This definition obliquely suggests that agricultural entrepreneurship is not restricted to producing raw commodities, but rather it also involves taking creative risks by converting raw materials into some form of innovatively processed or value-added products. In order to develop this capacity, it is fundamentally important that agrarian communities are equipped with the right combination of technical and practical agri-entrepreneurship skills. The acquisition of such skills involves a combination of approaches including formal education, training through extension programs, and non-formal capacity-building initiatives such as information dissemination (Kahan, 2012). The availability of a wide array of possible approaches to deliver agri-entrepreneurship programs is particularly important to an effective training program.

An agri-entrepreneurship initiative that provides job skills and workforce development capacity-building activities can subsequently translate to regional development success, which can lead to better income opportunities among individuals (USAID, 2006; Specker *et al.*, 2010; Lemmon, 2012). For example, Lemmon (2012) noted that in Iraq a 10% increase in funding for job-creation initiatives reduced violence by 10%. While the potential benefits of agricultural entrepreneurship are being recognized across the globe, there is paucity of on-the-ground implementation of entrepreneurship programs in politically fragile, war-torn, and transitioning agrarian economies (Brück *et al.*, 2012). The need for such programs in regions where a lack of economic opportunities and high unemployment exacerbate primary concerns for personal safety and security, particularly among youth and women, are highlighted in the works of Gunter (2014) in Middlle East and Northern African countries, Blattman *et al.* (2013) in Uganda, and Specker *et al.* (2010) in Burundi. In general, governmental and non-governmental agencies have found that infusing entrepreneurship into conflict-affected communities, including those engaged in small-scale agriculture, is valuable, and can be achieved through skill-building, job creation, and enterprise investments.

4.1.1 The capabilities approach

As non-governmental and governmental development agencies recognize the positive effects of productive entrepreneurship on economic growth and poverty alleviation, they increasingly promote entrepreneurship education and training programs in conflict areas among the rural poor. However, DeJaeghere and Baxter (2014) suggested that the current standard for donor entrepreneurship education lies within the neoliberal school of thought and neglects consideration of material and social constraints to entrepreneurial endeavors. For example, a female striving to become an agri-entrepreneur in a patriarchal society may find that a man refuses to sell land to her based upon her gender, precluding her from agribusiness start-up. To overcome this impediment, a neoliberal-focused training program would advise that she become more entrepreneurial, focusing on the skills trained rather than the cultural context in which the skills will be applied. Entrepreneurs in the neoliberal framework would bear the responsibility and risk associated with business success and failure in a changing market. The focus of this type of existing entrepreneurship training is on developing the individual through education, but omits consideration of social inequalities, individual and family values, and the need for a strong support system to facilitate entrepreneurship.

The main focus of the capabilities approach is to help people achieve well-being through enabling them to live a life they consider valuable. In the context of this chapter, we aim to help impoverished agrarian people in conflict regions live a life they consider valuable through the capabilities approach applied to entrepreneurial training programs. In contrast to a neoliberal framework, the capabilities approach considers the social and economic impediments to agri-entrepreneurs, such as cost-prohibitive start-up expenses and lack of community support, and helps people overcome these barriers as part of the training. In this approach, capabilities are built upon human and social capital and financial and physical assets (Oughton and Wheelock, 2003). The framework conceptualizes capabilities as valuable opportunities, and the freedom to choose to pursue those opportunities. The life one chooses to lead among these options is considered a person's functionings. The framework elucidates two consecutive points of transformation: (i) from capital (or endowments) to capabilities; and (ii) from capabilities to functions. For example, the capabilities approach outlines transforming a person's (i) existing access to farm land (capital) to (ii) having the knowledge and assets to create a successful agribusiness from that farm land (capabilities), and finally, being able to choose this as a means of livelihood, should they find it valuable. From this capabilities approach, we created a conceptual model (see Fig. 4.1 in Section 4.2) specific to agri-entrepreneurship in conflict regions, discussed in more detail in Section 4.2. First, we provide definitions of different types of entrepreneurs, as these types are possible outputs of the conceptual model, and each type may result in different training program outcomes.

4.1.2 Entrepreneurship typologies and training program outcomes

The relationship between entrepreneurship and improved livelihoods is complex and non-linear, and largely depends on the type of entrepreneur acting (Naudé, 2007). Baumol (1990) classified entrepreneurs based upon the strategies they employ: productive (contributing to economic growth), unproductive (rent-seeking), and destructive (illegal). Among these strategic types, entrepreneurs are often biased towards personal profit/gain as opposed to creating some positive social outcome (Naudé, 2007). The outcome for seeking profits can result in an economy that is largely productive or unproductive. The ability for entrepreneurship training to improve livelihoods and create a productive economy requires an investigation into the various types of entrepreneur that could result from increased access to financial, social, and human capital. Strategizing how to create entrepreneurs who improve livelihoods, rather than just increasing the supply of entrepreneurs in general, is an important aspect of the capabilities approach in developing entrepreneurial training programs.

The supply of entrepreneurs in conflict regions is not a limiting factor for achieving economic growth. In studying regions just after the conflict has ended, Naudé (2007) found that entrepreneurs are, in fact, ubiquitous. However, the supply of productive entrepreneurs, or what the 2007 Global Entrepreneurship Monitor (GEM) Report referred to as high potential growth entrepreneurs, is limited (Autio, 2007). Naudé (2007) suggested a fourth type of entrepreneur, that of the productive entrepreneur who has little to no effect on economic growth. This group falls under the informal entrepreneur type, or those who are entrepreneurs primarily to maintain their lifestyle (hereafter referred to as the informal entrepreneur). While Naudé (2007) did not explicitly suggest that the informal entrepreneur be added to Baumol's typology of entrepreneurs (productive, unproductive, destructive), he clearly denoted their unique impacts on conflict and post-conflict societies. Drawing on Baumol's typology, Naudé (2007) concluded that "efforts to increase the supply of entrepreneurship itself may be less important than efforts to change the allocation of entrepreneurial effort into productive entrepreneurship" (p. 4). Informal entrepreneurs are common among rural, poor, agrarian societies in conflict areas as people try to navigate failed markets and networks to make enough income to survive. As informal entrepreneurship is a potential outcome of training programs in conflict regions, we have chosen to include it in our typology.

Distinguishing the general characteristics of each type of entrepreneur is useful in order to develop strategies for desired outcomes of entrepreneurial trainings (e.g. economic growth, poverty alleviation, improved livelihoods). *Productive entrepreneurs* allocate resources more efficiently by creatively seizing opportunities to increase the productive use of resources, which can lead to more efficient, stable, and diverse economies (Naudé, 2007). Productive entrepreneurs can also interrupt the status quo of a society with radical innovation, advancing new technologies for society (Baumol, 1990). The *informal entrepreneur*, on the other hand, is productive, but by and large does not contribute to economic growth, only generating enough income to survive. This type of entrepreneur often has a microbusiness or "cottage industry," and is prevalent in rural areas that commonly lack infrastructure requisite to business expansion such as roadways and communications (Cañares, 2011). Informal entrepreneurs are often necessity entrepreneurs, that is, they have no alternative earning opportunities and are entrepreneurs as a means of survival. This can be contrasted with opportunity entrepreneurs that are choosing to be an entrepreneur because they perceive it to be a valuable opportunity (DeJaeghere and Baxter, 2014). Informal entrepreneurship offers societal benefits including alleviating, or at least, stagnating poverty, and the flexibility to adjust and benefit from new technological change or shifts in the consumer market for value added (Naudé, 2007). *Unproductive entrepreneurs* engage in rent-seeking activities, such as bribery, which increase personal profit but do not contribute to the economic growth of a region. *Destructive entrepreneurs*, whom the GEM refers to as conflict entrepreneurs, feed on and contribute to conflict, often prolonging and resurrecting violence, ultimately leading to further destabilization of the community (Brück *et al.*, 2012). We focus on a conceptual model (see Fig. 4.1 in Section 4.2) for facilitating the establishment of productive entrepreneurs engaged in agriculture as a means to enhancing livelihoods in conflict regions.

More recently, additional types of entrepreneur that fall outside the normal contexts of entrepreneurship have been identified, including intrapreneurs, habitual entrepreneurs, and social entrepreneurs. *Intrapreneurship* is entrepreneurship within existing firms, which may improve firm performance through innovative solutions to challenges (Antoncic and Hisrich, 2001). Intrapreneurs, then, are not self-employed, but can have similar positive impacts on development through improving existing firm resiliency and effectiveness. *Habitual entrepreneurs* are entrepreneurs whose firms have failed, but they try again, often increasing their chances of success. Habitual entrepreneurs may not be a distinct category of entrepreneur in and of themselves, but an attribute associated with successful entrepreneurs. Firm failure is often due to insufficient entrepreneurial ability. However, those that learn from those failures expand their entrepreneurial ability and are more likely to succeed (Naudé, 2008). Lastly, *social entrepreneurs* are entrepreneurs who seek to maximize social value rather than personal profit, and may play an important role in empowering and strengthening communities where governance has failed, such as in conflict regions (Acs and Kallas, 2007). The many definitions of social entrepreneurs can be characterized as those agents of change, often working in the nonprofit sector, that possess the skills of an entrepreneur (e.g. business acumen, risk-taking, innovative), but use their entrepreneurial ability to pursue opportunities that have positive social impact (e.g. health services for the poor). These forms of entrepreneurship, while not necessarily the intended product of entrepreneurship trainings, may have significant positive impact on livelihoods of the rural poor. As such, further investigation on how best to tailor training programs for these specific outputs may be useful.

Given that some types of entrepreneur contribute more to economic development and poverty alleviation than others (productive and informal entrepreneurs to some extent), one must consider that entrepreneurial training may free people to be any type of entrepreneur or not an entrepreneur at all. For example, DeJaeghere and Baxter (2014) found that some youth involved in a livelihood program in sub-Saharan Africa opted to spend earned income from their entrepreneurial endeavors on leisure activities or relationships, sometimes injuring their social networks or limiting their opportunities for further entrepreneurial endeavors. In addition, barriers to becoming entrepreneurs in conflict regions differ from those in stable regions. For example, the pervasive insecurity and corruption that occur in conflict regions predispose impoverished people to take fewer financial risks. Entrepreneurial trainings that consider the specific barriers to entrepreneurship and potential outcomes of increased entrepreneurial endeavors may better succeed at improving livelihoods in conflict-affected areas.

Drawing upon the capability model as cited in DeJaeghere and Baxter (2014) and Baumol's (1990) typology of entrepreneurs, we present a conceptual model for devising trainings aimed at improving the livelihoods of rural poor in conflict-affected regions by enhancing capabilities and mentoring the choice of functions (see Fig. 4.1 in Section 4.2). Specifically, we propose the development of an agricultural entrepreneurship training program to enhance livelihoods by fostering capabilities to become productive agri-entrepreneurs. In Section 4.2, the proposed conceptual model is discussed in more detail. To complement the framework, Section 4.3 discusses overcoming barriers associated with transforming from capital to capabilities, and Section 4.4 outlines the human value and agency considerations for transforming from capabilities to functionings. Section 4.5 presents how the conceptual model applies to an agri-entrepreneurship development program for rural youth in Mindanao, the Philippines, a region exposed to perennial conflict. From the Central Mindanao case study, Sections 4.6 and 4.7 define training program success as well as lessons learned and conclusions.

4.2 The Capabilities Framework Tailored for Agri-entrepreneurial Development in Conflict-affected Regions

The proposed conceptual model for framing the development of an agri-entrepreneurship training program (Fig. 4.1) is adapted from a recent iteration of philosopher and economist Amartya Sen's capabilities approach (DeJaeghere and Baxter, 2014). Two points of transformation are indicated from existing human, social, and financial capital to capabilities and from capabilities to an array of possible functions. Material and social constraints that limit capabilities must be overcome to achieve the first transformation (Table 4.1a), and the second transformation may be constrained by family values, cultural norms, and societal pressures (Table 4.1b). Possible functions or outputs of enhancing agri-entrepreneurial capabilities include becoming a productive, informal, unproductive, or destructive type of entrepreneur, or not becoming an entrepreneur and being a productive, unproductive, or destructive member of society.

Out of the range of potential outputs and outcomes of agri-entrepreneurial trainings in conflict regions, understanding the socio-economic context and incorporating it into the training material should help the program achieve the desired output (productive entrepreneur) and outcome (improved livelihoods). The zone of success in Fig. 4.1 is defined by those training outputs that lead to improved livelihood outcomes. This zone includes outputs beyond the productive entrepreneur, that of informal entrepreneurs and those who are not entrepreneurs but are productive members of society. While the primary desired output of many development efforts is to create productive entrepreneurs, this is not the only means of achieving improved livelihood outcomes. For example, informal entrepreneurship is widespread in many developing countries and is used as a means of survival in countries like Angola where the majority of the gross national product comprises informal markets (Naudé, 2007). Productive entrepreneurs have the special quality of tending to both improve livelihoods and generate economic growth. Informal entrepreneurs and members of society

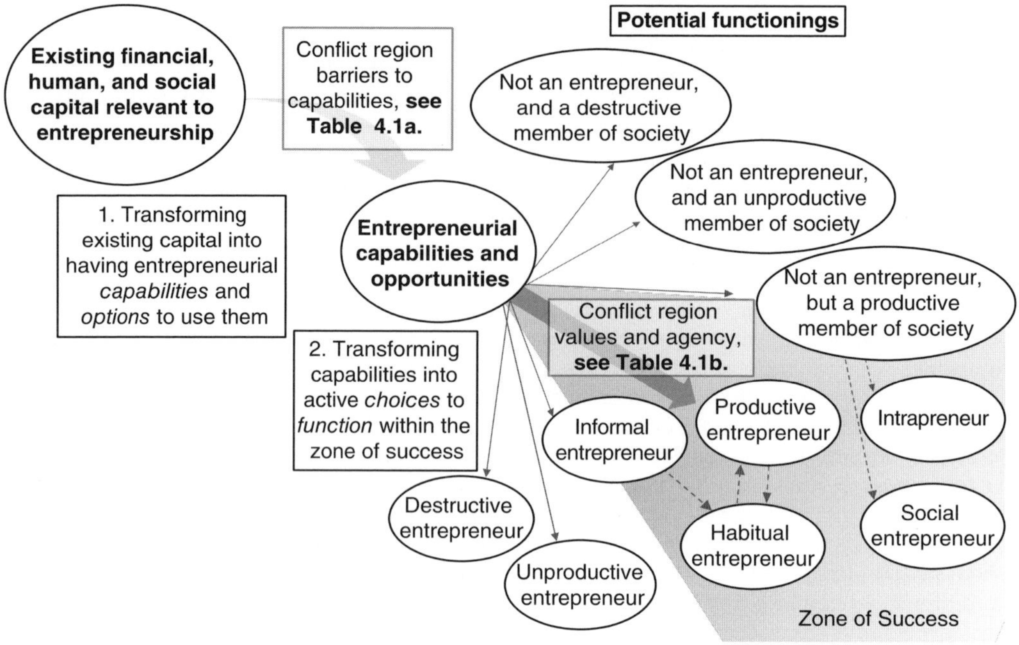

Fig. 4.1. The capability approach for agri-entrepreneurial training in conflict regions, adapted from DeJaeghere and Baxter (2014).

Table 4.1a. Conflict region barriers to transforming from existing capital to possessing entrepreneurial capabilities.

Conflict region barriers	Short-term solutions	Long-term solutions
Pervasive insecurity	Calculated and minimized risk; build from existing capital	Investment; community and institutional support
Unpredictable stability and schedule interruptions	Flexible schedule; stand-alone courses; involve household; budget for contingency plans	Improve communication ability; family support
Corruption	Community-driven development	Community and institutional support and checks; advisory board
Insufficient or underutilized human capital: language, education, and experience	Appropriate curriculum; one-on-one mentoring; internships	Expand agricultural extension program capacity; center of excellence to continue to provide expert advice and job opportunities; national recognized certificate
Insufficient or underutilized social capital: shifting landscape, distrust, and isolated communities	Network-building conferences and workshops	Center of excellence and agricultural extension to maintain and expand social network
Insufficient or underutilized financial capital: conflict and poverty cycle, low incentive to save and invest	Provide seed money	Saving and lending coaching, cooperatives

Table 4.1b. Conflict region individual values and agency constraining the transformation from entrepreneurial capabilities to functioning within the zone of success.

Conflict region values	Short-term solutions	Long-term solutions
Safety	Create a worry-free zone: provide meals, shelter, safe venue, reliable transportation, safe place to voice concerns, include household in planning for someone to take care of trainee's home responsibilities while at training	Mentorship program; community and institutional support
Group membership	Build a trust and responsibility network across ethnic and gender divisions	
Short-term investments	Coaching individual and household finances and investments	

who are productive but not entrepreneurs, however, may also alleviate poverty, strengthen the community, and be more resilient against conflict. Further, intrapreneurs may strengthen existing firm performance, habitual entrepreneurs may be a necessary condition for productive entrepreneurship, and social entrepreneurs may use their entrepreneurial acumen for positive social change outside of direct economic growth, including work in the non-profit sector. Given that these entrepreneur types contribute to economic growth or positive social change, we include them in the zone of success.

4.3 Transforming from Capital to Capabilities: Overcoming Conflict Region Barriers

Within the capabilities approach, developing functionings in a person necessitates considering the socio-cultural context in which that person operates and makes decisions. This larger context can promote or constrain the transformations from capital to capabilities. Conflict regions create constraints on this transformation that need to be overcome in order to develop a valued life through entrepreneurship training. Armed conflict poses significant barriers to entrepreneurship across individual, household, community, and society levels (Naudé; 2007; Brück *et al.*, 2012). On the societal level, conflicts can depress economic growth through increasing firm failure rates and decreasing foreign investment. In addition, governments in conflict regions often increase military spending at the expense of funding education, health services, and infrastructure, which are all underlying necessities for productive entrepreneurship (Brück *et al.*, 2012). On the individual, household, and community levels, the pervasive insecurity, unpredictable stability, corruption, and insufficient or underutilized human, social, and financial capital are the most pressing barriers to productive entrepreneurship in conflict regions (Fig. 4.1; Table 4.1a). While societal norms and government policies are important factors for entrepreneurial success in conflict regions, we focus on the individual and their community to facilitate agri-entrepreneurship. When conflict is prevalent, government and social systems are in flux and often unpredictable. As opposed to attempting to make systematic changes that better facilitate agri-entrepreneurship at a time when systems of power are already contentious, a better investment may be in the individual and their community as the starting point for developing more productive agri-entrepreneurs. In addition, development efforts at the community level may facilitate stability and act as a bottom-up approach to peace.

In starting with individuals and communities for promoting productive entrepreneurship in conflict regions, the specific socio-cultural context in which these individuals and communities exist may promote or constrain the transformations outlined in the capabilities approach. These individuals face specific barriers to transforming their existing human, financial, and social capital into agri-entrepreneurial capabilities and opportunities (i.e. completing the first transformation of the capabilities approach)

(Fig. 4.1; Table 4.1a). Through our work with rural agrarian youth in conflict regions, we have derived six barriers to developing effective agri-entrepreneurship training specific to conflict regions (pervasive insecurity, unpredictable schedules, corruption, insufficient or underutilized human, social, and financial capital). We then provide suggestions for overcoming these barriers.

4.3.1 Pervasive insecurity

Perhaps the most significant effect of conflict on entrepreneurial endeavors is the pervasive insecurity it creates among the poor. Insecurity may be both a result of and a cause of diminished human, social, and financial capital, and also of protracted corruption and conflict. Oughton and Wheelock (2003) stated that many economic theorists suggest that markets function most efficiently under insecure conditions because they incentivize people to avoid personal loss through seeking alternative opportunities for capital. However, they also pointed to literature that details the difference between risk, which can be calculated, and insecurity, which refers to the "processes that are presently taking a painful toll on people's lives, as well as to the psychological state of mind that is a consequence of this daily brutalizing experience" (Vail, 1999, p. 6). Insecurity and risk affect individual and household entrepreneurial choices differently, stimulating some and oppressing others. While economic insecurity can act as a catalyst for some to pursue entrepreneurial opportunities, it can also prevent many from risking financial investment (Oughton and Wheelock, 2003). The aversion to financial risk is especially pronounced among the poor who are faced with higher transaction costs, and among those living in rural areas where there are smaller markets and fewer profitable opportunities (Brück *et al.*, 2012). In addition, conflict affects not only income security, but also personal safety, supply of basic goods and services, and access to critical infrastructure such as hospitals and schools. It can threaten home permanence as many people are forced into diaspora as refugees or internally displaced people. Insecurity in social networks is manifested through burgeoning distrust brought about by corruption and unpredictable disruptions in communication that fracture business commitments. Furthermore, conflict can affect livelihood security through restricted access to education, training, and other forms of human capital (Cañares, 2011).

To help the entrepreneurs-in-training manage insecurity, risks associated with building and using capital should be calculated and minimized. Building financial, social, and human capital is the foundation for entrepreneurial endeavors as these forms of capital enable an entrepreneur to invest in materials to establish and grow a business, network to assist in the development and distribution of the product, and develop the skills to run an innovative and successful business. Risk is implied in entrepreneurship because new products and businesses are facing unknown demand among potential consumers. However, risk can be minimized and strategically taken. For financial capital, this could mean providing seed money for business start-up for the trainees. We found in our work with youth in Mindanao that training the entrepreneurs how to conduct a market study helped them to better understand their customers and how best to tailor their products to different markets. In this way, they minimize risk as they determine the demand for their products and product attributes before they market to retailers. Also, building social capital presents risks to those not yet confident in their abilities to communicate business ideas effectively and who have fear of commitment. Arranging internships may allow entrepreneurs-in-training to build social networks without the expectation of business expertise. Furthermore, allowing the entrepreneurs-in-training to practice presentations and communication styles within a local support network may help them feel more confident when they present themselves to potential business contacts. As many entrepreneurs fail on their road to success, framing failure as a learning opportunity may help the entrepreneurs-in-training to value taking calculated risk, necessary to the success of many entrepreneurial endeavors.

Over the long term, the entrepreneurs-in-training should gradually assume more of the investment into their entrepreneurial endeavors. For example, a matching funds system followed by saving and lending cooperatives could be used to fund new firms. Embedding the program within a community that values entrepreneurial endeavors and both trusts and can be trusted among the target population may also help the entrepreneurs-in-training to navigate the pervasive insecurity associated with entrepreneurship in conflict regions. Lastly, including the household in entrepreneurship decisions may mitigate risks as the household capabilities may be more than the sum of individual capital considered separately (Oughton and Wheelock, 2003). For example, entrepreneurs-in-training who have children may be able to have a household member provide child care while they build their business; they may be able to borrow money from a household member; or household members with diverse skills may be able to problem-solve challenges facing the firm in creative ways. These short- and long-term recommendations are location and culture dependent, and so tailoring the training program to meet the specific needs of the entrepreneurs-in-training is necessary.

4.3.2 Unpredictable schedules

As the stability within a region can shift daily, many members of the target population in conflict regions may have unpredictable schedules. They may have to evacuate their homes or cope with road closures without warning and ability to communicate these changes. Training programs should be prepared for such events by ensuring that agri-entrepreneurship curricula for the trainees are flexible and courses are stand-alone, that is, one does not need to take them at a particular time in a specific order. In our experience with youth in Mindanao, the stand-alone course schedule allowed the entrepreneurs-in-training to complete the training coursework in spite of tensions and conflict temporarily closing access to the training location. These youths joined in the next cohort of trainees to finish their entrepreneurial training courses.

4.3.3 Corruption

Corruption, such as bribes and misallocation of funds, is often an effect of impoverished and conflict-affected regions. The Asia Foundation, for example, found that Mindanao (Philippines) is so deeply corrupted that donor projects, even if well designed, often fall short of having truly transformative impacts, and may actually work to strengthen corrupted power structures (Parks *et al.*, 2013). One solution to corruption that has been suggested is to implement community participation in designing, planning, and implementing donor projects (Whalen, 2014). Including community stakeholders, local officials, and leaders in the decision-making process may empower the community to dismantle existing corruption and help with recruitment of future entrepreneurs-in-training. Inviting community feedback at project milestones before and after the commencement of training may improve overall program success over the long term. Having the community devise their own system of checks and balances and gradually replacing outside board members with local community members may help avoid corruption over the length of the project.

4.3.4 Human capital

Human capital, including experience, skills, and education, has been found to be a pivotal factor of entrepreneurial success in both developing and developed nations (Goedhuys and Sleuwaegen, 2000; Oughton and Wheelock, 2003; Naudé, 2008; Preisendörfer *et al.*, 2012). Human capital such as communication abilities, budget and business planning skills, and risk analysis capacity can help people obtain credit, effectively market their product, and form social networks (Gries and Naudé, 2009). The rural poor in conflict regions, however, often chronically lack the human capital requisite to productive entrepreneurship since many cannot afford education. Furthermore, conflict suppresses access to education, and those who are educated may migrate to urban areas. Entrepreneurship education has been suggested to be the panacea to fill gaps in human capital to engender economic growth and increase national security (Efe, 2014). In addition, education can have a spillover effect whereby those educated may pass on entrepreneurial knowledge to the less educated through publications, discussion, and openness to innovation (Acs and Kallas, 2007).

In order to implement effective entrepreneurship education and capacity-building programs in the form of training, the program design should be culturally relevant, at the appropriate education level, and built upon existing knowledge. A baseline assessment can help gauge the education level and existing knowledge, and allows trainees to voice certain topics they desire to learn. A comparable baseline assessment to partner organizations can assess capacity to deliver such a program. Local leaders within the partner organization can add relevant cultural examples and instruct in the native language of entrepreneurs-in-training to help ensure the program is relevant to the area and target population. In addition, experiential learning through internships and on-the-job training may help entrepreneurs-in-training to gain work experience, one of the most significant predictors of entrepreneurial success (Acs and Kallas, 2007; Sancho, 2010). Experiential learning that is culturally relevant allows the entrepreneurs-in-training to be engaged in the training, both as a receiver and contributor to the training program itself. This type of participatory approach aligns well with the capabilities approach that places value on the trainee having the freedom to make their own valued life choices.

The traditional methods for agricultural extension have been top-down and focused on the individual (Anandajayasekeram *et al.*, 2008). More recently, however, training programs have shifted to a participatory approach where the beneficiaries are collaborators in the knowledge base for the training curriculum, process for distributing the content, and eventual owners of both the process and the product (Cummings, 1997). A participatory program that seeks to empower its learners entails a shift from the traditional teacher-centered to learner-centered paradigm and from abstract to competency-based learning. Teacher-centered learning is oriented towards students learning passively as the teacher lectures as opposed to students actively participating in the course and applying the course concepts. In this learning-centered paradigm, students actively construct knowledge through self-discovery and are "co-producers of learning," and teachers shift from the standard lecturer to a more dynamic facilitator role (Saulnier *et al.*, 2008). Aligned with the learning-centered paradigm is a shift towards competency-based learning whereby students are asked to demonstrate mastery of a skill before moving onto the next step. This can be compared with the standard summative testing where a student may pass a course being totally deficient in one skill so long as the sum of their progress exceeds some set mark, where students have passed a set number of credit-hours in coursework but may still lack a skill, or where grading schemes are normalized and encourage competition. The desired performance outcomes and skills mastered drive the content and design of a competency-based curriculum (Voorhees, 2002). Competency-based learning in an agri-entrepreneurship development capacity-building program may be applied through asking students to demonstrate completing a cash-flow worksheet, developing a business plan, and actually starting a new venture. In addition, entrepreneurship literature suggests that work-based experience is critical to entrepreneurial development (Pittaway and Cope, 2007; Sancho, 2010) and that entrepreneurs learn through doing, including by mimicking, making mistakes, and problem-solving (Gibb, 1997; Cope and Watts, 2000). Real-world entrepreneurial experience, learner-centered instruction, and competency-based learning intersect in their ability to encourage independence, risk-taking in a safe-environment, and self-efficacy.

Learning-centered, competency-based, and participatory paradigms involve students in the process of creating their own knowledge, encourage experiential learning with real-world applications, and promote self-reliance and self-efficacy. The skills developed from these capacity-building models are essential to cultivating a practiced and spirited entrepreneur. They also align with the capacities-approach value of the freedom for entrepreneurs-in-training to make their own choices with their learned capabilities from the training program.

4.3.5 Social capital

Building networks and social capital also factor into sustainable entrepreneurial success (Sancho, 2010). Networks are built from both formal and informal contacts, including community leaders, family members, business team members, previous employers, friends, and colleagues. When entrepreneurs use their network to secure advice and obtain resources leading to successful outcomes, this access is considered social capital. Social networks provide alternative access to resources when conflict or other shocks prevent normal routes of transportation and communication. This has significant impacts especially among isolated rural areas with limited infrastructure and points of access. When conflict precludes access to the main thoroughfare, for example, isolated communities with little infrastructure may have no alternative means of entry and exit. Broad social networks, such as family or friends in neighboring areas, may provide information on new access routes such as back road networks that are safer or information that the tensions have calmed. In addition, personal relationships may smooth over the break of a business commitment due to conflict. For example, if business partners are friendly, they may be more empathetic to a missed delivery of product due to conflict closing normal transportation routes. Further, social networks may provide access to borrowed or gift resources that can assist with recovery from a conflict. For example, one can borrow a car from a neighbor to deliver products if a vehicle is difficult to access or broken down. Geographic location also affects conflict proximity and access to essential entrepreneurial resources such as financial capital, with rural areas suffering the most (Cañares, 2011). Social capital may help the rural entrepreneurs in conflict regions navigate these shifting landscapes. In addition, an entrepreneur's social relationships play a central role in business cluster formation, a concentration of businesses that can pool resources such as delivery trucks or manufacturing facilities, resulting in more productive and competitive businesses (Feldman *et al.*, 2005; Acs and Kallas, 2007). Increased social capital can facilitate the success of new agri-entrepreneurs amid conflict region setbacks.

Providing multiple opportunities for face-to-face interaction may facilitate a strong and trusted social network among entrepreneurs-in-training and their potential customers, manufacturers, distributors, and retailers. Hosting workshops and network-building conferences that are attractive to local stakeholders can provide this opportunity. Through life skills training, entrepreneurs-in-training should be prepared to communicate effectively with these stakeholders. As part of the participatory learning program, those trained in entrepreneurship curriculum can also host events in the community to spread the knowledge of both the curriculum and program. Trainees may also draw upon the network of the supporting institution in the local community already involved. Over the long term, a community center that maintains these social contacts and continues to provide opportunities for face-to-face interactions may help to foster profitable social capital.

4.3.6 Financial capital

Many studies in the literature identify the lack of access to financial capital as a constraint to pro-entrepreneurship development (Banerjee and Newman, 1993; Acs and Kallas, 2007; Autio, 2007). However, some have argued that human capital supersedes financial capital in importance for entrepreneurial success. For example, entrepreneurial education and skills can help people obtain access to credit, identify higher potential markets, and form social networks with more profitable relationships (Gries and Naudé, 2009). In addition, many financial capital woes are the result of underlying inadequate technical and managerial skills in the entrepreneurs (Lloyd-Ellis and Bernhardt, 2000). In such cases, developing human capital is necessary for increasing financial capital. Teaching participants how to present themselves to creditors, build and manage budgets, and understand investments and cash flow may help them overcome barriers to financial capital over the long term. However, in the short term, providing seed money for business start-up will allow entrepreneurs-in-training to actually practice starting a new firm, and should the firm fail, learn an exit strategy that minimizes financial burden. Hands-on experience with managing a budget and firm is important, as previously noted, because previous experience is one of the greatest predictors of small and mid-size start-up success (Sancho, 2010).

4.4 Transforming from Capabilities to Functionings: Navigating Values and Agency in Conflict Regions

The second transformation from capabilities to functionings entails human value and agency, or the capacity for individuals to make their own free and independent choices. An individual's personal values are shaped by culture and family life. These values affect how one may learn within and react to an entrepreneurial training program. Through our work with rural agrarian youth in conflict regions, we derived three values common in conflict regions that affect entrepreneurial skill development: safety, group membership, and short-term investing. Understanding how to tailor a training program to these values may help to increase the success of the program.

4.4.1 Safety

The threat to one's personal safety, as well as to the safety of peers and families, can be a constant worry to agri-entrepreneurs, making it difficult to focus in training. Insecurity of basic needs such as food and shelter will also impinge on learning abilities. People living in chronic poverty in conflict areas are often food and shelter insecure. Furthermore, they may risk encountering violence when travelling to the capacity-building activities. Training programs should be developed to consider the insecurities faced by entrepreneurs-in-training and attend to the insecurities so that the program participants are freed from the underlying conditions and can actively learn. As such, it is essential to provide meals, a safe venue, lodging, and reliable transportation to and from the training program before learning can take place. Having a counselor from the local community available so that entrepreneurs-in-training can voice concerns about their personal lives may also help to relieve worries. In addition, planning with the household to ensure that someone is attending to the home responsibilities of entrepreneurs-in-training in their absence may allow the participants to learn with more security.

4.4.2 Group membership

People in conflict areas often have a strong desire to be a part of a group as a safety mechanism. Idleness can contribute to involvement in groups that have negative influences on life choices, and entrepreneurial training may help the nascent entrepreneur to refocus their energies to more positive and productive choices. For example, DeJaeghere and Baxter (2014) found that entrepreneurship education that included life skills and future planning training for youth in sub-Saharan Africa helped them develop new relationships and more economical spending habits. It may be difficult to end these unhealthy relationships without an existing support network within a positive group. Providing entrepreneurs-in-training with a positive support network may free the participant from the risk of ending negative group membership without a safety net. Enabling healthy relationships, thus, begins with providing a positive group into which the participants can transition. With additional counseling, entrepreneurs-in-training may make the active choice to leave negative associations behind. Additionally, in areas where conflict divides ethnic groups and cultural norms prohibit effective female entrepreneurship, capacity-building programs that work across ethnic and gender divisions may help to strengthen the community overall.

4.4.3 Short-term investments

In regions with protracted conflict and chronic poverty, investment into the future tends to be weak. Short-term thinking abounds in these areas because returns on long-term investments are often too low to be attractive. In addition, given the abundant poverty depressing people's livelihoods, short-term investments into parties, weddings, and other cultural celebrations make day-to-day life valuable. Further, family medical emergencies and securing food and shelter are often immediate needs that preclude surplus time or money for further investment. If entrepreneurs are successful, however, they may be free to invest earned income in long-term investments as opposed to struggling to meet short-term needs. Here, an effective capacity-building program should influence short-term mentality with coaching and mentoring about household finances and investments.

Overall, strong trust networks embedded within institutional and community support may help individuals feel safe, associate with positive group members, and invest in long-term endeavors. A participatory program that empowers the entrepreneurs-in-training that complete the program with the responsibility to mentor the next cohort of entrepreneurs allows these successful participants to continue to be a part of an entrepreneurial community. In participating in the program as a trainee, they can then adapt the curriculum continuously as a trainer to make it stronger and more applicable to the needs of the next set of entrepreneurs-in-training. Including a mentorship component in the entrepreneurship program may engender lasting effects on these values by giving the individuals that completed the program the responsibility to mentor the next sets of entrepreneurs-in-training. Supporting this group in a permanent institution may help to maintain these pro-entrepreneurial mentalities and values over the long term.

4.5 Development of Agri-entrepreneurship Training Program for Out-of-school Youth in Rural Mindanao, Philippines: A Case Study of the Capability Approach

This section draws upon experience with a development project funded by the United States Agency for International Development (USAID) to better elucidate challenges in implementing a capacity-building program in conflict regions and methods to overcome those challenges. This project, the *U*niversity *P*artnership *L*inking *O*SY (Out-of-School Youth) to *A*gri-entrepreneurship and *D*evelopment to promote *J*ob *O*pportunities and *B*usiness *S*cale-up for Mindanao (UPLOAD JOBS for Mindanao), aimed to sustainably increase the institutional and human capacity of the rural workforce. Specifically, through the development of a training program, the project targeted the livelihood and income of OSY in Central Mindanao.

4.5.1 Central Mindanao: a conflict region

In the past decade, the Philippines has experienced steady economic growth. The World Bank (2015) reported that low and stable inflation made it possible for the Philippines to sustain continuous economic growth despite a weakening global economy. This set the Philippines onto a path that, according to the World Bank, "reduces poverty and creates more and better jobs."

Despite the national economic development in recent years and the government's efforts to improve the lives of Filipinos across the country, the geographic distribution of such growth has been uneven, with some parts of the country lagging due to armed, religious, and ethnic conflicts. For instance, vicious cycles of violence and poor governance continue to beleaguer several regions in Mindanao, resulting in high poverty. This poverty is due to massive destruction of the physical, human, and social capital of its residents. In 2013, the National Statistical Coordination Board reported that the poorest provinces with high incidences of poverty are mostly located in Mindanao (NSCB, 2013a). Of the 15 provinces with the highest poverty incidence, ten are Mindanao provinces: namely Bukidnon, Camiguin, Lanao del Norte, Lanao del Sur, Maguindanao, North Cotabato, Sarangani, Sultan Kudarat, Sulu, and Zamboanga del Norte.

While the poverty-stricken provinces are spread across Mindanao, the majority of them—North Cotabato, Sarangani, and Sultan Kudarat—are in the Central Mindanao area (Region XII). Central Mindanao has been most heavily affected by violent conflicts in the past two decades. The poverty gap ratio between the conflict-affected areas of Region XII and the rest of Mindanao continues to grow, increasing from 7.3% in 2003 to 7.6% in 2009 (NSCB, 2013b).

Decades of violent armed, political, religious, and ethnic conflicts have influenced the current conditions in Central Mindanao. The humanitarian crisis overwhelmed not only the national and local governments but international organizations as well. At the height of the conflict, trauma displaced a large portion of the population. Coincidentally, a number of those displaced are young adults who, in the process of displacement, miss a substantial number of school days and some are unable to go back to receive formal education. The young people trapped in this vicious cycle of violence and conflict have turned into security threats with an increasing number being recruited to join informal forces to bear arms and train for terrorist activities.

In 2012, the Philippine Center for Investigative Journalism (PCIJ) reported on the steep costs of the conflict in Central Mindanao to its youth population (Lingao, 2012). Conflict limits the capacity of the government to provide necessary services of its citizens (Weihe, 2004) such as education, health, and social services. Children entrapped in conflict areas in Central Mindanao learned that bearing arms was more important than education since arms were not only a means of defending themselves but also an easy means to earn a living (Lingao, 2012). The struggle experienced by the young people in Central Mindanao continues to disincentivize productivity and employment (Quitoriano and Francisco, 2004) (Fig. 4.2).

Cognizant of the vulnerability of and the potential from the youth sector in Central Mindanao, international donor and aid agencies have made youth development, particularly of out-of-school youth (OSY), a priority program. As Sommers (2006) asserted, youth exposed to and affected by violence and conflict "need a means to generate an income or a pathway towards employment." Since Central Mindanao has the highest prevalence of OSY, almost twice as many as the national level (E-Net Philippines, 2012), USAID committed to addressing the issues confronting the OSY in the region by equipping them with the capacity to be productive members of the community. Programs such as agri-entrepreneurship initiatives in the form of job skills training and business development programs are particularly relevant in conflict-stricken communities of Central Mindanao, Philippines as it may help dampen rural unrest and decrease engagement of the OSY in terrorist activities.

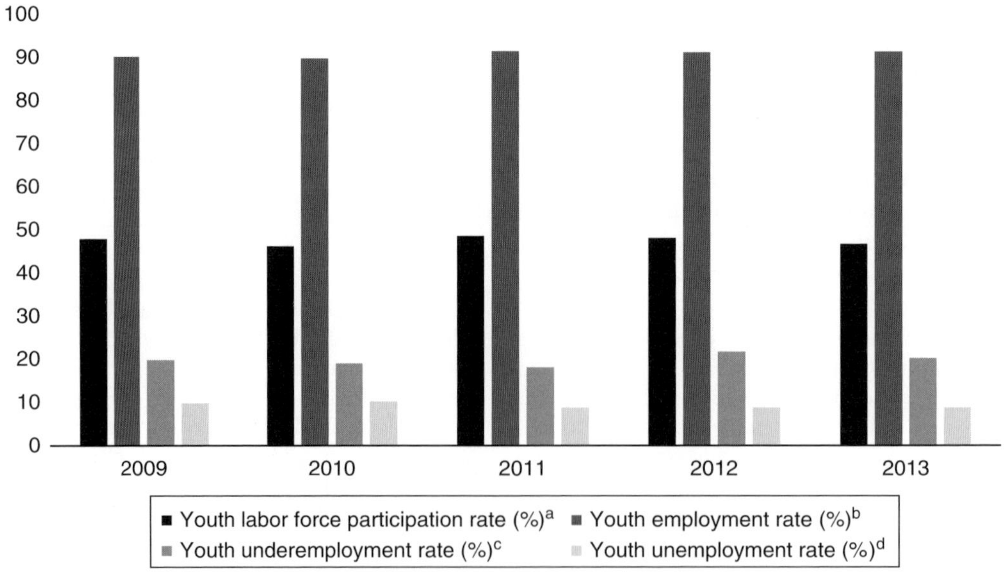

Notes:
[a]Youth labor force participation rate is the ratio of the youth labor force to the total youth household population multiplied by 100.
[b]Youth employment rate is the ratio of the employed youth to the youth labor force multiplied by 100.
[c]Youth underemployment rate is the ratio of the underemployed youth to the employed youth multiplied by 100.
[d]Youth unemployment rate is the ratio of the unemployed youth to the youth labor force multiplied by 100.

Fig. 4.2. Percentage change in youth labor and employment statistics in the agricultural and fishing sectors.

4.5.2 Agriculture and entrepreneurship in Mindanao

Central Mindanao is predominantly an agricultural region endowed with 25% of land area categorized as agricultural lands. Agriculture, hunting, forestry, and fishing sectors in the region account for nearly 29% of the regional economy (BAS, 2015). Of the region's total agricultural output, 68% is attributed to the crop subsector[1] and 15% is contributed by the fisheries subsector (Table 4.2).

The significance of the agriculture sector is very apparent in the labor market with about half of Central Mindanao's employment absorbed by the agricultural sector. This implies that farming is a major source of livelihood for many of the residents in Central Mindanao. The improvement of the rural livelihood and economy of the region, therefore, largely depends on sustaining the development of the agricultural sector, which is the largest industry to employ youth between 15 and 24 years old at the national level (Table 4.3).

While agriculture is still predominant in rural areas, interest in this sector has diminished among the younger generation (Bihis-Tolentino, 1995; Kruijssen, 2009; Briones, 2011). It is for this reason that incorporating the concept of entrepreneurship in agriculture is a strategic approach. With this form of intervention, OSY will see viability of agriculture in a new light. The need to develop initiatives to reach out to the OSY with an encouraging message that agriculture is a viable sector and enterprise is imperative. Propelling institutional and OSY human capacity-building among rural agricultural communities in the Central Mindanao area has never been more significant.

The common practice of overlooking the participation of youth in improving agricultural enterprises has shifted in the last decade. Recent years have seen an influx of development training programs to engage rural youth in agriculture-related enterprise. Many of the development programs focusing on rural areas have targeted the OSY as central components of agricultural rural enhancement programs. Enhancing knowledge and work skills in agriculture and business is a strategic intervention vital to accelerate rural agricultural economic development and reduce rural unemployment.

4.5.3 Training program design for out-of-school youth in Mindanao

The UPLOAD JOBS for Mindanao project, which includes design and implementation of a

Table 4.2. Top agricultural commodities in Central Mindanao in 2013 (cf. BAS, 2015).

Commodity name	Production growth rate (%)	Share to total agricultural output of the region (%)	Rank of the region in national production of the commodity
Palay	6.04	19.82	5th
Corn	8.68	15.00	2nd
Banana	2.61	8.56	3rd
Hog	0.16	8.33	7th
Pineapple	0.75	6.42	2nd
Yellowfin Tuna	4.18	5.60	1st
Skipjack	4.88	5.17	1st

Table 4.3. National youth employment in the agricultural sector (2009–2013) (all figures in thousands).

Major industry group	2009	2010	2011	2012	2013
All industries	6731	6816	7258	7322	7335
Agriculture, hunting, and forestry	2145	2063	2183	2185	2078
Fishing	328	331	322	320	298

capacity-building training program for out-of-school youth, demonstrates how differing types of existing capital can be transformed to nurture entrepreneurial capabilities and to provide options to use these capabilities. Starting with the implementation of the situational analysis (SA) technique, the project assessed the different types of capital existing in the area, particularly human capital, not only of OSY but also of potential trainers for entrepreneurial skills initiatives in Central Mindanao. The SA technique is a systematic method of collecting and evaluating existing information relevant to the issue and subject of interest such as comprehension level and interests towards different issues, including entrepreneurship. The SA technique employed field investigation tools in the form of OSY focus-group interviews and questionnaires developed and administered in selected communities in Central Mindanao. Through a random selection process, OSY from ten villages in Central Mindanao between the ages of 15 and 24 years participated in the field investigation, including participation by major ethnic groups (migrants, Maguindanaoans, and the Manobos) and both males and females. The survey questionnaire covered topics related to current job skills, interests in the field of farming and agriculture, expectations of job opportunities, training capacity, socio-demographics, and interest in training programs and capacity-building. This SA analysis sought to establish a baseline on the business and entrepreneurial skills of the OSY such as identifying the current skills and status of OSY in Central Mindanao.

Part of employing the SA technique, and as a component of the ground-truthing exercise, is assessment of the strengths and weaknesses of different stakeholders identified for potential inclusion as implementation partners. By conducting the SA, the implementing team was able to better understand the challenges to expect as well as to identify opportunities for intervention and options to address the needs of the OSY. For instance, educational-level disparities among potential participants as well as language barriers for program delivery were identified during the SA process. Further, the SA approach provided the development team with valuable information on what the target participants deemed were relevant skills needed to become a potentially productive entrepreneur.

To create an enabling environment whereby capabilities are transformed into active choices to function within the zone of success, the development team complemented the SA approach with a gap analysis (GA). The goal of the GA was to determine demand-driven OSY training program goals and to develop indicators for program success. Through this process, the development team was also able to assess the human capacity strengths and weaknesses of partner institutions in delivering demand-driven training programs. It was apparent from the GA approach that, as far as the desired goals, project stakeholders were aspiring for an improvement in their personal and societal economic growth. They agreed that, with appropriate skills, they saw entrepreneurship as a key to open that door for economic improvement.

A critical step in the GA involved a review of related formal programs on entrepreneurship in the USA and matching these with similar programs in the Philippines.[2] Specifically, this step included assessment and comparison of training certificate program modules and syllabi as well as pedagogical approaches and content delivery techniques between the Philippines and the USA. This particular step aided in designing the second stage of transformation to create a zone for success for the OSY. With this step, the development team was able to understand the pathways of intervention to ensure a certain level of success. For instance, the GA technique revealed that institutional and instructional capability in delivering training on developing business plans is very limited. Hence, one of the key interventions of the project was to involve experts who could train potential trainers in Central Mindanao to deliver a business plan course in which they turn from trainees to trainers.

To further effect capability transformation and ensure sustainability of the project interventions, the development team carefully reviewed existing syllabi of different entrepreneurship courses from various formal and informal learning institutions from the USA and the Philippines. The review of the syllabi was necessary for completing the design of the certificate program and to identify topics of relevance to OSY. The review was also important in developing materials such as modules for the OSY.

This particular step was one of the most relevant in that it helped in the identification of training courses and the overall design of a transformative agri-entrepreneurship training program for OSY. It was through this particular process that the development team was able to devise a suite of courses targeted to enhance the financial, human, and social capital of the OSY. The courses included: Introduction to Entrepreneurship; Entrepreneurial Marketing; Entrepreneurial Finance; and Business Plan and Model Development. With regular monitoring and evaluation of the implementation of courses during the project cycle, Product Development and Field Immersion courses were subsequently added to enable further capability transformation.

With the relevant information from the SA and GA, the development team designed and implemented a relevant and reputable agri-entrepreneurial training program targeting OSY in Central Mindanao. The process demonstrated that data analysis should be coupled with iterative participatory approaches that provide spaces for engagement to empower stakeholders. By doing this, the project team was able to build upon the existing value system and, at the same time, identify constraints that persistently or periodically impacted the success of the program such as inability to participate due to language barriers, lack of transportation, or household demands. Further, this allowed the project team to identify indicators of success and measure various components of the goals and aspirations of the stakeholders as identified in the SA and GA.

To facilitate sustainability of the capacity-building program beyond the project scope, post-SA and post-GA approach application the development team established a minimum set of standards to determine the best quality program that would be most beneficial to the OSY. This helped the development team understand that, to fully empower the OSY, providing the skills training may not be sufficient if it is not institutionally recognized and approved as an acceptable skills certification program. For this reason, and to fully implement successful transformation, through the local assistance of partner institutions, the development team applied for recognition by the Technical Education and Skills Development Authority (TESDA), a Philippine government agency tasked to manage, supervise, recognize, and certify technical education and skills development programs that "help develop Filipino workforce with world-class competence and positive work values." By doing this, the development team institutionalized a sustainability framework to ensure that quality capacity-building programs for OSY are consistently delivered even in the absence of technical and financial support from external agencies. The sustainability framework included elements such as a teach–coach–mentor strategy wherein those participants trained in the first year become trainers in the second year.

Moving the OSY to the zone of success, building sustainable capacity-building programs, and establishing social networks for OSY were the cornerstones for transformation in the project. To further the impact of the transformation beyond the OSY group into the wider community afflicted by conflict, the capacity-building team established a center of excellence called the Center for Agricultural and Farmland Entrepreneurship (CAFE). The CAFE functioned not only to provide sustainable income after donor funding ceased, but also to provide consultation services for OSY and other members of the community as they develop and engage in new ventures.

4.5.4 Lessons learned: practical experience and forward-looking steps to address barriers to transformation in conflict areas

Within the framework of the capability approach and in the context of barriers to transformation, the UPLOAD JOBS for Mindanao project encountered and addressed a varied suite of challenges, both natural and human-induced. From these challenges, one key lesson of relevance arising from the capacity-building program in a conflict region is the development of a safety strategy mainly due to unstable security issues such as bombing threats, road blockages, and demonstrations, among others. When implementing a development program in a conflict area, it is important to incorporate safety plans in the implementation process. By doing this, safety of the development team, institutional partners, and most importantly, of the target participants would have been taken into consideration and a plan would have been in place should an unprecedented event occur.

Safety issues have implications on any capacity-building program schedule. For instance, the unpredictable stability in certain parts of Central Mindanao impinged on training plans. During the scheduled training of the first group of target participants, some of the invited OSY were evacuated from their villages and, as such, the numbers of OSY were compromised for that scheduled training. The training was delayed, and the OSY were asked to attend the relevant trainings with the second cohort of participants when the security situation was relieved. With security concerns, delays have economic implications. It is, therefore, imperative to have some level of flexibility not only in the schedule but also with the budget.

Given the likelihood of conflict which restricts access of various partners to geographic areas during calamities or extreme weather conditions, capacity-building approaches should be revisited. In the UPLOAD JOBS for Mindanao project, extreme weather events prompted the development team to deliver a new method of training whereby future trainings were planned to be *à la carte*. In this training style, OSY could attend modules they had missed by joining trainings outside of their original cohort. This approach proved effective in a sense that OSY were not automatically rejected from the program for missing a scheduled training program. It provided the OSY with the opportunity to complete the capacity-building program despite personal interruptions. In relation to this challenge and lessons learned, it is necessary for the development team to consider a wide array of options to remain on time and on budget. For example, in the second year of the capacity-building implementation, a training was planned in Midsayap. OSY that were highly ranked in the business plan competition were set to conduct market studies with consumers in the local area to determine different ways they could enhance and market their products. Local faculty-trainers were also scheduled to practice conducting and coordinating market studies. This particular training had been planning-intensive due the stringent criteria for selecting market survey locations and the approvals needed to conduct surveys in these locations. Prior to the implementation of the capacity-building activity, the US Department of State issued a travel advisory prohibiting travel to the southern Philippines. Because of the level of investment in the activity, it was necessary to exercise flexibility and engage local partners to propose potential solutions such as relocating the activity into a safer location while remaining within the budget. With this incident, the training was relocated to Cagayan de Oro City, which is 9 hours away from the original location.

In terms of human capital barriers, the customization of the training program to the local cultural context helped build the entrepreneurial capability of the OSY. It should be noted that enhancing the capacity of local instructors, teachers, and mentors was critical in the customization of the training program. Further, the provision of business plan start-up funds and the availability of mobile technological assistance via the CAFE helped in the overall capacity development objective of increasing institutional and human capacity in the rural workforce as well as improving livelihood and incomes of OSY. Specifically, the program included hands-on exercises of developing different parts of a business plan such as the product development strategy, marketing and sales plan, financial statements, and scaling-up options. With this approach, the OSY developed a portfolio of materials including product description, financial statements, and contingency plans that they could always refer to as they implemented their business plan.

The application of the teach–coach–mentor approach provided the opportunity to develop confidence and leadership as well as nurture skills of OSY who showed interest in being a training recipient and a provider of capacity development to others in their communities. With this approach of becoming a mentor, the OSY develop non-cognitive skills such as resourcefulness, creativity, determination, critical thinking, and focus, the very skills that most productive and successful entrepreneurs possess. This particular strategy of delivering capacity development provided for practical and experiential learning as well as team-building experience.

The lack of adequate start-up finance is the primary reason OSY from agrarian-based conflict areas do not engage in entrepreneurial ventures, and so, the capacity development program incorporated financial literacy through lessons on simple accounting. To complement the lessons on accounting and financial literacy, a business plan competition was organized to allow participants to apply the lessons learned. Based on merit and relevance of the business plans presented, OSY were awarded start-up cash rewards. In the first round of competition, some of the OSY who put together business plans disbanded due to mobility and personal constraints of having regular communication to develop products and address business concerns. As such, in conflict areas such as Central Mindanao, proximity of partners in an entrepreneurial venture is very important, and it may help if more on-site workshops can be provided by local businesses and credit cooperatives through the agricultural loan officers.

Information and communication technologies (ICT) are now an essential component of any entrepreneurial venture. The situation in conflict areas is no different. The OSY in Central Mindanao need access to affordable ICT especially when they start their ventures and begin to expand. Most of the OSY have access to a cellular phone; however, due to financial constraints, most cannot afford the cost of loading their phones with enough credits to make a phone call or send text messages. Further, most OSY cannot afford a simple computer system, or internet service that they can use to connect to potential buyers or to research for new and innovative ideas to upgrade their products. As such, a capacity development program, such as the CAFE, needs to offer technical assistance to the OSY by means of access to computers and instructional help on how to improve the entrepreneurial venture. In the future, it may be more beneficial to OSY if satellite CAFE offices or a mobile CAFE office is established in more remote villages as most OSY face financial barriers and constraints on travel to the municipal center to access the CAFE services. Mobile CAFE offices can be patterned after the mobile library programs launched by the Philippine Department of Education, where library trucks travel around from village to village providing young children with access to books and materials that they cannot access in their small or non-existent local libraries.

The way young people perceive entrepreneurship is usually shaped by their personal environment and social relationships. Supportive entrepreneurship culture where there are well established social relationships is essential to nurture motivational factors that will help OSY become successful entrepreneurs. For this reason, a capacity-development program was organized to bring in the OSY participants from different villages in one location so that they could meet other OSY and establish friendships and networks to utilize in developing their entrepreneurial ideas primarily through information exchange and best-practice sharing. Through the mentor–mentee program of the CAFE, OSY extended their network beyond their peers by partnering with an established member of the community and the business sector. Although not many OSY participated in the mentor–mentee program, those who did professed that the experience was particularly helpful in understanding the entrepreneurial process, especially in turning challenges into opportunities.

As every conflict-affected area is different, no one-size-fits-all approach will succeed in the delivery of a successful capacity-development program. Exercising bottom-up strategies has proven a degree of success in the case study as well as having flexibility when unpredictable circumstances take place. The lessons learned through the Central Mindanao case study, although not prescriptive, are suggestive of possible items to consider when developing a capacity-enhancement program in areas of conflict.

4.6 Defining Program Success

This chapter uses the capability approach as a conceptual tool for designing capacity-building training programs for productive agri-entrepreneurs in conflict regions. The success of the capability approach is based on human freedom and functionings (Robeyns, 2005). While the primary intended output of this capacity-building training program is the productive entrepreneur, in light of the capability model, it is recognized that once people are given alternative options, they may choose to live out an alternative option other than entrepreneurship. When this choice leads to improved livelihoods, this can be considered a success. For example, given the access to a greater social network or a better understanding of investments, one may more easily find employment, return to school, or start a nonprofit venture. These choices, though not resulting in business start-ups, do result in productive members of society with improved livelihoods. Alternatively, one may choose to use the access to new financial, social, and human capital as a means to destructive ends. For example, one could start a business that funds the ongoing conflict or use earned profit for drugs and alcohol. These outcomes would be considered program failures as these behaviors can depress livelihoods overall. To avoid failures, the case study illustrated practical ways to overcome barriers and shape values specific to agri-entrepreneurship in conflict regions.

Success as defined by the capabilities framework will always be limited if basic needs are not met. Those engaging in entrepreneurship out of necessity are not doing so because it is a valued life option; they are striving to meet their basic needs to survive given their resources. Those engaged in necessity entrepreneurship may become relegated to the margins of society where it is more difficult to maintain influential and positive social relationships, making it difficult to access profitable earning opportunities. Bribery or other illegal or rent-seeking behaviors may then commence, driving down the economy further (DeJaeghere and Baxter, 2014). Although these behaviors were not observed in the case study area, Cañares (2011) pointed out that lines between necessity and opportunity entrepreneurship are often blurred as conflict can both suppress and create entrepreneurial opportunities simultaneously, and many people are driven by both opportunity and necessity. As the capability approach succeeds when people are free to live valued lives, the capacity-building program will seek to develop entrepreneurs who value the choice of being an entrepreneur, and although they may also be striving to meet their basic needs, do not fall in the poverty cycle as anecdotally observed with a number of participants in the UPLOAD JOBS for Mindanao project. Breaking this cycle would entail producing entrepreneurs that are well connected, with access to enough financial capital to fund profitable investments, and the knowledge to manage a thriving business.

4.7 Conclusions

This chapter has discussed the capabilities framework as a means to conceptualize the creation of successful agri-entrepreneurial programs. In applying this framework to conflict regions, specific barriers to entrepreneurial capabilities have been defined. In addition, values that influence people's choices of what they do, or do not do, as a result of the development efforts have been discussed. The success of the capability approach is based on human freedom and functionings. The primary desired output of the model put forth in this chapter is productive entrepreneurship. However, this is not the only means of achieving improved livelihoods that can result from increasing social, financial, and human capital. Informal entrepreneurs and people who are not entrepreneurs, but are productive members of society, may also enhance well-being. Also in the entrepreneurial vein are intrapreneurs, habitual entrepreneurs, and social entrepreneurs. Further investigation into the effects of these entrepreneurs on livelihoods may help foster entrepreneurship among people not drawn to the standard entrepreneurial model. The freedom to live a valued life as an entrepreneur may also improve regional economic growth, strengthen the community, and quell further conflict.

Development organizations seeking to improve livelihoods among the rural poor in conflict-affected agrarian regions might consider agri-entrepreneurship training as a means to do so. However, careful thought should be taken in designing a program that fosters the type of entrepreneur associated with those outcomes. Productive entrepreneurship can both improve livelihoods and increase economic growth. Unproductive and destructive entrepreneurs, on the hand, may depress economic growth and protract conflict, extending chronic poverty overall. Conflict regions pose barriers to entrepreneurship, and without thoughtful consideration into overcoming them, limit the ability of organizations to develop productive entrepreneurs.

A case study in Central Mindanao, Philippines with rural out-of-school youth illustrated the positive impact of entrepreneurial training and barriers that were overcome, avoided, or remained impediments. When designing training programs for areas affected by conflict, targeting populations that are ethnically diverse, or implementing a program in a region with a high incidence of poverty, transition from paper to practice can be difficult. Creating time for evaluation and reflection at any planned training milestone is essential to navigating this transition to transformation. As the social, economic, political, and geographic context of the target population changes, so too should the training program flux.

The development of the design and validation framework for capacity-building programs for out-of-school youth indicated that a good proportion of the OSY are still connected to agriculture. This implies that, when equipped with the necessary skills through capacity-building activities, and introduced to relevant agricultural entrepreneurial opportunities through business-plan competitions and seed funding, the OSY desire to be involved in agriculture-related entrepreneurial activities will be nurtured and transformation towards the zone of success may occur. Further analysis of the OSY situation and the existing institutional human capacity revealed that additional efforts through development of sustainable OSY-centered training programs are necessary if community-wide economic growth is to be achieved.

4.8 Acknowledgments

The authors extend their sincerest gratitude to the United States Agency for International Development (USAID) by funding the UPLOAD JOBS for Mindanao project through the Higher Education for Development (HED). RJM and MMP thank Dr Elma Neyra for her diligent efforts to help OSY succeed, and Emilie Bayona and Dawn Hope Sulit of Southern Christian College for their untiring determination to reach as many people in their community as possible. The authors also thank Dr Catherine Chan for her contributions in conceptualizing and implementing the UPLOAD JOBS for Mindanao project and this chapter. A special thanks to Brent Sipes and Tina Lee for their efforts in editing and coordinating the development of this chapter.

Notes

[1] Central Mindanao produced 1.35 and 1.31 million metric tons of rice and corn, respectively, in 2013.

[2] The GA considered comparing only US-based and Philippine-based entrepreneurship programs because the educational systems in the two countries are very similar. The similarity of the way the programs are organized in the two countries made the matching exercise relatively easier and comparable. It should be noted that only formal learning institutions were considered, due to the availability of relevant information.

References

Acs, Z. and Kallas, K. (2007) *State of Literature on Small to Medium-Size Enterprises and Entrepreneurship in Low-Income Communities. 2007-02.* Max Planck Institute of Economics, Entrepreneurship, Growth and Public Policy Group, Jena, Germany.

Alsos, G.A., Carter, S. and Ljunggren, E. (2011) *The Handbook of Research on Entrepreneurship in Agriculture and Rural Development.* Edward Elgar Publishing, Cheltenham and Northmpton, UK.

Anandajayasekeram, P., Puskur, R., Workneh, S. and Hoekstra, D. (2008) *Concepts and Practices in Agricultural Extension in Developing Countries: A Source Book.* International Food Policy Institute and International Livestock Research Institute, Washington, DC.

Antoncic, B. and Hisrich, R.D. (2001) Intrapreneurship: construct refinement and cross-cultural validation. *Journal of Business Venturing* 16, 495–527.

Autio, E. (2007) *2007 Global Report on High-Growth Entrepreneurship*. Global Entrepreneurship Monitor, London.

Banerjee, A. and Newman, A. (1993) Occupational choice and the process of development. *Journal of Political Economy* 101(2), 274–298.

BAS (2015) *Regional Profile: SOCCSKSARGEN*. Philippine Statistics Authority, Quezon City, Philippines.

Baumol, W.J. (1990) Entrepreneurship: productive, unproductive, and destructive. *Journal of Political Economy* 98(5), 893–921.

Bihis-Tolentino, J. (1995) Institutional strengthening and programme management related to rural youth programmes. In: *Expert Consultation on Extension Rural Youth Programmes and Sustainable Development*. FAO, Rome, pp. 109–122.

Blattman, C., Fiala, N. and Martinez, S. (2013) Generating skilled self-employment in developing countries: experimental evidence from Uganda. *Quarterly Journal of Economics* 129(2), 697–752.

Briones, R. (2011) *Workforce Development Initiatives for Out-of-School Youth – What Works? A Participatory Research with Youth and Communities in Southern Philippines*. Education Quality and Access for Learning and Livelihood Skills (EQuALLS) Phase 2, Manila, Philippines.

Brück, T., Naudé, W. and Verwimp, P. (2012) Business under fire: entrepreneurship and violent conflict in developing countries. *Journal of Conflict Resolution* 57(1), 3–19.

Cañares, M.P. (2011) In violence as in peace: violent conflict and rural entrepreneurship in the Philippines. *Journal of Small Business and Entrepreneurship* 24(2), 253–264.

Cope, J. and Watts, G. (2000) Learning by doing: an exploration of experience, critical incidents and reflection in entrepreneurial learning. *International Journal of Entrepreneurial Behaviour and Research* 6(3), 104–124.

Cummings, F.H. (1997) Role of participation in the evaluation and implementation of development projects. *Knowledge and Policy* 10(1), 24–33.

DeJaeghere, J. and Baxter, A. (2014) Entrepreneurship education for youth: a capabilities approach as an alternative framework to neoliberalism's individualizing risks. *Progress in Development Studies* 14(1), 61–76.

Efe, A.J. (2014) Entrepreneurship education: a panacea for unemployment, poverty reduction and national insecurity in developing and underdeveloped countries. *American International Journal of Contemporary Research* 4(3), 124–136.

E-Net Philippines (2012) *Out of School Children and Youth in our Midst*. Available at: http://www.enetphil.org/index.php?option=com_content&view=article&id=72%3Aout-of-school-children-and-youth-in-our-midst&catid=55&Itemid=1 (accessed 3 November 2012).

Feldman, M., Francis, J. and Bercovitz, J. (2005) Creating a cluster while building a firm: entrepreneurs and the formation of industrial clusters. *Regional Studies* 39(1), 129–141.

Gibb, A.A. (1997) Small firms' training and competitiveness: building on the small business as a learning organisation. *International Small Business Journal* 15(3), 13–29.

Goedhuys, M. and Sleuwaegen, L. (2000) Entrepreneurship and growth of entrepreneurial firms in Côte d'Ivoire. *Journal of Development Studies* 36(3), 122–145.

Goodhand, J. (2001) *Violent Conflict, Poverty, and Chronic Poverty*. Working Paper No. 6. Chronic Poverty Research Centre, London.

Gries, T. and Naudé, W. (2009) Entrepreneurship and regional economic growth: towards a general theory of start ups. *Innovation – The European Journal of Social Science Research* 22(3), 309–328.

Gunter, F. (2014) *Entrepreneurship and Political Stability in the Middle East and Northern Africa*. ASSA/MEEA, Boston, Massachusetts.

Kahan, D. (2012) *Entrepreneurship in Farming*. Food and Agriculture Organization of the United Nations, Rome.

Kruijssen, F. (2009) *Youth Engagement in Agricultural Research Development with a Focus on Sub-Sahara Africa*. Wageningen: Wageningen International, Wageningen, Netherlands.

Lemmon, G. (2012) *Entrepreneurship in Postconflict Zones*. Council on Foreign Relations, New York.

Lingao, E. (2012) *Mindanao: The Hidden Costs of War*. Available at: http://pcij.org/stories/mindanao-the-hidden-costs-of-war/ (accessed 8 June 2015).

Lloyd-Ellis, H. and Bernhardt, D. (2000) Enterprise, inequality and economic development. *Review of Economic Studies* 67, 147–168.

Naudé, W. (2007) *Peace, Prosperity, and Pro-Growth Entrepreneurship*. UNU-WIDER Discussion Paper 2007/02. UNU-WIDER, United Nations University, Helsinki, Finland.

Naudé, W. (2008) *Entrepreneurship in Economic Development*. UNU-WIDER Research Paper 2008/20. UNU-WIDER, United Nations University, Helsinki, Finland.

NSCB (2013a) *2012 Full Year Official Poverty Statistics of the Philippines*. National Statistical Coordination Board, Makati City (Metro Manila), Philippines.

NSCB (2013b) *MDG Watch*. National Statistical Coordination Board, Cotabato, Philippines.

Odeh, L.E. (2010) A comparative analysis of Global North and Global South economies. *Journal of Sustainable Development in Africa* 12(3), 338–348.

Oughton, E. and Wheelock, J. (2003) A capabilities approach to sustainable household livelihoods. *Review of Social Economy* LXI(1), 1–22.

Parks, T., Colletta, N. and Oppenheim, B. (2013) *The Contested Corners of Asia: Subnational Conflict and International Development Assistance*. The Asia Foundation, San Francisco, California.

Pittaway, L. and Cope, J. (2007) Simulating entrepreneurial learning: integrating experiential and collaborative approaches to learning. *Management Learning* 38(2), 211–233.

Preisendörfer, P., Bitz, A. and Bezuidenhout, F.J. (2012) Business start ups and their prospects of success in South African townships. *South African Review of Sociology* 43(3), 1–23.

Quitoriano, E. and Francisco, T. (2004) *Their War, Our Struggle – Stories of Children in Central Mindanao*. Save the Children UK, Quezon City, Philippines.

Robeyns, I. (2005) The capability approach: a theoretical survey. *Journal of Human Development* 6(1), 93–114.

Sancho, F. (2010) *Agricultural and Rural Entrepreneurship: Concepts for Modeling Development*. Inter-American Institute for Cooperation in Agriculture (IICA). Available at: http://repiica.iica.int/docs/B2032i/B2032i.pdf (accessed 8 June 2015).

Saulnier, B.M., Landry, J.P. and Wagner, T.A. (2008) From teaching to learning: learner-centered teaching and assessment in information systems education. *Journal of Information Systems Education* 19(2), 169–174.

Sommers, M. (2006) *Youth and Conflict – a Brief Review of Available Literature*. USAID and Equip3/Education Development Center, Washington, DC. Available at: http://www.crin.org/docs/edu_youth_conflict.pdf (accessed 8 June 2015).

Specker, L., Briscoe, I. and Gasana, J.M. (2010) *Early Economic Recovery in Fragile States Case Study Burundi: Operational Challenges*. Conflict Research Unit, Netherlands Institute of International Relations Clingendael, Wassenaar, Netherlands. Available at: http://www.clingendael.nl/sites/default/files/20100100_cru_paper_lspecker_burundi.pdf (accessed 15 July 2015).

UNEP (2013) *Smallholders, Food Security, and the Environment*. United Nations Environment Programme/International Fund for Agricultural Development (IFAD), Rome. Available at: http://www.unep.org/pdf/SmallholderReport_WEB.pdf (accessed 15 July 2015).

USAID (2006) *Poverty Reduction in Conflict and Fragile States: Perspectives from the Household Level*. Summary of Proceedings from a Conference held 8–9 November 2006, *Washington*, DC. USAID, Washington, DC. Available at: http://pdf.usaid.gov/pdf_docs/PNADK232.pdf (accessed 15 July 2015).

Vail, J. (1999) Insecure times: conceptualising insecurity and security. In: Vail, J., Wheelock, J. and Hill, M. (eds) *Insecure Times: Living with Insecurity in Contemporary Society*. Routledge, London, pp. 1–20.

Voorhees, R.A. (2002) Competency-based learning models: a necessary future. *New Directions for Institutional Research* 2001(110), 5–13.

Weihe, T. (2004) *Cooperatives in Conflict and Failed States*. US Overseas Cooperative Development Council, Washington, D.C.

Whalen, R. (2014) Delivering foreign aid to Mindanao: challenges and opportunities. In: Hopmann, P.T. and Zartman, I.W. (eds) *Mindanao: Understanding Conflict 2014*. Johns Hopkins School of Advanced International Studies, Washington, DC, pp. 173–182.

World Bank (2015) *Philippines: High and Inclusive Growth Can Eradicate Poverty and Boost Shared Prosperity within a Generation – World Bank*. World Bank, Manila, Philippines. Available at: http://www.worldbank.org/en/news/press-release/2015/01/13/philippines-high-and-inclusive-growth-can-eradicate-poverty-and-boost-shared-prosperity-within-a-generation (accessed 8 December 2016).

5 Measuring Youth Entrepreneurship Attributes: The Case of an Out-of-school Youth Training Program in Mindanao, Philippines

Cynthia Lai,[1]* Catherine Chan,[2] Domenico Dentoni[3] and Elma Neyra[4]
[1]University of Hawai'i at Mānoa, Honolulu, HI, USA; [2]University of Hawai'i at Mānoa, Honolulu, HI, USA; [3]Wageningen University, Wageningen, Netherlands; [4]Southern Christian College, Midsayap, North Cotabato, Philippines

5.1 Introduction

The implementation of youth entrepreneurship training programs is motivated by the realization that fostering entrepreneurship, defined in this chapter as starting a new business (Kelley *et al.*, 2012), can help in addressing youth unemployment when no other alternatives exist (Rosa, 2006; Geldhof *et al.*, 2014; UNCDF, 2014). The Millennium Development Goals, established in the Millennium Summit of the United Nations in 2000, promoted entrepreneurship as one of the major platforms to support sustainable social and economic development for youth (UNDP, 2004; Brock, 2008; UNSD, 2008). Especially in regions of conflict (Brück *et al.*, 2013) or affected by weak government institutions (Dentoni and Krussmann, 2015), discrimination (Rindova *et al.*, 2009) and unequal access to limited capital resources (Langevang *et al.*, 2012), entrepreneurship has attracted the attention of youth training programs to support and stimulate local endogenous growth. However, little is still known about youth entrepreneurs in these marginalized contexts that could improve the implementation of such training programs (Langevang *et al.*, 2012; Evangelista, 2013). Inherent social and institutional differences (Abbasi and Al-Marmah, 2000; Burnside and Dollar, 2000) coupled with diverse applications of entrepreneurship (Garravan and O'Cinneide, 1997) and fieldwork challenges (Cohen *et al.*, 2011) limit our knowledge to support the effective design, implementation and performance outcomes of entrepreneurship training programs in these regions. Therefore, developing adaptable tools that collect, measure, and inform a baseline of entrepreneurship attributes of any program's potential entrepreneurs prior to program implementation is imperative.

5.2 Defining the Potential Entrepreneur

While there are multiple and sometimes contrasting ways of defining the entrepreneur (Shane and Venkataraman, 2000; Zahra and Dess, 2001; Steyaert and Katz, 2004; Filion, 2011), this chapter relies on the definitions outlined by the Global Entrepreneurship Monitor (GEM) (Kelley *et al.*, 2012). To date, GEM, established at the London School of Business and Babson College in 1999, still represents the largest and most comprehensive effort to

*laicynth@hawaii.edu

 Enabling Agri-entrepreneurship and Innovation: Empirical Evidence and Solutions for Conflict Regions and Transitioning Economies (eds C. Chan *et al.*)

measure entrepreneurship (GERA, 2015). GEM acknowledges that the pursuit of entrepreneurship is a dynamic process that ultimately entails the creation of a new business venture (Kelley *et al.*, 2012). Therefore, GEM has categorized and defined the entrepreneur into three interdependent phases of measurement: (i) potential entrepreneurs, who are those seeking opportunities in their scope of activities and believing that they have the capabilities to start businesses (Krueger and Brazeal, 1994); (ii) nascent entrepreneurs, who are in the first 3 months of starting a new business; and (iii) new entrepreneurs, who are former nascent entrepreneurs with an established start-up for more than 3 months, but less than 3-and-a-half years (Kelley *et al.*, 2012). GEM measures individual entrepreneurship attributes and activities at each phase to gain a comprehensive understanding of why, in different countries, some populations are more 'entrepreneurial' than others. For the purpose of this chapter, we focus exclusively on measurement applications of potential entrepreneurs.

5.3 Attributes of Potential Entrepreneurs

GEM and other scholarly sources indicate that there are multiple attributes characterizing the potential entrepreneur (Hornaday and Aboud, 1971; Timmons, 1978; Gartner, 1985, 1989; Mueller and Thomas, 2001). These attributes have contributed to our knowledge and understanding of key attributes that drive the 'successful' entrepreneur (Krueger and Brazeal, 1994; Shane and Venkataraman, 2000; UNCDF, 2014).

5.3.1 Entrepreneurship characteristics

Entrepreneurship characteristics have been of particular interest for research defining the entrepreneur (Gartner, 1989; Solomon and Winslow, 1988; Shane, 1997). The Panel Study of Entrepreneurship Dynamics (PSED), founded by the University of Michigan in 1996 and the precursor to GEM, still represents one of the most recognized national and longitudinal studies of the entrepreneur (PSED, 2007). Developed in the USA, the PSED has been replicated in different countries such as Argentina, Canada, Greece, the Netherlands, Norway, and Sweden to collect and report population demographics, and entrepreneurship characteristics (i.e. former entrepreneurship experience, family entrepreneurship experience, entrepreneurship motivations and intentions, and access to social, financial and human capital) to evaluate a population's potential for entrepreneurship (PSED, 2007). Related studies have also used these characteristics to differentiate and compare entrepreneurs from non-entrepreneurs within a population (Davidsson and Honig, 2003; Schoon and Duckworth, 2012). As it relates to youth, GEM's global report on youth entrepreneurship, *Future potential: a GEM perspective on youth entrepreneurship* (Schøtt *et al.*, 2015), places a particular emphasis on measuring youth's former involvement in entrepreneurship, entrepreneurship motives, and human, social, and financial capital characteristics in order to better understand why and how youth get involved in entrepreneurship (Schøtt *et al.*, 2015).

5.3.2 Personality traits

The field of psychology has contributed much theoretical and empirical research to support and validate the use of personality traits to characterize the entrepreneur (Baum *et al.*, 2014). The personality trait approach is one of the classical and early approaches to entrepreneurship and thus most extensively applied and explored in this field (Baum *et al.*, 2014). As a result, numerous personality traits have been identified and measured in different countries such as Finland (Littunen, 2000), Hong Kong (Koh, 1996), China (Taormina and Lao, 2007), Sweden (Hansemark, 2003) and Germany (Caliendo and Kritikos, 2008) to characterize the entrepreneur. Results from these studies have advanced our knowledge on personality traits and the impacts of institutional, geographical, and cultural context on individual entrepreneurship potential and outcomes (Thomas and Mueller, 2000; Rauch and Frese, 2007). Personality traits have only recently been applied and measured for youth, mainly of students in formal institutional settings and controlled environments where large sample sizes are more readily available and attainable for this demographic (Mueller and Thomas, 2001; Hansemark, 2003; Ahmed *et al.*, 2010). For example, Mueller and Thomas (2001) used a large data set (n=1800) of third- and fourth-year students in 25 universities across nine countries to explore and compare their entrepreneurship personality traits of autonomy and innovativeness. Results from these and other analyses (Littunen, 2000; Hansemark, 2003; Gürol and Atsan, 2006; Ahmed *et al.*, 2010) revealed differences in personality traits for different countries that require additional research (Mueller and Thomas, 2001).

5.4 Study Aims and Significance

A notable gap persists in adapting or developing new measures of entrepreneurship potential for youth (Geldhof *et al.*, 2014) in developing contexts (Langevang *et al.*, 2012). To be specific, the majority of existing literature and empirical applications involving the potential entrepreneur originate in the US and European context (Shane, 1997; Mueller and Thomas, 2001). These are regions where access to research personnel, resources, and funding support are more readily available. A burgeoning literature on entrepreneurial characteristics, processes, and practices in developing countries (Bruton *et al.*, 2013) and indigenous communities (Peredo and Chrisman, 2006) have only come to light in recent years.

This chapter focuses on developing a survey that measures entrepreneurship attributes of potential youth entrepreneurs by adapting the extant entrepreneurship theory to the case study context of out-of-school youth in Mindanao, Philippines. The adapted survey measures a hybrid set of entrepreneurship attributes: specifically, youth demographics, entrepreneurship characteristics, and personality traits. In addition, the survey was developed to accommodate the challenges and limitations of collecting data on youth that reside in an international and marginalized context (e.g. language barriers, literacy level, limited resources). Case-study findings will contribute to the limited empirical data available regarding out-of-school youth potential entrepreneurs in Mindanao. Findings from this chapter aim to encourage other youth training programs to administer this adapted survey to enhance data availability, empirical comparisons, and program design that support potential youth entrepreneurs in regions of fragile and conflict contexts.

5.5 Case Study Area: Mindanao, Philippines

Mindanao represents the southernmost island and agricultural breadbasket of the country, with a rich production of palay, corn, abaca, banana, coconut, coffee, mango, pineapple, sugarcane, cacao, oil palm, camote, cassava and rubber (MDA, 2015). It is the eighth most populous island in the world, consisting of indigenous people, small-scale famers, landless workers and fishers who have persisted through generations of deep-seated local and ethnic conflict (Bloom *et al.*, 2012; MDA, 2015). The majority of the population resides on less than US$1.25/day (HED, 2011; Bloom *et al.*, 2012). Weak government institutions, discrimination, unequal access to limited capital resources and the presence of illegal markets have limited the island's sustainable growth and development (Co, 2004; Schiavo-Campo and Judd, 2005; Quitoriano, 2009; Evangelista, 2013; USAID, 2016). Among the causes of rural poverty and high unemployment in this region are the decline in the productivity and profitability of farming following the Green Revolution, outcomes of destruction by natural disasters, armed conflict, limited off-farm income-generating opportunities, poor access to markets, technical isolation, lack of effective infrastructure and access to finance (Co, 2004; Bloom *et al.*, 2012). Lack of effective policy that takes into consideration the sensitive environment, culture, societal norms, and ethnic diversities of these vulnerable populations has led to Mindanao having ten of the top 16 poorest provinces in the country (Bloom *et al.*, 2012).

5.5.1 Out-of-school youth (OSY)

Of particular concern in Mindanao is the rising number of OSY amongst the population (HED, 2011). OSY are defined as any youth who are not currently enrolled in school, have not completed college or any post-secondary education, and are unemployed (PSA, 2001). OSY who are unable to find sustainable or reliable incomes, and have limited professional skills and education, often end up contributing to the pre-existing high unemployment (HED, 2011). They often leave their barangay (or village) or family to find employment in the cities and overseas, or on the streets where they get lured into counterproductive activities that are often illegal but offer direct profit returns (Rajendran *et al.*, 2006; HED, 2011). In December 2015, Mayor Japal Guiani Jr of Cotabato City in Mindanao declared public concern for OSY as a major target for recruitment by the Islamic State of Iraq and Syria (ISIS). They were being offered an immediate 20,000 PHP (or US$423.14) for their commitment and were promised additional profitable returns (Hegina, 2015). It is for these reasons and the poverty and conflict that continues to plague the island's development that donor organizations, developmental projects, and international agencies are putting immediate investments into the OSY of Mindanao to train and foster positive contributors of society (HED, 2011; UFE, 2014). It is their mission to integrate OSY into the formal labor market and stimulate sustainable economic activity in these fragile and conflict contexts (Evangelista, 2013; UFE, 2014). To date, no data related to potential youth entrepreneurs has been collected in Mindanao, a conflict region.

5.6 Methods

The adaptation of the survey tool for OSY in Mindanao started with a literature review on entrepreneurship across the fields of business, psychology and behavioral psychology. This review identified the key entrepreneurship characteristics and personality traits featured by potential entrepreneurs.

5.6.1 Measuring OSY entrepreneurship characteristics

To identify empirical measures regarding youth entrepreneurship characteristics, measures from the PSED (2003) "Identification of Entrepreneurs Questionnaire" were extracted and adapted, specifically items related to individual demographics and interests in starting a business, business experience, and reasons for starting a new business (PSED, 2007).

5.6.2 Measuring OSY personality traits

The survey tool focuses on four key personality traits: risk-taking propensity, need for achievement, autonomy, and innovativeness (Ahmed, 1985; Low and MacMillan, 1988; Gartner, 1990; Johnson, 1990; Birley and Westhead, 1994; Mueller and Thomas, 2001; Ruvio and Shoham, 2011). These personality traits were selected based on their frequency and recognition (or statistical significance) for defining and characterizing the entrepreneur. For example, in a study by Stewart *et al.* (1999), need for achievement, risk-taking propensity, and innovativeness were used as determinants for entrepreneurs. In another study, by Gürol and Atsan (2006), need for achievement, autonomy, risk-taking propensity, and innovativeness were four of the six personality traits used to measure and define the entrepreneurship profile of Turkish university students. Table 5.1 lists the four personality traits, their definitions, and publication sources. Research in the field of psychology has further clarified that these four personality traits, assumed to be stable over time, are expected to promote and further motivate an individual with entrepreneurship intentions into action (Brockhaus, 1982; Rauch and Frese, 2007; Caliendo and Kritikos, 2008).

Question items to measure OSY personality traits were selected based on Rotter's (1966) original 29 item I-E scale (Thomas and Mueller, 2000), the Jackson Personality Inventory (Jackson, 1994; Thomas and Mueller, 2000; Zhao *et al.*, 2010), the Kogan–Wallach Choice Dilemma Questionnaire (Kogan and Wallach, 1964; Brockhaus, 1980) and the entrepreneurial self-assessment scale (Technonet Asia, 1981; Koh, 1996). These standardized assessment tools measure and assess individual personality traits related to entrepreneurship.

Table 5.1. Review of personality traits selected for this study including their definition and publication sources.

Personality trait	Definition	Sources
Risk-taking propensity	Personal willingness to assume risk (commit to and accept) associated with being self-employed	Kogan and Wallach, 1964; Teoh and Foo, 1997; Cromie, 2000; Entrialgo *et al.*, 2000; Mueller and Thomas, 2000
Need for achievement	Hardworking, determined, shows initiative, and strives in a competitive environment in the pursuit of excellence	McClelland, 1961; Entrialgo *et al.*, 2000; Littunen, 2000; Utsch and Rauch, 2000; Stewart *et al.*, 2003
Autonomy	Belief that one has complete control of one's personal welfare, takes full responsibility of one's successes and failures, and is not dependent on others	Rotter, 1966; Gilad, 1982; Riipinen, 1994; Koh, 1996; Engle *et al.*, 1997; Hansemark, 1998; Leone and Burns, 2000; Utsch and Rauch, 2000
Innovativeness	Personal preference for seeking and pursuing novel activity and is creative and resourceful.	Schumpeter, 1965; Zacharakis, 1997; Entrialgo *et al.*, 2000; Utsch and Rauch, 2000; Stewart *et al.*, 2003

5.6.3 Adapting a survey tool for OSY potential entrepreneurs

The adapted survey tool was much shorter (35 questions) compared with referenced personality trait surveys and the PSED questionnaire due to the process of adaptation. Question items had to account for OSY socio-economic context, literacy level, and program-specific conditions in Mindanao.

Question items from the PSED (2003) questionnaire were modified to measure OSY demographics and entrepreneurship characteristics. To be specific, the demographic measure of total household income was modified to measure OSY "individual annual income," a continuous response-variable. Additional demographic measures included categorical question items measuring OSY personal use of loans (No/Yes), religion (Christian/ Other), access to developmental training (No/Yes), and household farm status (Owner/Leaser/Laborer/None). In addition, youth were asked to give the name of their residential community and years of residence. Items related to individual interests in starting a business and business experience were modified from their more open-ended response to represent categorical choice items of No or Yes. Question items of "What are the one or two main opportunities that prompted you to start this new business?" and "Why do you want to start this business?" were modified to explore OSY entrepreneurship intentions. To be specific, the question item "What are the one or two main opportunities that prompted you to start this new business?" was expanded to explore individual attributes that influence entrepreneurship engagement, such as individual family entrepreneurship experience (i.e. family owns or manages a business) and personal entrepreneurship networks (i.e. knows an entrepreneur personally) (Djankov *et al.*, 2005; Duval-Couetil *et al.*, 2010). These items represented categorical choice responses of 'No' coded as 0 and 'Yes' coded as 1. The response for the question item, "Why do you want to start this business?" was coded as 0 for "Other" and 1 for "I have always aspired to / I see it as an opportunity." Survey responses were modified to represent categorical choice items in order to account for the limited staff and resources available for survey-response translation and to limit confusion in evaluating OSY survey responses in this cross-cultural context.

OSY personality traits were measured using 16 question items that were selected from existing measurement tools and adapted to ask respondents about their perceived level of agreement measuring their risk-taking propensity, innovativeness, need for achievement, and autonomy based on a five-point likert scale (1 "Strongly disagree" to 5 "Strongly Agree"). A likert scale was used so that a standardized composite score of means, or quantitative measure, could be determined for each personality trait in order to assess and compare results (Boone and Boone, 2012).

Finally, 12 entrepreneurs, professionals and experts in the field of entrepreneurship and business in the USA and Mindanao were asked to review the survey tool for basic comprehension and suitability of survey measures to characterize the potential youth entrepreneur. As a result, the adapted survey tool comprised 12 measures of demographics, seven measures of entrepreneurship characteristics, and 16 likert items of personality traits (four survey measures for each of the four personality traits). Table 5.2 provides a description of the 35 survey measures and the units used for this chapter's analysis.

5.6.4 Data collection: administering the adapted survey tool

The adapted survey was administered and data collected as part of a project titled *U*niversity *P*artnership *L*inking *O*SY to *A*gri-entrepreneurship and *D*evelopment to promote *J*ob *O*pportunities and *B*usiness *S*cale-up (UPLOAD JOBS) for Mindanao, Philippines in 2012–2015 (UPLOAD JOBS, 2012; USAID, 2014). This particular training program was funded by the United States Agency for International Development (USAID) through Higher Education for Development (HED). The project was implemented in partnership with the University of Hawai'i at Mānoa (UHM) as the leading institution, in collaboration with its local counterpart, Southern Christian College (SCC) (HED, 2011).

Two weeks prior to program implementation, a mandatory program announcement was held. Youth were administered the adapted survey by the project team (amongst seven individuals) to collect data on OSY entrepreneurship attributes prior to program implementation. The adapted survey was administered in the municipalities of Alamada, Carmen, Libungan, Midsayap, Aleosan, and Pigcawayan in Cotabato Province and Esperanza in Sultan Kudarat Province between December 2012 to August 2014. These seven municipalities were selected based on SCC's existing involvement and relations with local barangays (also known as villages) within each municipality. A total of 463 potential entrepreneurs, defined as any OSY who expressed interest in starting a new income generating and/or business venture, participated in the survey.

5.6.5 Modifying the adapted survey tool

Since the adapted survey had never been administered or applied to OSY in Mindanao the initial (or pilot) survey had some discrepancies after its administration in December 2012 (Cohort 1, n=101). To be specific, while experts in the field of entrepreneurship and local counterparts conducted a review of the survey prior to administration, Cohort 1 OSY experienced difficulties understanding survey questions measuring the four personality traits. In addition, OSY and local project staff found the initial survey to be too long, given the allotted time for administration and completion. As part of the process of adapting the survey tool to measure OSY entrepreneurship attributes in Mindanao, feedback was considered and survey questions were modified.

In order to shorten the adapted survey length, question items to measure each personality trait were reduced from four to three. In addition, preliminary results of Cohort 1 responses indicated that the personality trait of need for achievement represented the highest total and mean composite score. It also represented the smallest standard deviation compared with the other three personality traits. This suggested that, of the four personality traits, this trait was prevalent across the sample population and would be the least effective in differentiating the entrepreneurship potential of OSY. As a result, the personality trait was removed in order to minimize survey length while maintaining a variety of personality traits (three remaining). With support and feedback from local counterparts, additional revisions were made to the remaining survey measures as described in Table 5.3. A final revised survey was administered to the remaining OSY cohorts (Cohorts 2–4, n=289) from September 2013 to August 2014.

Table 5.2. Description of the survey measures and units used for analysis to inform an out-of-school youth (OSY) entrepreneurship-training program of their demographics, entrepreneurship characteristics, and personality traits prior to program implementation. This survey tool was administered in seven select municipalities in the provinces of Cotabato and Sultan Kudarat in Mindanao, Philippines (2012–2014) to 463 OSY.

Measure	Description and unit
Demographics	
Gender	Categorical. 0 is Male and 1 is Female
Age	Continuous
Income	Continuous, individual income per year (at time of survey) *(in Philippines pesos/PHP)*
Loans	Categorical, whether or not the respondent has an existing loan. 0 is No and 1 is Yes. *(Formal and/or informal)*
Education	Categorical, the highest level of educational attainment. 0 is below High School and 1 is High School Diploma achieved
Employment	Categorical, individual employment status. 0 if Unemployed and 1 if Self-Employed or Employed
Community that you reside	Name of community
Migrant status	Continuous, number of years resided in current community
Head of household	Categorical, whether or not the respondent is the head of household and has dependents. 0 is No and 1 is Yes
Religion	Categorical. 0 is Other and 1 is Christian
Access to training	Categorical, whether or not the respondent has participated in a developmental training program in the past year. 0 is No and 1 is Yes
My family owns their own farm	Categorical, family farm, contextualized to agricultural context, that an individual is directly exposed to
Entrepreneurship Resource Characteristics	
My family owns their own business	Categorical, family business ownership experience that an individual is directly exposed to. 0 is No and 1 is Yes
Entrepreneurship plans	Categorical, whether or not the respondent has plans to start a new business within the next year. 0 is No and 1 is Yes
Entrepreneurship experience	Categorical, whether or not the respondent has business ownership or management experience (of the past year). 0 is No and 1 is Yes
Entrepreneurship network	Categorical, whether or not the respondent knows at least one business entrepreneur in their social network. 0 is No and 1 is Yes
Entrepreneurship aspiration	Categorical, whether or not the respondent has entrepreneurial aspirations in their individual development. 0 is No and 1 is Yes
Entrepreneurship interest	Categorical, individual interest in starting a business. 0 is No and 1 is Yes
Entrepreneurship intention	Categorical, reason for interest in starting a business. 0 is Other and 1 is "I have always aspired to/ I see it as an opportunity"
Personality Traits of Entrepreneurship	
Autonomy	Four different likert items measuring individual belief that they have complete control of their personal welfare, take full responsibility of their successes and failures, and are not dependent on others; each with five choice items of: 1 as Strongly Disagree, 2 as Disagree, 3 as Neutral, 4 as Agree and 5 as Strongly Agree
Need for achievement	Four different likert items measuring individual hard-work ethic, determination, and initiative in the pursuit of excellence; each with five choice items of: 1 as Strongly Disagree, 2 as Disagree, 3 as Neutral, 4 as Agree and 5 as Strongly Agree
Innovativeness	Four different likert items measuring individual preference for seeking and pursuing novel activity and is creative and resourceful; each with five choice items of: 1 as Strongly Disagree, 2 as Disagree, 3 as Neutral, 4 as Agree and 5 as Strongly Agree
Risk-taking propensity	Four different likert items measuring individual willingness to assume risk (commit to and accept) associated with being self-employed; each with five choice items of: 1 as Strongly Disagree, 2 as Disagree, 3 as Neutral, 4 as Agree and 5 as Strongly Agree

Table 5.3. List of modifications made to the initial survey for Cohort 1 (December 2012) to establish a 'Final Survey' for Cohorts 2–4 (September 2013 to August 2014) data collection in Mindanao, Philippines. Modifications were made based on feedback from out-of-school youth (OSY), local counterparts and preliminary analyses of data collected from Cohort 1 (*italic font* indicates specific suggestions by local counterparts).

Personality trait	Cohort 1: Initial Survey (*n*=101)	Cohorts 2–4: Final Survey (*n*=289)
Autonomy	20. I usually work hard to improve on my past performance	I work hard to improve on my past performance
	21. I prefer to be my own boss	I prefer to be my own boss
	22. Making money is primarily a matter of good fortune	*Counterparts suggested removing this question as OSY are constantly trying to earn money to survive and may not understand this question*
	23. I believe that I have control over the future of my life	I have control over the future of my life
Need for achievement	24. I enjoy competition and winning	Removed — OSY responses all had agree–strongly agree, suggesting minimal sample differentiation
	25. For my job, I perform above and beyond expectations, there is always something more to be done or improved	Removed — OSY responses all had agree–strongly agree, suggesting minimal sample differentiation
	26. If I'm told something is impossible to do, I often can't resist seeing if it's true. I don't mind failing if I learn something in that process	Removed — OSY responses all had agree–strongly agree, suggesting minimal sample differentiation
	27. Any criticism is good criticism	Removed — OSY responses all had agree–strongly agree, suggesting minimal sample differentiation
Innovativeness	28. I prefer to step outside of my comfort zone to explore and try new things	I prefer to step outside of my comfort zone to explore and try new things
	29. I have the ability to anticipate and troubleshoot problems	People get excited by my ideas
	30. Change is a precursor to development and growth	I am constantly looking for the next challenge in my life
	31. I always like to have the latest information and technology	*Counterparts suggested we remove this question as OSY do not often have the money to have the latest technology nor access to information*
Risk-taking propensity	32. If forced to choose between them, I would rather "be safe than sorry"	I often do whatever it takes to win
	33. I have confidence in my ability to recover from my mistakes, no matter how big	I am a risk-taker
	34. When facing a decision with uncertain consequences, the potential benefits (not losses) are my greatest concern	When facing a challenging decision, I am more focused on what I will GAIN than lose
	35. I would choose a twenty-thousand-pesos bonus over a four-thousand-pesos annual raise, even if I had about a one-in-three chance of winning the bonus	*Counterparts suggested we remove this as OSY do not have a good concept of money and may not understand the question*

5.7 Results and Discussion

Data for this study was organized and evaluated using SPSS Statistical Software Version 23 to provide relevant descriptive statistics of the OSY population in Mindanao. A total of 463 surveys were administered; however, based on program age criteria of 18–24 years, only 390 surveys were considered. Finally, four surveys were removed due to missing cases, leaving a total of 386 surveys for analysis.

Tables 5.4–5.7 represent a summary of the results of 386 OSY demographics and entrepreneurship attributes that participated in the adapted survey in Mindanao, Philippines. These OSY represent potential entrepreneurs (defined in Section 5.2), expressing an interest in pursuing entrepreneurship and participating in the upcoming entrepreneurship-training program.

OSY considered for this analysis had an average age of 20 years. In this sample population, 38.9% of respondents represented females. Approximately 87.8% of OSY indicated completing a high school education and 37.9% of the sample indicated being employed or self-employed. When asked to elaborate on their length of (self-)employment, the majority of OSY stated that they had held this employment status for less than a year. These demographics support the training program's target population of OSY: youth (18–24 years) that are not currently enrolled in school, have not completed college or any post-secondary education, and are unemployed (or have unstable employment).

Table 5.4. Socio-economic profiles of out-of-school youth (OSY) (*n*=386) that participated in the entrepreneurship training program announcement, met program age criteria (18–24 years), and completed the survey in one of the seven selected municipalities in Mindanao, Philippines (2012–2014).

Demographics	Frequency (%)	Mean (average)
Gender, female	38.9	
Age		20 (20.49)
Education, high school diploma	87.8	
Religion, Christian	79.9	
(Self-)employed	37.9	
Head of household	18.2	
Family owns their own farm	41.2	
Family size		7 (6.69)
Resided in their village for over 10 years	81.8	
Individual annual income		19,110.45 PHP (US$411.27)
Has applied for a loan in the past 3 years	22	
Participated in a government-funded developmental training program in the past year	63.3	

Table 5.5. Frequency statistics of out-of-school youth (OSY) Cohorts 1–4 (*n*=386) entrepreneurship resource characteristics that participated in the survey administered from December 2012 to August 2014 in seven selected municipalities in the provinces of Cotabato and Sultan Kudarat in Mindanao, Philippines.

Entrepreneurship characteristics	Frequency (%)
Aspires to be an entrepreneur	93
Interested in starting a business	95.1
Interested in starting a business because it is an opportunity	36.8
Expected to start a business within the next year	87.4
In the past 12 months has owned or managed a business	20
Knows at least one entrepreneur personally	70.2
Family owns their own business	23.7

Table 5.6. Descriptive statistics (mean, mode, standard deviation, and range) of Cohort 1 (n=101) out-of-school youth (OSY) personality traits collected from the initial survey administered in seven selected municipalities in the province of Cotabato in Mindanao, Philippines, in December 2012. Respondents were asked about their perceived level of agreement for each five-point likert item (1 Strongly disagree, to 5 Strongly agree) given statements measuring their autonomy, need for achievement, innovativeness, and risk-taking propensity.

Survey question/ likert item	Mean	Mode	Standard deviation	Min.	Max.
Autonomy					
20. I usually work hard to improve on my past performance	4.06	4	0.998	1	5
21. I prefer to be my own boss	3.75	4	1.043	1	5
22. Making money is primarily a matter of good fortune	3.56	4	1.153	1	5
23. I believe that I have control over the future of my life	4.33	5	0.861	1	5
Total	*15.7/20*				
Need for Achievement					
24. I enjoy competition and winning	3.84	4	0.967	1	5
25. For my job, I perform above and beyond expectations, there is always something more to be done or improved	4.16	4	0.821	1	5
26. If I'm told something is impossible to do, I often can't resist seeing if it's true. I don't mind failing if I learn something in that process	4.12	4	0.909	1	5
27. Any criticism is good criticism	4.16	4	0.987	1	5
Total	*16.28/20*				
Innovativeness					
28. I prefer to step outside of my comfort zone to explore and try new things	4.03	4	1.072	1	5
29. I have the ability to anticipate and troubleshoot problems	3.69	4	1.027	1	5
30. Change is a precursor to development and growth	4.19	4	0.987	1	5
31. I always like to have the latest information and technology	4.17	4	0.861	1	5
Total	*16.08/20*				
Risk-taking propensity					
32. If forced to choose between them, I would rather "be safe than sorry"	3.75	4	1.220	1	5
33. I have confidence in my ability to recover from my mistakes, no matter how big	4.04	4	0.916	1	5
34. When facing a decision with uncertain consequences, the potential benefits (not losses) are my greatest concern	3.49	4	1.092	1	5
35. I would choose a 20,000-pesos bonus over a 4000-pesos annual raise, even if I had about a one-in-three chance of winning the bonus	3.18	4	1.252	1	5
Total	*14.46/20*				

Household demographics of these OSY supported the shifting agricultural background of the population to off-farm activities, indicating that 41.2% of the population have a family-owned farm. Average family size of OSY was about seven individuals, with approximately 18.2% of OSY representing the head of household. Approximately 82% of the study population indicated having lived in their current barangay for more than 10 years, a particular

Table 5.7. Descriptive statistics (mean, mode, standard deviation, and range) of Cohorts 2–4 (*n*=285) out-of-school youth (OSY) personality traits collected from the final survey administered in seven municipalities in Mindanao, Philippines, from September 2013 to August 2014. Respondents were asked about their perceived level of agreement for each five-point likert item (1 Strongly disagree, to 5 Strongly agree) given statements measuring their autonomy, innovativeness, and risk-taking propensity.

Survey question/ likert item	Mean	Mode	Standard deviation	Min.	Max.
Autonomy					
19. I work hard to improve on my past performance	4.48	5	0.700	1	5
20. I prefer to be my own boss	4.07	4	0.990	1	5
21. I have control over the future of my life	4.45	5	0.885	1	5
Total	*13/15*				
Innovativeness					
22. I prefer to step outside of my comfort zone to explore and try new things	4.34	4	0.759	1	5
23. People get excited by my ideas	3.98	4	0.729	1	5
24. I am constantly looking for the next challenge in my life	4.40	5	0.723	1	5
Total	*12.72/15*				
Risk-taking propensity					
25. I am a risk-taker	4.11	4	0.871	1	5
26. I often do whatever it takes to win	4.37	5	0.811	1	5
27. When facing a challenging decision, I am more focused on what I will GAIN than lose	4.19	4	0.886	1	5
Total	*12.67/15*				

advantage to the entrepreneurship training program whose underlying objective was to catalyze and support local endogenous growth. To be specific, it is assumed that OSY who have resided in their barangay for over 10 years are more likely to be committed, resourceful and have greater social capital in pursuing an entrepreneurship venture than an OSY who is mobile or new to the region.

Despite the region's unequal and limited access to capital resources, OSY in this region have access to developmental training programs, with 63.3% OSY indicating former training participation in the past year. In addition, while the number of OSY that have applied for a loan in the past 3 years remains low (22%), this number indicates that OSY do have access to loans within the area. However, further clarification from local counterparts indicated that the majority of these loans, if not all of them, were informal either from a family member, friend or loan shark (informal financiers that offer loans at extremely high interest rates). Finally, the survey question measuring OSY annual incomes indicated that 79.2% OSY were generating incomes, with an average of US$1.13/ day, below the United Nations Millennium Development Goals indicator of US$1.25/day (UNSD, 2008).

The findings of entrepreneurship characteristics indicate a positive outlook of OSY entrepreneurship potential in terms of their aspirations, interests, intentions, and social networks in Mindanao (Table 5.5). To be specific, 93% of OSY aspired to be an entrepreneur and over 95% indicated an interest in starting a business. Yet, out of the OSY interested in starting a business, only 36.8% expressed this interest in relation to the pursuit of a specific business opportunity. When OSY elaborated on the 'Other' reasons for starting a business, their responses were predominantly related to the necessity to do so (i.e. no other source of income generation or employment, need an additional source of income to support the family or relatives).

Next, 87.4% of OSY stated that they were expecting to start a business within the next year, implying the population's enthusiasm towards entrepreneurship despite only 20% having owned or managed a business in the past 12 months. While OSY may lack personal experience in entrepreneurship, the majority of OSY have some form of personal exposure to entrepreneurship. To be specific, 70.2% of OSY know at least one entrepreneur personally and 23.7% are part of a family that owns their own business. This type of personal exposure allows for a source of communication, information, support, and mentorship for potential entrepreneurs. In addition, the presence of this network around OSY represents an advantage to entrepreneurship training programs aiming to foster youth (self-)employability in the region.

Tables 5.6 and 5.7 represent a summary of survey results measuring their personality traits for Cohort 1 (n=101) and Cohorts 2–4 (n=285), respectively. Survey results were organized by survey administered: Cohort 1 was administered the initial survey and Cohorts 2–4 the final (or modified) survey. Results were then organized by personality trait, survey question and SPSS outputs. Data analysis was completed using SPSS Statistical Software Version 23 to determine the mean, mode, standard deviation, and range for each five-point likert item, given OSY perceived level of agreement of 1 (Strongly disagree) to 5 (Strongly agree). For this analysis, the means of the item responses were summed to create a composite score for each personality trait for comparison.

Results for Cohort 1 (n=101) are based on the initial (or pilot) survey of 35 question items administered to OSY in December 2012. According to these results, OSY responded positively to the likert items related to the personality traits of autonomy, need for achievement, innovativeness, and risk-taking propensity. OSY responses ranged from a minimum of 1 Strongly disagree to 5 (Strongly agree) across likert items. Comparing mean composite scores out of 20 for each personality trait, Cohort 1 scored highest for the personality trait of need for achievement with a mean total score of 16.28, followed by innovativeness with 16.08, autonomy with 15.7, and risk-taking propensity with 14.46. For each likert item, OSY responded most frequently (mode) with a perceived level agreement of 4 (Agree), with the exception of question/likert item 23 of Autonomy, "I believe that I have control over the future of my life," which had a mode of 5 (Strongly agree). Standard deviations (SD) of OSY responses are greater than 0 and relatively small (close to the mean), with the greatest variation in personality traits of autonomy (survey question 22 with SD of 1.153) and risk-taking propensity (survey question 35 with SD of 1.252). Feedback from local counterparts indicated that the variability in these two questions might be due to the minimal understanding of OSY as it relates to money in the context of these question items (Table 5.3). The two question items were revised for the final survey.

Results for Cohorts 2–4 are based on the final (or modified) survey of 27 question items administered to 289 OSY in September 2013 to August 2014. Due to missing cases, four surveys were removed leaving a total of 285 surveys for analysis (n=285, 4 missing). Based on survey results, OSY responded positively to the likert items related to the personality traits of autonomy, innovativeness, and risk-taking propensity. OSY responses ranged from a minimum of 1 (Strongly disagree) to 5 (Strongly agree) across likert items. Comparing mean composite scores of each personality trait out of 15, Cohorts 2–4 scored highest for the personality trait of autonomy with a mean total score of 13, followed by innovativeness with a mean total score of 12.72, and risk-taking propensity with a mean total score of 12.67. OSY responded most frequently (mode) with a perceived level agreement of 4 (Agree) and 5 (Strongly agree) across personality items. To be specific, 4 (Agree) represented the mode for survey questions 20 of autonomy, 22 and 23 of innovativeness, and 25 and 27 of risk-taking propensity. Next, 5 (Strongly agree) represented the mode for survey questions 19 and 21 of autonomy, 24 of innovativeness, and 26 of risk-taking propensity, respectively. Standard deviations (SD) of OSY responses are greater than 0 and relatively small (close to the mean) with the greatest variation in the personality trait of autonomy (survey question 20 with a SD of 0.990) and risk-taking propensity (survey question 27 with a SD of 0.886). The final survey reflected lower SD than the initial survey, suggesting that the revisions were successful in enhancing survey measures and subsequent OSY responses.

5.8 Conclusion

This chapter outlines the process used to develop an adapted survey tool that collects information and measures key entrepreneurship attributes of OSY potential entrepreneurs in Mindanao. The survey, adapted from the extant entrepreneurship literature to the empirical context of this study, resulted in a data set (*n*=386) of demographics, entrepreneurship characteristics and personality traits of OSY potential entrepreneurs between the ages of 18 and 24 years, in Mindanao.

Demographic and entrepreneurship characteristics of OSY collected prior to program implementation were successful and assisted the project team to customize their training program efforts accordingly. To be specific, despite enthusiasm (95% interested in starting a business) and 87.4% of the OSY population expecting to start a business in the next year, only 20% of OSY had actual business and management experience. As a result, the training program catered to a more beginner-level audience that focused on introducing the significant topics of entrepreneurship (i.e. business planning, marketing, finance, production, and business expansion), how to prepare and present a business plan, and individual empowerment. The low number of OSY (under 20%) who were head of household represented a concern for participant program commitment. To be specific, as dependents within a household and often of large family sizes, parents relied on OSY for daily household activities. Therefore, training programs were shortened in length and OSY invited for training 2 weeks in advance so that they could organize their schedules accordingly. Next, since the majority of OSY still resided at home with their parents they did not have complete autonomy over decision making. As a result, the training program was customized for a younger generation of potential entrepreneurs who resided at home.

Positive personality traits of autonomy, need for achievement, innovativeness, and risk-taking propensity reflected the potential of OSY as entrepreneurs in Mindanao. In addition are the advantages that the program's training and skills development could provide for these OSY and their entrepreneurship ventures. Future studies should investigate the notable gap between theoretical entrepreneurial intentions and OSY need to pursue a business. To be specific, OSY showed motivation to start a business in the next year predominantly for reasons of necessity. These findings confirm the need to distinguish "necessity entrepreneurship" from "opportunity-based entrepreneurship" (Rosa, 2006) as different drivers for new venture creation, particularly in the marginalized context of Mindanao. Moreover, mean annual incomes of 19,110.45 PHP (US$411.27) of these OSY indicate the significance of entrepreneurship training programs to support youth, particularly when counterproductive activities such as ISIS are offering OSY an immediate 20,000 PHP (US$423.14) for their commitment alone (Hegina, 2015).

The survey developed in this chapter was adapted to offer an instrument that advances our knowledge and empirical information about youth potential entrepreneurs. The survey was adapted to accommodate youth and program-specific conditions in Mindanao, overcoming the challenges of collecting data on youth in a marginalized context of limited resources. By measuring key entrepreneurial attributes of OSY in Mindanao, this chapter draws implications for future entrepreneurship training programs to replicate the adapted survey. Future training programs have potential to generate additional data sets of OSY entrepreneurship characteristics and personality traits to strengthen their programs and advance empirical research to support youth entrepreneurship. Future studies by chapter authors will test the reliability and effectiveness of the adapted survey as a screening tool to identify youth with superior entrepreneurship potential to participate in entrepreneurship training. Results of this study tend to support the performance outcomes of the case study entrepreneurship-training program in Mindanao.

References

Abbasi, G.Y. and Al-Mharmah, H. (2000) Project management practice by the public sector in a developing country. *International Journal of Project Management* 18(2), 105–109.

Ahmed, S.U. (1985) N-Ach, risk-taking propensity, locus of control and entrepreneurship. *Personality and Individual Differences* 6(6), 781–782.

Ahmed, I., Nawaz, M.M., Ahmad, Z., Shaukat, M.Z., Usman, A., Rehman, W.U. and Ahmed, N. (2010) Determinants of students' entrepreneurial career intentions: evidence from business graduates. *European Journal of Social Sciences* 15(2), 14–22.

Baum, J.R., Frese, M. and Baron, R.A. (2014) *The Psychology of Entrepreneurship*. Psychology Press, Mahwah, New Jersey.

Birley, S. and Westhead, P. (1994) A taxonomy of business start-up reasons and their impact on firm growth and size. *Journal of Business Venturing* 9(1), 7–31.

Bloom, G., Karlin, A., Morgan, K., Holden, T. and Grosberg, M. (2012) *Lonely Planet Philippines*. Lonely Planet Press.

Boone, H.N. and Boone, D.A. (2012) Analyzing likert data. *Journal of Extension* 50(2), 1–5.

Brock, D.D. (2008) *Social Entrepreneurship: Using the New Millennium Development Goals to Make an Impact on Society*. SSRN Scholarly Paper No. ID 1344417. Social Science Research Network, Rochester, New York.

Brockhaus, R.H. (1980) Risk taking propensity of entrepreneurs. *Academy of Management Journal* 23(3), 509–520.

Brockhaus, R.H. (1982) The psychology of the entrepreneur. In: Kent, C., Sexton, D. and Vesper, K. (eds) *Encyclopedia of Entrepreneurship*. Prentice Hall, Eaglewood Cliffs, New Jersey, pp. 39–57.

Brück, T., Naudé, W. and Verwimp, P. (2013) Business under fire: entrepreneurship and violent conflict in developing countries. *Journal of Conflict Resolution* 57(1), 3–19.

Bruton, G.D., Ketchen, D.J. and Ireland, R.D. (2013). Entrepreneurship as a solution to poverty. *Journal of Business Venturing* 28(6), 683–689.

Burnside, C. and Dollar, D. (2000) Aid, policies, and growth. *American Economic Review* 90(4), 847–868.

Caliendo, M. and Kritikos, A.S. (2008) Is entrepreneurial success predictable? An ex-ante analysis of the character-based approach. *Kyklos* 61(2), 189–214.

Co, M.J. (2004) The formal institutional framework of entrepreneurship in the Philippines: lessons for developing countries. *The Journal of Entrepreneurship* 13(2), 185–203.

Cohen, N. and Arieli, T. (2011) Field research in conflict environments: methodological challenges and snowball sampling. *Journal of Peace Research* 48(4), 423–435.

Cromie, S. (2000) Assessing entrepreneurial inclinations: some approaches and empirical evidence. *European Journal of Work and Organizational Psychology* 9(1), 7–30.

Davidsson, P. and Honig, B. (2003) The role of social and human capital among nascent entrepreneurs. *Journal of Business Venturing* 18(3), 301–331.

Dentoni, D. and Krussmann, F. (2015) Value network analysis of Malawian legume systems: implications for institutional entrepreneurship. Paper presented at the workshop on 'Complex-systems dynamics principles applied to food systems', June 6th–8th, 2015, Food and Agriculture Organization (FAO) Meeting Urban Food Needs Program. FAO, Rome.

Djankov, S., Qian, Y., Roland, G. and Zhuravskaya, E. (2005) Who are China's entrepreneurs? *American Economic Review* 96(2), 348–352.

Duval-Couetil, N., Reed-Rhoads, T. and Haghighi, S. (2010) Development of an assessment instrument to examine outcomes of entrepreneurship education on engineering students. In: *Proceedings ASEE/IEEE Frontiers in Education Conference, Washington, DC, October 2010*. Institute of Electrical and Electronics Engineers (IEEE), New York, pp. T4D-1–6.

Engle, D.E., Mah, J.J. and Sadri, G. (1997) An empirical comparison of entrepreneurs and employees: implications for innovation. *Creativity Research Journal* 10(1), 45–49.

Entrialgo, M., Fernández, E. and Vázquez, C.J. (2000) Characteristics of managers as determinants of entrepreneurial orientation: some Spanish evidence. *Enterprise and Innovation Management Studies* 1(2), 187–205.

Evangelista, R. (2013) *Entrepreneurship in the Philippines: opportunities and challenges for inclusive growth*. Center for International Private Enterprise, Washington, DC. Available at: http://www.cipe.org/publications/detail/entrepreneurship-philippines-opportunities-and-challenges-inclusive-growth (accessed 1 October 2015).

Filion, L.J. (2011) Defining the entrepreneur. In: Dana, L.-P. (ed.) *World Encyclopedia of Entrepreneurship*. Edward Elgar Publishing, Cheltenham, UK, p. 41.

Gartner, W.B. (1985) A conceptual framework for describing the phenomenon of new venture creation. *Academy of Management Review* 10(4), 696–706.

Gartner, W.B. (1989) Some suggestions for research on entrepreneurial traits and characteristics. *Entrepreneurship Theory and Practice* 14(1), 27–38.

Gartner, W.B. (1990) What are we talking about when we talk about entrepreneurship? *Journal of Business Venturing* 5(1), 15–28.

Geldhof, G.J., Porter, T., Weiner, M.B., Malin, H., Bronk, K.C., Agans, J.P. and Lerner, R.M. (2014) Fostering youth entrepreneurship: preliminary findings from the young entrepreneurs study. *Journal of Research on Adolescence* 24(3), 431–446.

Gilad, B. (1982) On encouraging entrepreneurship an interdisciplinary analysis. *Journal of Behavioral Economics* 11(1), 132–163.

GERA (2015) *GEM Global Entrepreneurship Monitor*. Global Entrepreneurship Research Association, London. Available at: http://www.gemconsortium.org/ (accessed 12 October 2015).

Gürol, Y. and Atsan, N. (2006) Entrepreneurial characteristics amongst university students: some insights for entrepreneurship education and training in Turkey. *Education+ Training* 48(1), 25–38.

Hansemark, O.C. (1998) The effects of an entrepreneurship programme on need for achievement and locus of control of reinforcement. *International Journal of Entrepreneurial Behavior and Research* 4(1), 28–50.

Hansemark, O.C. (2003) Need for achievement, locus of control and the prediction of business start-ups: a longitudinal study. *Journal of Economic Psychology* 24(3), 301–319.

HED (2011) *Philippines: Job Opportunities for Business Scale-up (JOBS) for Mindanao 2011*. Higher Education Development (American Council on Education), Washington, DC. Available at: http://archive.hedprogram.org/funding/rfa/archive/2011-JOBS-Mindanao.html (accessed 21 November 2015).

Hegina, A.J. (2015) ISIS recruiting minors, out-of-school youth — Cotabato City mayor. *Inquirer.net (Philippine Daily Inquirer)*. Available at: http://newsinfo.inquirer.net/745374/isis-recruiting-minors-out-of-school-youth-cotabato-city-mayor (accessed 9 January 2015).

Hornaday, J. and Aboud, J. (1971) Characteristics and role demands of entrepreneurship. *Personal Psychology* 24, 141–153.

Jackson, D.N. (1994) *Jackson Personality Inventory: Revised manual*. Sigma Assessment Systems, Port Huron, Michigan.

Johnson, B.R. (1990) Toward a multidimensional model of entrepreneurship: the case of achievement motivation and the entrepreneur. *Entrepreneurship Theory and Practice* 14(3), 39–54.

Kelley, D., Singer, S. and Herrington, M. (2012) *The Global Entrepreneurship Monitor: 2011 Global Report*. The Global Entrepreneurship Research Association (GERA) and Babson College, Wellesley, Massachusetts.

Kogan, N. and Wallach, M.A. (1964) *Risk Taking: a Study in Cognition and Personality*. Holt, Rinehart & Winston, New York.

Koh, H. (1996) Testing hypotheses of entrepreneurial characteristics: a study of Hong Kong MBA students. *Journal of Managerial Psychology* 11(3), 12–25.

Krueger, N.F. and Brazeal, D.V. (1994) Entrepreneurial potential and potential entrepreneurs. *Entrepreneurship Theory and Practice* 18, 91–104.

Langevang, T., Namatovu, R. and Dawa, S. (2012) Beyond necessity and opportunity entrepreneurship: motivations and aspirations of young entrepreneurs in Uganda. *International Development Planning Review* 34(4), 439–460.

Leone, C. and Burns, J. (2000) The measurement of locus of control: assessing more than meets the eye? *The Journal of Psychology* 134(1), 63–76.

Littunen, H. (2000) Entrepreneurship and the characteristics of the entrepreneurial personality. *International Journal of Entrepreneurial Behavior and Research* 6(6), 295–310.

Low, M.B. and MacMillan, I.C. (1988) Entrepreneurship: past research and future challenges. *Journal of Management* 14(2), 139–161.

McClelland, D.C. (1961) *Achieving Society*. Van Nostrand Reinhold, Princeton, New Jersey.

MDA (2015) *Mindanao at a glance*. Mindanao Development Authority, Philippines. Available at: http://www.minda.gov.ph/ (accessed 9 October 2015).

Mueller, S.L. and Thomas, A.S. (2001) Culture and entrepreneurial potential: a nine country study of locus of control and innovativeness. *Journal of Business Venturing* 16(1), 51–75.

Peredo, A.M. and Chrisman, J.J. (2006) Toward a theory of community-based enterprise. *Academy of Management Review* 31(2), 309–328.
PSA (2001) *Out-of-School Youth.* Philippine Statistics Authority, Quezon City, Philippines. Available at: https://psa.gov.ph/content/out-school-youth (accessed 9 October 2015).
PSED (2003) *Identification of Entrepreneurs Questionnaires.* Panel Study of Entrepreneurship Dynamics, Michigan University, Ann Arbor, Michigan. Available at: http://www.psed.isr.umich.edu/psed/data (accessed 4 June 2012).
PSED (2007) *Background: History of Research Program, Research Rationale and Design.* Panel Study of Entrepreneurship Dynamics, Michigan University, Ann Arbor, Michigan. Available at: http://www.psed.isr.umich.edu/psed/home (accessed 20 November 2015).
Quitoriano, E.L. (2009) *Land, Foreign Aid and the Rural Poor in Mindanao.* Focus on the Global South Research Report. Belgian Alliance of North–South Movements (11.11.11) and Focus on the Global South, Quezon City, Philippines.
Rajendran, S., Veronesi, D., Mohammad, N. and Mala, A. (2006) *The impact of armed conflict on male youth in Mindanao, Philippines.* The World Bank Social Development Papers: Conflict Prevention and Reconstruction (Paper No. 35/July 2006). World Bank, Washington, DC.
Rauch, A. and Frese, M. (2007) Let's put the person back into entrepreneurship research: A meta-analysis on the relationship between business owners' personality traits, business creation, and success. *European Journal of Work and Organizational Psychology* 16(4), 353–385.
Riipinen, M. (1994) Extrinsic occupational needs and the relationship between need for achievement and locus of control. *The Journal of Psychology* 128(5), 577–587.
Rindova, V., Barry, D. and Ketchen, D.J. (2009) Entrepreneuring as emancipation. *Academy of Management Review* 34(3), 477–491.
Rosa, P.J. (2006) Entrepreneurial motivation in developing countries: what does "necessity" and "opportunity" entrepreneurship really mean? *Frontiers of Entrepreneurship Research* 26(20), 4.
Rotter, J.B. (1966) Generalized expectancies for internal versus external control of reinforcement. *Psychological Monographs: General and Applied* 80(1), 1.
Ruvio, A.A. and Shoham, A. (2011) A multilevel study of nascent social ventures. *International Small Business Journal* 29(5), 562–579.
Schiavo-Campo, S. and Judd, M. (2005) *The Mindanao Conflict in the Philippines: Roots, Costs, and Potential Peace Dividend.* Conflict Prevention and Reconstruction Paper No. 24. World Bank, Washington, DC. Available at: http://siteresources.worldbank.org/INTCPR/214578-1111996036679/20482477/WP24_Web.pdf (accessed 9 December 2016).
Schoon, I. and Duckworth, K. (2012) Who becomes an entrepreneur? Early life experiences as predictors of entrepreneurship. *Developmental Psychology* 48(6), 1719–1726.
Schøtt, T., Kew, P. and Cheraghi, M. (2015) *Future Potential: A GEM Perspective on Youth Entrepreneurship.* Global Entrepreneurship Monitor, London.
Schumpeter, J.A. (1965) *Geschichte der ökonomischen Analyse I und II.* Vandenhoek & Ruprecht, Göttingen.
Shane, S.A. (1997) Who is publishing the entrepreneurship research? *Journal of Management* 23(1), 83.
Shane, S. and Venkataraman, S. (2000) The promise of entrepreneurship as a field of research. *Academy of Management Review* 25(1), 217–226.
Solomon, G.T. and Winslow, E.K. (1988) Toward a descriptive profile of the entrepreneur. *The Journal of Creative Behavior* 22(3), 162–171.
Stewart, W.H., Watson, W.E., Carland, J.C. and Carland, J.W. (1999) A proclivity for entrepreneurship: A comparison of entrepreneurs, small business owners, and corporate managers. *Journal of Business Venturing* 14(2), 189–214.
Stewart Jr, W.H., Carland, J.C., Carland, J.W., Watson, W.E. and Sweo, R. (2003) Entrepreneurial dispositions and goal orientations: a comparative exploration of United States and Russian entrepreneurs. *Journal of Small Business Management* 41(1), 27–46.
Steyaert, C. and Katz, J. (2004) Reclaiming the space of entrepreneurship in society: geographical, discursive and social dimensions. *Entrepreneurship and Regional Development* 16(3), 179–196.
Taormina, R.J. and Lao, S. (2007) Measuring Chinese entrepreneurial motivation: personality and environmental influences. *International Journal of Entrepreneurial Behavior and Research* 13(4), 200–221.
Technonet Asia (1981) *Entrepreneurs Handbook.* University of Philippines Institute for Small Scale Industries, Singapore.

Teoh, H.Y. and Foo, S.L. (1997) Moderating effects of tolerance for ambiguity and risktaking propensity on the role conflict-perceived performance relationship: evidence from Singaporean entrepreneurs. *Journal of Business Venturing* 12(1), 67–81.

Thomas, A.S. and Mueller, S.L. (2000) A case for comparative entrepreneurship: assessing the relevance of culture. *Journal of International Business Studies*, 287–301.

Timmons, J.A. (1978) Characteristics and role demands of entrepreneurship. *American Journal of Small Business* 3(1), 5–17.

UFE (2014) *Mindanao*. Underfunded Emergency Window, UN Central Emergency Response Fund, New York. Available at: http://www.philippines.hvu.nl/mindanao1.htm (accessed 20 December 2014).

UNCDF (2014) *Financial Inclusion of Youth*. United Nations Capital Development Fund, New York. Available at: http://bit.ly/1Nc7x5D (accessed 20 January 2015).

UNDP (2004) *Unleashing entrepreneurship: making business work for the poor.* Report to the Secretary-General of the United Nations by the Commission on the Private Sector and Development. United Nations Development Program, New York.

UNSD (2008) *Millennium Development Goal Indicators*. United Nations Statistics Division, New York. Available at: http://unstats.un.org/unsd/mdg/Metadata.aspx?SeriesId=580 (accessed 10 February 2016).

UPLOAD JOBS (2012) University Partnership Linking OSY to Agri-Entrepreneurship and Development to promote Job Opportunities and Business Scale-Up (UPLOAD JOBS) for Mindanao. Available at: http://manoa.hawaii.edu/ctahr/uploadjobs/ (accessed 1 October 2012).

USAID (2014) *Mindanao Youth for Development*. USAID, Manila, Philippines. Available at: http://www.usaid.gov/philippines/education/mydev (accessed 9 January 2015).

USAID (2016) *Our Work in Mindanao*. USAID, Manila, Philippines. Available at: https://www.usaid.gov/philippines/cross-cutting/mindanao (accessed 9 January 2015).

Utsch, A. and Rauch, A. (2000) Innovativeness and initiative as mediators between achievement orientation and venture performance. *European Journal of Work and Organizational Psychology* 9(1), 45–62.

Zacharakis, A. (1997) Entrepreneurial entry into foreign markets. *Entrepreneurship: Theory and Practice*, 23–40.

Zahra, S. and Dess, G.G. (2001) Entrepreneurship as a field of research: encouraging dialogue and debate. *Academy of Management Review* 26(1), 8–10.

Zhao, H., Seibert, S.E., and Lumpkin, G.T. (2010) The relationship of personality to entrepreneurial intentions and performance: a meta-analytic review. *Journal of Management* 36(2), 381–404.

6 Coping Strategies for Youth Entrepreneurs in Conflict Areas

Tina S. Lee,[1]* Katherine A. Wilson,[1] Catherine Chan,[1]
Jovelyn Bantilan[2] and Emilie Bayona[2]
[1]*University of Hawai'i at Mānoa, Honolulu, Hawaii;*
[2]*Southern Christian College, Midsayap, Philippines*

6.1 Introduction

Entrepreneurship is considered a pathway to global peace and economic development that increases localized economic freedom for marginalized populations to escape the poverty trap (Strong, 2010). Entrepreneurship has been shown to have a positive impact on job creation, economic growth, and innovation creation (Van Praag and Versloot, 2008). In countries or regions that have dire need to increase jobs and growth, entrepreneurship has been recognized as a solution. International development aid funds have designed and implemented market development initiatives to help crisis and post-crisis areas by increasing the participation of small enterprises in the economy (SEEP, 2007). These initiatives work at various levels, from individuals to policy institutions, leading to an increase in micro and small-scale entrepreneurship.

It is critical to consider that, during violent conflict in a developing country, citizens are stressed in ways that alter their ability to participate freely in the market, or alter the market itself. The conflict environment and local economy may not facilitate entrepreneurship as readily, as strong institutions and infrastructure that support stable markets are lacking. Entrepreneurs living in conflict areas need to be able to cope within their own lives, and as an entrepreneur, in order to have a sustainable enterprise. Within their own lives, people need to make decisions on a short-term and long-term basis that allow their households to survive during times of conflict and poverty. If they choose to become self-employed or entrepreneurial, they have a different suite of problems they must cope with on a daily basis. Ultimately, their ability to sustain their household can depend on what type of strategies they utilize. The impact of violent conflict and the way households cope with conflict is still not well researched or understood (Ciarli *et al.*, 2010; Brück *et al.*, 2013). While stabilizing market development can help crisis and transitional economies, market opportunities and challenges in these economies are not well understood (SEEP, 2007). Due to the security issues involved in working in conflict areas, the strategies that households and individuals utilize to cope with the stresses from conflict and rebuild their local economy are not well researched.

Coping strategies, in this chapter, refer to the ex-post measures taken to cope with short-term or long-term disturbances and shocks caused by economic changes, political changes, natural disasters, or violent conflict. In regions

*tslee07@gmail.com

with violent conflict, entrepreneurship in itself can be considered as a coping strategy for households, as opposed to a means of accumulating wealth; other forms of income generation like migration or remittances are sometimes preferred over entrepreneurship (Ciarli *et al.*, 2010). Although the literature does not commonly discuss coping strategies specifically for youth, or specifically for entrepreneurs (Brück *et al.*, 2013), this chapter will use the "coping strategy" lens to understand strategies implemented by youth entrepreneurs aged 18–24 years old.

There is little literature on the ways youth cope in conflict regions and how they generate income and build social capital in an environment of uncertainty and unpredictability. When industries are disrupted, jobs are likely to decrease, providing fewer opportunities for youth to engage in their own income-generating activities. Globally, even in non-conflict regions, youth unemployment remains one of the greatest challenges to overcome (Anyanwu, 2013). Entrepreneurs face challenges regardless of location but these challenges are more pronounced in conflict regions and for the youth entrepreneurs. These challenges need to be addressed and resolved, because these regions and sectors can benefit the most from entrepreneurial activities.

This chapter examines the challenges faced by entrepreneurs in conflict regions, and ways that entrepreneurs cope in conflict regions, with a focus on youth entrepreneurs. It provides a case study from Mindanao, Philippines conducted through a qualitative/quantitative survey on youth entrepreneurs examining their own coping strategies utilized as youth and as entrepreneurs.

6.1.1 Economic challenges faced by entrepreneurs in conflict regions

Conflict can cause entrepreneurship to be deeply hindered by reducing markets and opportunities to invest and increasing uncertainty and costs (Brück *et al.*, 2013). However, even with these negative effects, statistics show that entrepreneurial activities often increase during times of conflict because people need alternative income as wage jobs decrease (Brück *et al.*, 2013). How conflict will affect a business depends on the entrepreneurial activity, the characteristics of the conflict, and their relationship (Brück *et al.*, 2013).

The challenges that entrepreneurs face are strongly correlated to the supply and demand of their market. Every business has to make sure a consistent source of input resources for their product is available and there is an adequate demand for their product. Entrepreneurs have to coordinate the delivery of input resources as well as the delivery of the product to the market. Conflict can change variables in input resources, construction of product, the market size, and the market location, thereby forcing entrepreneurs to adapt. If they cannot adapt to these challenges, they are forced to leave the market. It can be challenging to adapt quickly to these changes, especially when resources tend to increase in scarcity during conflict. If these challenges on the supply and demand are analyzed and possible solutions for quick adaptation proposed, it could help these entrepreneurs succeed and have a positive impact on the economy and their livelihood.

The supply side of business is crucial in order to get the product assembled. Markets in conflict regions can negatively affect the input resources needed to make the product; for example, theft or looting can remove needed resources (Brück *et al.*, 2013). Input resources could become more expensive during times of conflict and entrepreneurs might have to find substitute input resources. If entrepreneurs can overcome those supply challenges, they still have to face market challenges. Conflict can negatively affect accessibility to a market if roads are closed or considered too dangerous to travel on (Brück *et al.*, 2013), or the buyers cannot get to the market because of road closures, causing a decrease in business. However, conflict can also indirectly or directly benefit entrepreneurs (Bennett, 2010). Some entrepreneurs benefit from products that increase in demand during times of conflict; some may benefit from decreased market entry for new entrepreneurs due to conflict (Bennett, 2010). Entrepreneurial activities could also increase locally if another competing market were shut down by conflict impacts (Brück *et al.*, 2013). Demand can decrease if the product is not a necessity for households as their income in times of conflict decreases. The impacts that conflict have on the supply and demand and markets of an entrepreneurial business is variable but, if entrepreneurs are aware of the potential impacts, they can better adapt quickly if they occur.

6.1.2 Unemployed youth and youth entrepreneurs in conflict regions

Focusing on youth when discussing entrepreneurship in conflict regions is pertinent because those aged 15–24 years make up the bulk of human capital in developing countries and sometimes up to 85% of the total population (Nagarajan, 2005). Youth in developing countries who are employed often have informal jobs or are underpaid (ILO, 2013). In the developing world, 37.8% of employed youth are poor, with 17.7% in extreme poverty and 20.1% in moderate poverty (ILO, 2015). These alarming statistics worsen in conflict regions, which have higher consequences, as youth have to resort to undesirable employment for income generation. Unemployed youth in these regions become a target for militia recruiters and child traffickers (Nagarajan, 2005). Case studies in Africa showed that the majority of youth involved in conflict activities were unemployed or jobless youth (Zakaria, 2006).

The World Bank defines youth unemployment as the labor force between the ages of 15 to 24 that is without work but is seeking and available for employment (World Bank, 2014). An important distinction to make is that improving unemployment rates may not improve income levels. If unemployment is low, it does not necessarily mean there is no poverty; many are employed with poor wages and therefore are still at the poverty line (ILO, 2015). In low-income countries, 72.5% of youth can be categorized as having "irregular employment" (International Labour Organization, 2015). Often youth with informal jobs are underpaid so the solution is not just to find them employment, but to find them employment with fair and steady wages.

One solution to youth unemployment in developing countries is self-employment or entrepreneurship (Anyanwu, 2013; ILO, 2013). The 2014 G20 Leaders' Summit recognized this and stated that a strategy to reduce youth unemployment is to encourage entrepreneurship (Mazzarol, 2014). However, while entrepreneurship thrives in conflict regions, collecting data is difficult. Youth entrepreneurs in conflict areas need to manage risk, as well as face challenges associated with their age. Likewise, some entrepreneurial activities can become unproductive if they are illegal, and conflict can sometimes encourage these activities (Brück *et al.*, 2013). The goal to strive for is to introduce youth to legal entrepreneurial options, while preparing them for the challenges of being an entrepreneur in a conflict region.

The initial investment cost of starting a business is a hurdle that all entrepreneurs face but is especially prohibiting for youth, who often do not have any personal wealth or credit, nor source of income. Starting and operating any enterprise requires funding up-front. As small entrepreneurial firms have less created value proportionate to their costs of inputs compared with large firms (Van Praag and Versloot, 2008), they are prone to instability at their initiation. Youth in developing countries do not get formal loans through banks and rely on their family or friends for financial support (SEEP, 2013). Informal entrepreneurs tend to be more prevalent in conflict regions (Branzei and Abdelnour, 2010) and youth are often informal entrepreneurs. These informal establishments tend to make less money but have higher costs than formal entrepreneurial activities (Branzei and Abdelnour, 2010). Youth are not only facing financial investment challenges but also they are not earning as much and their input costs are higher. In order for entrepreneurship to be a successful option for unemployed youth in conflict regions, a solution that is easily attainable by these youth needs to be created. The solution needs to provide enticing employment opportunities to counter being recruited by militia or engaging in other unproductive activities (Richards and Chauveau, 2007). This will not only help the youth, their families, and the economy but also keep militia recruits lower, therefore promoting more peace in the region instead of the youth feeding the conflict (Richards and Chauveau, 2007).

Collecting any data in conflict regions is difficult (Verpoorten, 2009; Brück *et al.*, 2013) and the impacts that conflict causes on businesses is not well researched in academia (Brück *et al.*, 2011). However, thriving businesses and burgeoning entrepreneurship can play a critical role in local and national economies. The impacts and contingencies that entrepreneurship has on the needy in developing countries remains unanswered (Branzei and Abdelnour, 2010).

6.2 Youth Coping Strategies in Conflict Regions

Studying coping strategies employed by youth entrepreneurs to overcome the barriers procured by conflict and the barriers of being a youth will help training methods for entrepreneurship development. Households in developing countries face high income variability and have developed numerous coping mechanisms, including savings and informal insurance (Dercon, 2002). Formal insurance and credit options are rarely utilized to offset income variability (Dercon, 2002; Skoufias, 2003). There is a lack of knowledge about household actions against income shocks caused by conflict (Verpoorten, 2009); however, some studies have been done on coping strategies implemented by households in times of economic crises, conflict, or natural disasters that interrupt their income or employment. Economic crises can decrease job availability, decrease the probability of finding a new job, decrease the level of earnings, and increase unemployment (Skoufias, 2003). One coping strategy resorted to by households is selling off their valuable assets (Skoufias, 2003; Ghorpade, 2013), but this is a very short-term solution and has no long-term gain. Another coping strategy that has negative outcomes is spending less on tangible items such as food (Skoufias, 2003). Both of these have a negative impact on the future generation. The youth in these homes and these coping mechanisms (ex-post) may not allow households ever to escape the poverty trap, transmitting poverty from one generation to the next (Skoufias, 2003). Individuals or households can also cope by finding secure income sources or migrating and therefore reducing risk (Skoufias, 2003). However, most of these coping strategies do not apply to youth specifically, and limited studies have been done on how youth cope with income instability and unemployment in conflict regions. They do not have valuable assets to sell, they do not have control over their family's finances, and due to the costs associated with migration, this may not be an option either. Entrepreneurship can offer a solution for youth who need to gain income during times of conflict to support themselves or to help support their family.

6.2.1 Risk-reducing strategies and conflict avoidance

Risk-reducing strategies are strategies to reduce exposure to conflict. This could include relocation to safer areas and sale of visible valuable assets (Ghorpade, 2013).

A case study of psychology research on youth in urban areas in the USA may draw parallels on the risk-reducing strategies that youth use to stay safe in conflict areas. Community-based participatory research on youth from violent urban North American neighborhoods suggested that youth have a highly nuanced sense of the safety of their neighborhoods, knowing when to avoid certain places to minimize violence and stress (Teitelman *et al.*, 2010). Instead of navigating into unsafe areas, youth functioned in 'micro' environments within neighborhoods and maintained high levels of vigilance (Teitelman *et al.*, 2010). Safe areas for youth may be considered extremely unsafe to the outsider who does not have the residents' highly nuanced security knowledge.

Unfortunately, constantly having to avoid and navigate around unsafe areas is not ideal for an entrepreneur who has to identify ways to reduce costs along a supply chain. Entrepreneurs are encouraged to get to know their suppliers and scrutinize their pricing, quality, and business practices. Localized danger can disrupt an entrepreneur's access to supply routes and resources and discourage youth from engaging in entrepreneurship. Youth entrepreneurs in conflict areas may be forced to work with a hyper-local supply chain, rather than taking a broader, regional, or even global scope, and not be as selective with their suppliers.

6.2.2 Benefiting from conflict

In some cases, entrepreneurs can benefit from conflict, in either a destructive or constructive manner. Entrepreneurial activity may profit from conflict by providing supplies or services to fuel the conflict. There are many examples globally of destructive entrepreneurs who fuel the economy and supplies of rebel groups, thus further contributing to continued violence.

Entrepreneurs can also participate in peace-building initiatives by providing a product or service that is in need during wartime and improves the welfare of citizens affected by the conflict (Brück *et al.*, 2013). The ability for entrepreneurs to benefit from conflict may depend heavily on their geographic location and whether their region is considered to be in-conflict, post-conflict, or does not have conflict. A study of self-employment in Colombia found that communities that are impacted by high homicide and displacement rates at the origin of conflict faced reduced self-employment rates, while communities that were receiving an influx of displaced people increased self-employment (Bozzoli *et al.*, 2011). This study suggested that in conflict areas, self-employment is not being reduced, but is being relocated. While entrepreneurs may benefit from conflict in the short term, the literature still suggests that, in the long term, the negative impacts of conflict overshadow these entrepreneurial opportunities (Brück *et al.*, 2013).

6.2.3 Coping through household and social networks

In conflict areas, social and family networks provide services and support that people do not have access to through government and local institutions. A study of high-risk areas in the Philippines found that, instead of borrowing money through a bank or credit union, household and friend networks provided informal loans and gifts (Lund and Fafchamps, 1997). In these types of risk-sharing relationships, network quality matters: households with more numerous and prosperous friends will receive more informal gifts and loans, particularly in times of hardship (Lund and Fafchamps, 1997). A similar study from Afghanistan also found that credit sources for entrepreneurs do not differ from non-entrepreneurs; i.e. loans from friends and family (Ciarli *et al.*, 2010). These social insurance networks may also serve as low-risk strategies for entrepreneurs to fund their enterprises.

For youth, the household network may also serve as an emotional support and guidance system to cope with conflict. Youth in violent US neighborhoods identified their mothers, aunts, and grandmothers as major assets for guidance and emotional support (Teitelman *et al.*, 2010). This shows that having a robust social network of family and friends is emotionally and financially beneficial and can help with the success of a youth's entrepreneurial endeavor.

6.3 Case Study Introduction

Mindanao, the Philippines' second largest island, has harbored centuries of conflict deeply rooted in its history. Modern-day conflict stems from the migration of Christian and Catholic settlers to the predominantly Muslim sultanates of Maguindanao and Sulu (Schiavo-Campo and Judd, 2005). Today, violence is more complex, involving clans that maintain vibrant shadow economies of illegal items, including guns and ammunition (Vellema *et al.*, 2011). Pockets within the Autonomous Region of Muslim Mindanao, Central Mindanao, and the Sulu Archipelago are led by political sub-groups, or rebel groups, which are increasingly organized and adapted into society so that conflict-related businesses now fuel local livelihoods and economies (Vellema *et al.*, 2011).

Youth are often the victims of the negative impacts of conflict, and can be led to resort to undesirable employment or income generation. The Mindanao population comprises 30% youth aged 15–24, of which 17.2% are unemployed (Philippines National Statistics Office, 2011). In 2011, 40% of unemployed people in the Philippines were youth (ILO, 2012). Specifically within the Autonomous Region of Muslim Mindanao, the youth unemployment rate is a staggering 42%, compared with the national average of 6% (USAID, 2007). In Central Mindanao, only 52% of its young adults, ages 15–24, register to secondary schools (USAID, 2007).

Thirty-five per cent of Overseas Filipino Workers are youth (ILO, 2012); even youth that receive higher education increasingly move abroad to work as unskilled laborers to supplement their family income. Unemployed youth who are not able to migrate or receive an education become targets for militia recruiters and child traffickers (Nagarajan, 2005). Case studies in Africa show that the majority of youth involved in conflicts are unemployed or jobless (Zakaria, 2006). Even if unemployed youth do not partake in counterproductive activities, they are still a burden on their family because they are dependent on their family's income (Nagarajan, 2005). Household savings in the Philippines are low because there is a high population of 0–14-year-olds who have not entered the workplace and therefore are dependent on their family's income (Mapa and Bersales, 2008). It is of value to research how youth can become employed or create income and be a beneficial contribution to society instead of taking part in counterproductive activities such as theft, militia, and financial burdens.

The case study described in this chapter looks at the impacts of conflict on youth and youth entrepreneurs in Mindanao, Philippines, which suffers from a long history of conflict and also a high youth unemployment rate. Qualitative research was conducted as part of the UPLOAD JOBS for Mindanao project (see Chapter 3), a university partnership between the University of Hawai'i at Mānoa/College of Tropical Agriculture and Human Resources/Department of Natural Resources and Environmental Management and the Southern Christian College in the provinces of North Cotabato and Sultan Kudarat. The project ran from July 2012 to July 2015, and qualitative interviews were conducted at the end of the project in 2015. The program chose youth ranging from ages 18 to 24 to participate in entrepreneurship training so that they could have the skills and resources to successfully start their own business and be in control of their income and be a productive part of society. The goal of the project was to create opportunities for unemployed youth and teach them positive income generation activities instead of counterproductive activities, aiming to show youth that they had the ability to be self-employed or entrepreneurs, thereby benefiting themselves, their family, and society.

The UPLOAD JOBS for Mindanao Project implemented an Agri-Entrepreneurship Extension Training program for 200 out-of-school youth in four separate cohorts, modeled to be a condensed course of standard entrepreneurship training at the high-school and college level. Courses including Introduction to Entrepreneurship, Entrepreneurial Marketing, Finance, Business Plan and Development, Production Plans, and New Ventures were taught. In a predominantly agricultural region, the training program built skills in agri-entrepreneurship, yet most of the basic skills courses were taught for entrepreneurship topics, with agricultural topics as supplemental courses. The entire program took place over a period of 9 days: 4 days for the first part, 5 days for the second part. During the first segment, the out-of-school youth formed teams of four to five people and came up with a business concept. After the first segment, teams had 1 month to prepare their business plans. During the second segment, teams were required to give a business plan pitch to a panel of judges. Based on their scores, teams were given a monetary award to secure supplies to start their business, anywhere between US$1000 to US$50. The monetary award was meant to pay for basic machinery and equipment, supplies, and paid labor associated with producing their product.

In between cohort trainings, the program staff implemented activities to build social skills and provide opportunities for the youth to network with each other, and continue their active participation. Opportunities included a 2-day entrepreneurship training course with the community, leadership skills training, sales training, and other social activities. Twelve youth even became agri-entrepreneurship trainers and assisted with all of the cohort trainings. Youth also participated in an annual trade fair/colloquium where they displayed their products to the community and gave presentations on processes to create their product.

With a monetary business plan award, business teams commenced their businesses in their geographic regions. Within a few months, one or two team members typically stopped attending meetings or participating in the business, even selling their products. A few months later, there would be one remaining person running the enterprise. That one individual typically reported earnings, and in some cases even cited their business as a stable source of income. One of the greatest challenges throughout the program was monitoring youth businesses and encouraging them to continue running their business. Even in cases where the youth had earnings, were able to set their own hours and schedule, and finally had some of their own spending money, there would be an obstacle that made them leave their business behind or start selling 'by order only'.

Youth in conflict and poverty regions, despite being out-of-school and unemployed, have informal obligations to their social network, family, and religious organizations. Youth in the project often had informal arrangements with family and friends to provide services and many had to ask their parents' permission to attend the training program. Central Mindanao is an agricultural region, and youth were also expected to work on their family's farms. In the UPLOAD JOBS program, female and male youth participants had different obligations, as did youth that were Christian and those that were Muslim. Women were encouraged to marry by their families; once they were married, the program often stopped hearing from them.

> *A young mother and food-cart entrepreneur lives in the town center of Midsayap in Central Mindanao. At 23 years old, she was the only respondent that had a family and children of her own. Through the UPLOAD JOBS program, the respondent started a fried lumpia and barbeque food cart and was making a profit selling food outside hospitals and school cafeterias. Due to her drive and planning, the UPLOAD JOBS program offered to partner her to incubate her business. Unfortunately, shortly after the interview, the respondent had to stop her business at her husband's request.*

Youth were also obligated to contribute financially to their families. Often, youth entrepreneurs desired to go to school, as schooling was considered a necessity in the region, so they paid for schooling with the profits from their business. Occasionally women, particularly Bangsamoro women, were sent by their families to live and work in the Middle East or East Asia to send money home to their families. Some youth found full-time employment in a big city nearby such as Davao City or even Manila. In more than a few instances, the youth's family stepped in saying the youth had obligations at home and could not continue their business. Despite best efforts to train the youth and provide seed funding, entrepreneurship often was not their path, and at the end of the program, there were around ten businesses that continued their daily operations.

6.4 Methodology

6.4.1 Qualitative (household) survey methodology

At the end of the 3-year program from February 2015 to May 2015, program staff conducted face-to-face surveys with the youth participants. The survey was designed to get insight on the coping strategies that youth utilized to navigate conflict and the strategies they utilized as entrepreneurs in a conflict area. The survey asked about their business, income, and experience, and many aspects of starting a business were assessed including where they obtained the financial means it took to start a business. These surveys provided insight into the factors that an entrepreneur considers when starting a business in a conflict area. The survey was designed with a combination of straightforward questions and more introspective questions to derive their opinion on their environment and surroundings.

Eighteen individuals were selected based on their continuous and active participation with the UPLOAD JOBS for the Mindanao project. Since it is difficult to access interviewees in a conflict area like Central Mindanao, the 18 individuals were interviewed while they were present at UPLOAD JOBS for Mindanao project activities. Due to their active participation in the project, it can be assumed that these individuals have a higher interest in entrepreneurship training, and fewer obligations at home. Surveys took between 45 minutes and 1 hour to administer. Two individuals administered each survey: one English-speaking project staff member coupled with one Philippines-based local language-speaking project staff member who did the interview. The local language interviewer had close relationships with the youth and was able to supplement the interview with clarifying and probing questions. Prior to starting the survey, the respondent agreed voluntarily to complete the survey and was told that they had the option not to answer questions. Surveys were conducted individually and in a private setting.

Research presented in this chapter follows interviews with 18 out-of-school youth (OSY) entrepreneurs from Central Mindanao to answer the following questions. What type of coping strategies do OSY use to navigate conflict in their day-to-day lives? What type of coping strategies do agri-entrepreneurs incorporate? What implications does this have for entrepreneurship training programs in conflict areas?

6.5 Results

In this section, it should be noted that names of survey respondents have been changed to pseudonyms to respect their privacy. At the time the research was conducted, one Philippine peso (PHP) was equivalent to US$0.02.

6.5.1 Participant demographics and profiles

Of the 18 individuals who completed the survey, eight were male and ten were female between the ages of 18–24 years old. The participants had higher levels of education compared with many out-of-school youth in the region: ten had finished high school and eight had done some form of college or vocational education.

In 2000, the average household size in the Philippines was five people (Hogeschool Utrecht, 2000). All 18 individuals had at least two family members working in their household, with the majority having two or three working household members. Household sizes varied from three people to more than 12 (Table 6.1). The average household size was about seven people.

Six of the 18 respondents had no income when they started this training. After training, only two reported they had no income. Incomes ranged from not earning at all to an income of 84,000 PHP/year. The average income of all working respondents was 23,878 PHP/year, or about US$477.56/year. The average per capita poverty threshold for the Philippines in 2012 was almost 19,000 PHP/year and the average family income in 2009 for Northern Mindanao was 165,000 PHP (Philippine Statistics Authority, 2012). Table 6.2 demonstrates the importance of youth income on household income. The majority of the 18 individuals' households made less than the average family in Northern Mindanao. With youth bringing in their own income, this could alleviate some financial stress they may place on their households. Eight individuals reported they are expected to contribute income to the household if they are earning money; ten said they

Table 6.1. Household size and employment of rural out-of-school youth entrepreneurs in Central Mindanao, Philippines. Data taken from household surveys completed between February and May in 2015.

Question posed (*Total number of respondents = 18*)	Number of responses	Percentage (%)
Total number of household members (n=18)		
0 to 2	0	0.0
3 to 5	7	38.9
6 to 8	5	27.8
9 to 11	4	22.2
12 or more	2	11.1
Total number of working household members (n=18)		
0 to 1	0	0.0
2 to 3	12	66.7
4 to 5	5	27.8
6 to 7	0	0.0
8	1	5.5

Table 6.2. Personal and household income characteristics of rural out-of-school youth entrepreneurs in Central Mindanao, Philippines. Data taken from household surveys completed between February and May in 2015.

Question posed (*Total number of respondents varies per question*)	Number of responses	Percentage (%)
Household income (Philippine pesos per year) (n=15)		
0 to 50,000	6	40.0
51,000 to 100,000	3	20.0
101,000 to 150,000	4	26.7
151,000 to 200,000	2	13.3
Individual income (Philippine pesos per year) (n=17)		
0	2	11.8
1 to 10,000	3	17.6
11,000 to 40,000	6	35.3
41,000 to 80,000	2	11.8
81,000 to 100,000	4	23.5
Does your family expect you to contribute to the household income when you have earned income? (n=18)		
Yes	8	44.4
No	10	55.5
Does your family expect you to contribute to the household income when you have earned no income? (n=18)		
Yes	3	16.6
No	15	83.3

were not expected to contribute. Only three reported that they were expected to contribute income to the household if they were not earning an income and 15 said they were not expected to contribute.

Before the training program, ten of the 18 interviewed were unemployed and eight were employed, including self-employment. Fourteen of the 18 were employed after the program, with three employment statuses unknown. A total of eight out of the ten originally unemployed respondents were employed after the program. On average, income characteristics of the same 18 individuals improved after their participation in the training program. For some of the respondents, their reported income included a combination of part-time employment, self-employment, and occasional employment.

6.5.2 Employment profile and wages

Respondents were asked to respond to questions based on formal, informal, and entrepreneurial work that brings in monetary income. In SOCCSKSARGEN, the region in the Philippines that makes up Central Mindanao, the minimum wage in October 2015 was 275 PHP/day non-agriculture, 257 PHP/day agriculture, and 255 PHP/day retail/service establishment (DOLE, 2015). It is expected that youth from ages 18 to 24 make the minimum wage or less, as they tend to be prone to a youth wage discount (Grimshaw, 2014). Respondents had a combination of employment profiles: some balanced part-time jobs with managing their enterprises, some did not work at all, and some worked in full-time jobs. Forty-seven per cent of respondents worked below 8 hours a day, and 53% worked above 8 hours a day. While 47% worked most of the year (9 to 12 months), the other 53% worked zero to 8 months out of the year. Most of the respondents made between 100 and 200 pesos per day, with a few outliers who either didn't make any money, or made 300 to 500 pesos per day. The individual making 300 pesos per day was an entrepreneur selling balut, a popular Filipino street snack, in a far-away neighborhood known for conflict. Eighty-one per cent responded that their pay rate was higher than the minimum they were willing to accept. This shows that the majority were happy with their wages.

Youth employment and wages are variable between full-time employment, part-time employment, self-employment, school, and lack of employment.

> *Jessica works full-time at her uncle's roadside burger restaurant 11 hours per day, 12 months of the year. Although she went through the UPLOAD JOBS program, she does not have any type of enterprise. Her restaurant job pays 150 PHP per day, and she is saving up to own her own restaurant.*

Many respondents who were enrolled in school paid for their tuition and other school fees through self-employment. In fact, they continued their self-employment "by order only" in order to pay for schooling. As out-of-school youth, they are considered to be non-traditional students that dabble in school when they have money and quit when they do not have money. Across the board, when asked what they needed to do to achieve their desired careers, respondents identified education, either formal school or informal skills training, as a necessary step.

In order to determine how their income was allocated, respondents were asked questions to see how their income was spent every month. Respondents said that on average, 34% of earnings was for family, 46% for personal consumption, and 20% was saved. This varied from respondent to respondent depending on their personal and familial obligations. One respondent said that they gave 80% of their income to their family, while another respondent saved 70% of their monthly income. The Philippines has one of the lowest saving rates in East Asia (Mapa and Bersales, 2008). For the youth in this study, the majority of their earned income went towards personal consumption (46%) and the least amount of income to savings (20%). Only one respondent reported saving 70%, one 50%, and the majority, 12 individuals, reported saving less than 20%.

> *Ahmed, a Bangsamoro youth from the Municipality of Esperanza in the Province of Sultan Kudarat province, lives in a household of ten members, five of them working. In total, their household makes 100,000 PHP per year. As a student, Ahmed works eight hours a day driving a tricycle and helping his family grow rice, which is not paid monetarily. For the small jobs where he is paid, 80% of it goes to his family, and 20% is used for his own personal needs.*

The minimum and maximum reported percentages for each question are reported in Table 6.3. The mode is also reported in Table 6.3. Based on the data, a lower percentage of the youth's income went to family and personal savings than using their income for personal consumption.

Youth unemployment causes a burden on families because the family has one more person for which they are financially responsible. In the Philippines, if youth need money they often ask family or friends for informal loans to cope with income instability (Lund and Fafchamps, 1997). In this case study, this proved to be the case with 15 respondents saying if they needed money, they would ask their family and two would ask their friends (Table 6.4). Nine respondents out of the 15 respondents said they were responsible for paying their family back in full, while five respondents said they were not responsible for paying back the money, and one respondent said they had to pay 50% back (Table 6.4). In addition, youth are not able to qualify for commercial loans due to lack of assets for collateral to the bank or credit union, thus they are forced to go to their friends and relatives. Certain households

Table 6.3. Personal income allocation by rural out-of-school youth entrepreneurs in Central Mindanao, Philippines. Data taken from household surveys completed between February and May in 2015.

Question posed	Minimum	Mode	Maximum
What percentage of your earnings is for family?	0	20	80
What percentage of your earnings is for personal consumption?	20	50	90
What percentage of your earnings is for savings?	0	10	70

Table 6.4. Money borrowing practices by rural out-of-school youth entrepreneurs in Central Mindanao, Philippines. Data taken from household surveys completed between February and July in 2015.

Question posed (*Total number of respondents varies per question*)	Number of responses	Percentage (%)
If you needed money, where could you go to get it? (n=18)		
Family	15	83.3
Friends	2	11.1
Nowhere	1	5.6
If you received a personal loan, what percent do you have to pay back? (n=15)		
0%	5	33.3
50%	1	6.7
100%	9	60.0
What was/is the interest rate? (n=15)		
**This was a fill-in-the-blank question*		
0%	15	100.0

required youth who made income to contribute financially to the household. Eight out of the 18 respondents reported that if they were making some income, they were expected to contribute to the household income (Table 6.2). Only one individual made a contradictory answer and said zero per cent of their income was for family but they were expected to "give something" to the household income.

6.5.3 Living in a conflict region

Only three of the 17 respondents considered themselves to be living in conflict regions, although all respondents admitted to living near an unsafe area.

> *Paula, who works part-time at the local restaurant and recently enrolled in a local college, also makes cotton twine hammocks to supplement her income. She sells her hammocks at evacuation centers when her barangay gets evacuated due to militant activity in the area. Paula's target customers in evacuation centers are mothers, so she makes child and baby-sized hammocks for their children.*

> *Dawn works part-time as a veterinary assistant and has also just enrolled at the local college. Dawn learned how to grow mushrooms from a government livelihood development program using banana leaves as a substrate. Through the UPLOAD JOBS program, she started to grow mushrooms full-time and even hired her mother to assist. However, her barangay is evacuated every few months for weeks a time, making it difficult to tend to her mushroom plot. When her family is evacuated, she returns during the day to harvest her mushrooms for sale. Currently, her source for substrate is in a conflict area where there are people carrying guns, so her supply of substrate is also disrupted during times of conflict. Likewise, transportation costs to and from the evacuation center are high, and it is questionable whether Dawn will continue her enterprise.*

Conflict often interrupts employment and these out-of-school youth must adapt to these interruptions. The majority of the youth interviewed found alternative, productive activities to keep busy during times of unemployment and found another job or helped work at home (Table 6.5).

In general, youth were shown to be productive during times of unemployment. Of the 14 respondents who answered the question, six reported that they worked on schoolwork or went to school during times of unemployment, four said they helped with household chores, two stated they looked for another job and two reported they did nothing (Table 6.5). The 18 youth were also asked what they did if their income was disrupted. Of the 14 who answered, ten said they would find another job or work at home to get income from family, two stated they would borrow money, and two said they would do nothing. Some youth, however, demonstrated a lack of ambition and motivation

Table 6.5. Out-of-school youth entrepreneur reactions to income disruption and unemployment. Data taken from the household survey completed between February and May in 2015.

Question posed *(Total number of respondents varies per question)*	Frequency	Percentage (%)
What activities do you carry out when you have no employment? (n=14)		
Studying or school	6	42.9
Help around house	4	28.6
Find another job	2	14.3
No activities	2	14.3
What do you do if your income is disrupted? (n=14)		
Find another job or work at home to get income from family	10	71.4
Borrow money	2	14.3
Nothing	2	14.3
If you work on a farm, is it: (n=11)		
A formal arrangement	1	9.1
Based on availability for pay	3	27.3
Family obligation without pay	5	45.5
Other	2	18.2

to stay employed or make money. One of the respondents reported they did not work, they were not expected to contribute to the household income, and they borrowed money from family when they need money.

6.5.4 Doing business in a conflict region

Those who did not consider themselves to live in a conflict region were probed about the security of nearby commercial centers and access roads in order to have respondents consider the impact of conflict on their enterprises and employment. In addition, they were asked about whether they would consider doing business in conflict areas if there was a possibility to increase their income. Three of the 18 respondents said they would take a risk doing business in a conflict region if it would increase their profit. One respondent actually traveled to conflict areas as part of his business model, whereas the other two said they would if necessary.

> *James lives in the town proper of Midsayap in North Cotabato province. As it is a bustling town center close to different schools, he assists his mother part time with her sari-sari store. James was one of the youth respondents with a higher income making 300 PHP per day. Through the UPLOAD JOBS for Mindanao project, he obtained a grant to purchase 6 trays of undeveloped balut eggs. After preparation at home, the respondent takes his balut by scooter to a nearby neighborhood to sell his balut for a 50% higher price. He has an arrangement with the barangay captain, so he is not concerned about his own safety.*

6.5.5 Conflict avoidance

The remaining 11 of the 18 respondents said they did not live in a conflict area in Mindanao and would not do business or go to a conflict area even if it meant increasing income or profit. One of the respondents even said they would not sell in a conflict region even if it meant doubling their profits. One respondent stated that if their neighborhood started having conflict, they would sell in another 'non-conflict' neighborhood. Another respondent said that Bangsamoro/Muslims sold in the unsafe regions, so there was competition in unsafe markets. Each of these responses shows that these individuals had adapted different methods to avoid the risks of doing business in conflict-ridden areas.

6.5.6 Financial support from family and social network

Seventeen of the 18 respondents lived with one of their parents or a guardian; there was one female respondent who lived with her husband and child. Of the 17, 100% of respondents said that they would go to their mother, father, other close family member, or friends for money if they needed it, either for their enterprise or for personal use. While most of the respondents had to pay the loan back completely, a few respondents said that there was no time limit and they did not have to pay it back. All of the respondents said that when they borrowed money, they did not have to pay an interest rate (Table 6.4).

This social norm benefited these youth because their small incomes went further if they did not have to pay back an extra amount. No interest also means that youth may be more encouraged to ask for money to help and grow their enterprises if they know they do not have to pay interest. It also leaves them with more money in the future to further their business. This helps because a major barrier for these youth is financing their business venture. In developing countries, it is common for youth to get loans or money from family and friends instead of taking a loan out from a bank. In a 2002 case study in the West Bank/Gaza conflict region, many youth got loans from family or private options to start their business, which matches this case study's results (Nagarajan, 2005). None of the 18 respondents listed a bank as where they would get money if they needed it. Only four had previously taken out a loan from a bank. This is probably because families do not charge interest. Lund and Fafchamps (1997) completed a study in the Philippines and found that 80% of informal loans from family and friends had zero interest rate.

Since the UPLOAD JOBS program provided them with start-up funds for their enterprises, it is not clear if the respondents needed to borrow money for personal needs only, or for their enterprise. However, the youth responses implied that many youth depended on their family and social networks for financial support. Their ability to succeed as entrepreneurs may also be limited by how robust and wealthy their network is to provide them with interest-free loans. On various occasions prior to the interviews, representatives from local credit unions and banks were invited by UPLOAD JOBS to give presentations to the youth on opening a savings account, taking out a small business loan, and microfinance. Interestingly, respondents said that even if they really needed it, they would still not take out a formal loan or open a bank account.

6.6 Conclusion

The interviews conducted by the UPLOAD JOBS for Mindanao project team were first-hand accounts of the strategies that youth utilize in their personal lives and with their enterprise to cope with conflict in Central Mindanao, Philippines. The youth interviewed had a highly nuanced knowledge of security with an understanding that the impacts of conflict in Central Mindanao are dependent on the exact neighborhood and location. Most of the youth entrepreneur respondents avoided conflict in their surrounding areas even though it had a cost to their livelihoods; a few sought out opportunities within conflict areas to increase their profit margins. Although at the international scale occurrence of violent conflict is often reported on a large geographic scale, the impacts of conflict on-the-ground are highly nuanced and are dependent on place, person, village, etc.

In Mindanao, the youth that were interviewed did not consider themselves to be living in an area within a conflict region, yet were aware of conflict nearby, and avoided it. They had adapted to outside forces influencing the market and still were able to run their businesses. One respondent did mention that her supply chain was disrupted by conflict. She grew mushrooms and used banana leaves as substrate. She could not get the necessary substrate sometimes because she did not like going to areas where people were actively carrying guns. She also had to cope with evacuating her barangay for weeks at a time, leaving her mushroom plots unattended. This example coincides with literature on how conflict can disrupt an entrepreneur's supply chain and how they are forced to adapt. Training programs can be implemented to teach youth to expect disruptions and give them advice on how to adapt. Without support or advice, it can be easy for youth to get discouraged by these disruptions and fail to continue their business.

Youth in conflict regions can have numerous obligations, often because of their role in their family or social network. They also receive benefits from their household, social networks, and even local political structures. Of the 18 youth interviewed, 17 lived in multigenerational households and many had an obligation, either financial or laborial, to their household. The multi-faceted networks can be highly supportive of a burgeoning entrepreneur through money, beneficial partnerships, informal exchanges, marketing, and other exchanges. Of the 17 respondents living in multi-generational homes, all of them responded that they could go to a mother, father, brother, sister, etc. to get money when they needed it, without paying any interest. They also mentioned lack of confidence with official bank loans.

While being a part of a social network helps youth entrepreneurs cope with conflict, their networks also may burden them with different obligations that affect their ability to create livelihood opportunities, build savings, get higher education, and especially engage in new opportunities like entrepreneurship. Throughout the UPLOAD JOBS for the Mindanao training program, project staff reported youth with functioning agri-enterprises who left their enterprises due to family obligations to move and work abroad. The project staff also reported young females who got married, and thereafter making their own livelihoods was no longer necessary or allowed. Twelve of the 18 individuals interviewed stated that they had to receive their parents' approval before attending the training program. Seven out of the 18 stated that they had to receive their parents' approval to start their own business. This suggests that entrepreneurship programs working with youth should integrate family and friends into the curriculum to better relate to youth and to get the family to support the youth's entrepreneurial endeavors. Working with families of trainees may increase the percentage of youth who stay in the program. Having a strong support network can also help the youth cope with supply-and-demand challenges as well as risk-aversion coping mechanisms. In order for entrepreneurship training programs to be successful, the family needs to be involved and persuaded to be supportive of the youth's endeavor, not to discourage it.

Another coping strategy that the youth entrepreneurs utilize is to have multiple formal and informal income streams. Among the youth that were interviewed, some used income to support themselves and their families, others were paying for education and other professional development opportunities. The youth interviewed had a combination of non-monetary labor exchanges (e.g. babysitting, farm labor, laundry), entrepreneurial activities (e.g. choreography, agri-entrepreneurship) and/or informal jobs and work exchanges. Only a few of the youth had formal jobs. This provides an example of income security: if one source of income is disrupted, youth can fall back on another source till the other disruption diminishes or they find a way to adapt to that disruption.

Market development initiatives, like UPLOAD JOBS for Mindanao's agri-entrepreneurship training program, need to be better adapted to the post-crisis context (SEEP, 2007). Market development programs that work with youth in conflict areas should integrate family and friends into the curriculum to better relate to youth. Entrepreneurial programs serving regions affected by conflict should make an effort to learn about how conflict impacts the people they work with, and may alter their ability to be entrepreneurs. This is often difficult given the lack of real-time information during, and even post, conflict periods, but critical for the success of the training initiative. A basic supply chain analysis could be a useful tool for young agri-entrepreneurs who live in conflict areas to identify how to navigate conflict within their business.

Entrepreneurship and business development can contribute to peace building and post-crisis management efforts. From the program design process, entrepreneurship training programs should consider the impacts of continuous conflict and strategies to mitigate the impacts of conflict for entrepreneurs. Programs need to be tailored to take into account market disruption and constraints on target markets, whether for youth or adults.

References

Anyanwu, J.C. (2013) Characteristics and macroeconomic determinants of youth employment in Africa. *African Development Review* 25(2), 107–129.

Bennett, J. (2010) Informal firms in developing countries: entrepreneurial stepping stone or consolation prize? *Small Business Economics* 34(1), 53–63.

Bozzoli, C., Brück, T. and Wald, N. (2011) *Self-employment and Conflict in Columbia*. German Institute for Economic Research, Berlin.

Branzei, O. and Abdelnour, S. (2010) Another day, another dollar: enterprise resilience under terrorism in developing countries. *Journal of International Business Studies* 41(5), 804–825.

Brück, T., Naudé, W. and Verwimp, P. (2011) Small business, entrepreneurship and violent conflict in developing countries. *Journal of Small Business & Entrepreneurship* 24(2), 161–178.

Brück, T., Naudé, W. and Verwimp, P. (2013) Entrepreneurship and violent conflict in developing countries. WIDER Working Paper 2013/028. UNU-WIDER, Helsinki.

Ciarli, T., Parto, S. and Savona, M. (2010) *Conflict and Entrepreneurial Activity in Afghanistan*. UNU-WIDER, United Nations University, Helsinki.

Dercon, S. (2002) Income risk, coping strategies, and safety nets. *The World Bank Research Observer* 17(2), 141–166.

DOLE (2015) *National Wages and Productivity Commission*. Department of Labor and Employment, Manila, Philippines. Available at: http://www.nwpc.dole.gov.ph/rtwpb.html (accessed January 2016).

Ghorpade, Y. (2013) *Coping Strategies in Natural Disasters and Under Conflict: a Review of Household Responses and Notes for Public Policy*. HiCN WP 136. TAMNEAC, Households in Conflict Network, University of Sussex, Brighton.

Grimshaw, D. (2014) *At work but earning less: trends in decent pay and minimum wages for young people*. Employment Policy Department, Employment Working Paper no. 162. International Labour Office, Geneva.

Hogeschool Utrecht (2000) *The Philippines*. Hogeschool, Utrecht, Netherlands. Available at: http://www.philippines.hvu.nl/facts2.htm (accessed 1 January, 2016).

ILO (2012) *Fact Sheet on Youth Employment in the Philippines*. International Labour Organization, MDG Achievement Fund, Makati City, Philippines.

ILO (2013) *Global Employment Trends for Youth 2013: A Generation at Risk*. International Labour Organization, Geneva.

ILO (2015) *Global Employment Trends for Youth 2015*. International Labour Organization, Geneva.

Lund, S. and Fafchamps, M. (1997) *Risk-Sharing Networks in Rural Philippines*. Stanford University, Stanford, California.

Mapa, D.S. and Bersales, L.S. (2008) Population dynamics and household saving: evidence from the Philippines. *The Philippine Statistician* 57(1–4), 1–27.

Mazzarol, T. (2014) *6 ways governments can encourage entrepreneurship*. World Economic Forum. Available at: https://agenda.weforum.org/2014/12/6-ways-governments-can-encourage-entrepreneurship/ (accessed 9 December 2016).

Nagarajan, G. (2005) *Microfinance, youth and conflict: emerging lessons and issues*. USAID, Washington, DC. Available at: http://pdf.usaid.gov/pdf_docs/Pnade303.pdf (accessed 9 December 2016).

Philippine Statistics Authority (2012) *Family Income Statistics*. PSA, Manila, Philippines. Available at: https://psa.gov.ph/content/2012-fies-statistical-tables (accessed 12 December 2016).

Philippines National Statistics Office (2011) *Labor Force Survey 2006–2012: Philippines*. National Statistics Office (NSO), Manila, Philippines.

Richards, P. and Chauveau, J.P. (2007) *Land, agricultural change and conflict in West Africa: regional issues from Sierra Leone, Liberia, and Cote d'Ivoire*. CSAO and SWAC, Paris.

Schiavo-Campo, S. and Judd, M.P. (2005) *The Mindanao conflict in the Philippines: roots, costs, and potential peace dividend*. Paper No. 24, Social Development Papers, Conflict Prevention and Reconstruction. World Bank, Washington, DC.

SEEP (2007) *Market Development in Crisis-Affected Environments: Emerging Lessons for Achieving Pro-Poor Economic Reconstruction*. SEEP Network, Washington, DC.

SEEP (2013) *Understanding Youth and their Financial Needs*. SEEP Network, Arlington, Virginia.

Skoufias, E. (2003) Economic crises and natural disasters: coping strategies and policy implications. *World Development* 31(7), 1087–1102.

Strong, M. (2010) Peace through access to entrepreneurial capitalism for all. *Journal of Business Ethics* 89, 529–538.

Teitelman, A., McDonald, C.C., Wiebe, D.J., Thomas, N., Guerra, T. and Kassam-Adams, N. (2010) Youth's strategies for staying safe and coping. *Journal of Community Psychology* 38(7), 874–885.
USAID (2007) *Jobs for the 21st Century: Synthesis Paper.* United States Agency for International Development, Washington, DC. Available at: http://pdf.usaid.gov/pdf_docs/Pnadj499.pdf (accessed April 2015).
Van Praag, M. and Versloot, P. (2008) The economic benefits and costs of entrepreneurship: a review of the research. *Foundations and Trends in Entrepreneurship* 4, 65–154.
Vellema, S., Borras, S.M. and Lara, F. (2011) The agrarian roots of contemporary violent conflict in Mindanao, Southern Philippines. *Journal of Agrarian Change* 11(3), 298–320.
Verpoorten, M. (2009) Household coping in war – and peacetime: cattle sales in Rwanda, 1991–2001. *Journal of Development Economics* 88(1), 67–86.
World Bank (2014) Unemployment, youth total (% of total labor force ages 15–24). World Bank, Washington, DC. Available at: http://data.worldbank.org/indicator/SL.UEM.1524.ZS (accessed 9 December 2016).
Zakaria, Y. (2006) Youth, conflict, security, and development. In: *The Reality of Aid 2006 Report, Part II: Africa*. Reality of Aid Network, Quezon City, Philippines, pp. 47–60.

7 Allowing Entrepreneurs to Save Profits is Important to Motivation, Sustainability, and Resilience: Can All Cultures Support This?

James R. Hollyer*
University of Guam, Mangilao, Guam

7.1 Introduction

Successful international development—skill development that creates societal transformation rather than a transfer of wealth—requires a keen understanding of the culture where one works (Hall, 1976). As leadership and business guru Peter Drucker said, "Culture eats strategy for breakfast." All the great strategy and plans in the world will not create change or transformation if they do not, in some way, benefit the existing culture (or those in power). Anything that threatens a culture's norms will be resisted by those who benefit from the status quo (Morris *et al*., 2011). In some cases, sole entrepreneurs striking out on their own to make their "fortune" are such a threat in some countries' cultures. Entrepreneurs are especially threatening in cultures where collective behaviors frown on individual success because it disturbs the status quo or the culture's version of "sustainable" or "equilibrium" (Hall, 1976). Yet, a huge untapped potential of creativity and hard work exists just waiting to be given to the world's 1.8 billion young people between the ages of 10 and 24 years. In 2009, Namibian youth commented on their hidden potential, saying "I understand entrepreneurship is an opportunity for me to create a job for myself and for other people as well" (April, 2009). In this chapter, a background on some of the cultural frameworks that can impact individual entrepreneurial behavior will be discussed and compared. Working for someone and being an entrepreneur working for oneself (or family) have very different risk and reward profiles. The chapter will discuss the entrepreneur's needing to retain profits if they are going to maintain their desire to stay in business. For business development efforts, it is critical to have a firm grasp and develop deep knowledge on the "mechanics" or "DNA" of a culture from its many perspectives to increase the chance that entrepreneurial teachings will have meaning and effect in the target culture.

7.2 Getting to Know Cultural Frameworks

In the past 40 years, a great number of studies have identified how thought processes, expectations, cultural norms, and upbringing of one culture are not always readily understood or transferable to another culture, especially when it comes to business practices (Axtell 1993, 1995; Warner-Søderholm, 2013). In *Beyond Culture*, Hall (1976) pointed out that a square

*hollyer@hawaii.edu

peg cannot be fitted into a round hole when it comes to business practices. An entrepreneurial educator needs to spend time with a knowledgeable cultural translator to learn how a culture operates and then identify places where ideas or opportunities might succeed. Hall evaluated the amounts of "said" and "unsaid" information that were brought into a conversation and relationship when two different cultures met on the field of business. Hall defined "high-context" cultures as ones where much content "goes without saying" but is intrinsically known by players because of years of social programming (Table 7.1). At the opposite end of the spectrum, Hall defined "low-context" cultures that leave little unsaid or unwritten, "just to be crystal clear" (Table 7.1).

Hall used Japan as an example of a high-context culture and the USA as an example of a low-context culture. Other researchers have identified additional cultures as high- or low-context over the years (Table. 7.2).

Interestingly, of nearly 200 countries in the world, few developing countries appear on cultural context lists. Clues of how to work successfully in developing countries to assist them in developing their entrepreneur class within their cultural context are nonexistent. Thus, the adage of "when in Rome, do as the Romans do" is simply not that easy to implement for an outsider in the cultural context of many developing countries. Cultural context can be thought of as personal computer operating systems—Apple and Microsoft Windows. Even though the human interface is similar, the software run on the different operating systems is fundamentally different. While outwardly appearing similar, special software, knowledge, and experience are needed to bridge the two systems. The same is true with teaching entrepreneurial practices in a dissimilar society. A translation bridge needs to be constructed. Entrepreneurial practices forged and proven in one country's culture might not be transferable in the location where the training is taking place, as there may be no or few cultural "receptors" for the practices. Meyer (2014) updated the type of knowledge that would assist international practitioners allowing more success in their work through her book, *The Culture Map: Breaking Through the Invisible Boundaries of Global Business*.

Adding to Hall's (1976) high/low-context contribution to cultural framing is another dimension of understanding what operates in a culture by identifying how "collective" or "individual" a society is. Is it about the societal "we" as in collective countries, or is it about the "me" (and my immediate family) as in the more individual countries? This matters especially if the instructor is teaching about individual business behaviors in a collective society: little information is likely to be retained or implemented, because the information goes against the grain of the culture. The entrepreneur training will have little relevance to real life even if some of the ideas are "interesting." Table 7.3 illustrates one of many interpretations

Table 7.1. Cultural context plays an important role in how and why entrepreneurs save their profits (based upon Hall and Hall, 1990).

	Response or Behavior	
Aspect	High-context country	Low-context country
Directness	Implicit assumptions and understanding of the unspoken	Clear and direct communication with little assumed
Non-verbal communication	Very important and frequently used	Less important and often ignored
Emotion	Reserved and internalized	External and expressed
Attitude towards group	Ingroup and outgroup clearly identified	Pragmatic and elastic
Loyalty	Strong and important	Friable and less important
Commitment to relationships	Substantial and long-term	Less important and limited

of the basic tenets of collectivistic and individualistic societies.

These attributes, with variations, are mostly hardwired into the people of a society. Asking an entrepreneurial instructor from one type of society to "get it" and a learner from the other type of society to "get it" will take time, dedicated focus, and willingness to wear someone else's shoes until they feel like their own. And even if someone understands some of a culture's tenets from an intellectual perspective, one may not be able to implement it with an acceptable level of efficacy and/or may completely abandon one's original cultural perspective. Not all high-context cultures are equivalent, nor are all low-context cultures similar. For example, in the Marshall Islands, there are a number of stores owned by Chinese entrepreneurs from the Republic of China and from the People's Republic of China, but only a few are owned by Marshallese. While all three cultures are high-context, the former two operate the majority of the stores. Clearly, there is more to the notion of context and how it relates to business entry and success than deep cultural roots and cooperative relationships. Success also has to do with a culture's acceptance and/or support of individuals who want to succeed *individually*.

Table 7.2. Ranking of countries according to their use of cultural context from high to low (Hall and Hall, 1990; Nishimura *et al.*, 2008).

Higher Context
Japanese
Finnish
Indian
Arabian
Greek
Spanish
Italian
British
French
American and Canadian
Scandinavian
German
Lower Context

Another key aspect in the attributes of collective and individual societies is what is called a "bias for action." The concept was introduced by Peters and Waterman (2006) to identify those people who get things done within a society. Certainly, in both types of societies, entrepreneurs and businesses are working. In training and discussion about entrepreneur development, vis-à-vis the two cultural types, it is important to identify a specific culture's willingness to support those who want to get things done—rapidly—versus those who have a tendency to want to "talk about it." Talking may perhaps be a bias to maintain the status

Table 7.3. Societal values/behaviors by type (from Leake and Black, 2005).

Collectivistic	Individualistic
Interdependence	Independence
Obligations to others	Individual rights
Rely on group	Self-sufficiency
Adhere to traditional values	True to own values and beliefs
Maintain traditional practices	Continuously improve practices (progress)
Fulfill roles within group	Pursue individual goals/interests
Group achievement	Individual achievement
Competition between groups	Competition between individuals
Group or hierarchical decision making	Self-determination and individual choice
Shame/guilt due to failing group	Shame/guilt due to individual failure
Living with kin	Independent living
Take care of own	Seek help if needed
Property shared within group	Strong individual property rights
Elders transmit knowledge (often oral)	Individuals seek knowledge (often textual)
Objects valued for social uses	Objects valued for technological uses

quo or to collect some sort of cultural points for being seen in the room during the discussion. In some cultural contexts, discussion may be viewed as equal to solid progression achieved by action-oriented entrepreneurs. Perhaps it is not so much a "yes" or "no" as to whether a culture has a "bias for action," but rather how tightly the "cultural handbrake" is applied on action-oriented groups or individuals as a way to control progress or success.

> Culture unifies people's behaviour, but it may also create barriers between people, thus nowadays, innovation faces the consequences of culture for various reasons. People's beliefs and behaviour can contribute or block the process of developing and implementing new ideas.
>
> (Kaasa and Vadi, 2008)

In many ways, the "bias for action" boils down to how a culture measures success. For example, is it the number of new women and minority-owned businesses that were supported, or the number of those businesses that were thwarted by using intrinsic cultural mechanisms?

The cultural context, collectivism or individualism, and bias for action form the basis of a culture's DNA, or core operating system, that should be identified prior to engaging in meaningful conversations about entrepreneurial development. To add to the difficulty of understanding or decoding behaviors is actually communicating (and behaving) the culturally correct way while teaching in locations that might not be culturally similar to the instructor. The instructor must know and understand what is actually "allowed" by a culture. For example, can girls or women be entrepreneurs in the culture? Are entrepreneurs celebrated for being willing to stand out from their peers and take risks, i.e. being a "tall nail," "tall poppy," or "escaped crab"? In collective societies, these metaphors for "standing out" are bad and should be avoided at all cost (Hall, 1976). Yet, in individualized societies, they can be highly prized behaviors and may even be rewarded through peer recognition such as a "Small Business Person of the Year" award. Another question to ask, especially in collective societies, is: "What is the ratio of 'free riders' to entrepreneurs?" In an ideal collective society where everyone puts something of value into the pot, sharing is a means to support common needs from a variety of resource holders. The actual exchange or contribution practices of a culture, church, village, chief, family or other entity that might see themselves entitled to the profits from an entrepreneur can vary. The fundamental question and the major point of this chapter is: "Can entrepreneurs retain a portion of their profits for personal use?" Following is a discussion to explain more of the framework behind entrepreneurial decision making and the financial needs of the average entrepreneur.

7.3 Entrepreneurs Must Profit from Their Efforts

Regardless of the size of the entrepreneurial endeavor, maintaining control over and reinvesting profits from a business is critical to the long-term health and success of that entrepreneurial enterprise. "Saving for business purposes should foster entrepreneurial reinvestment" (Beck *et al.*, 2014). Investing in new inputs and paying off loans with profits is a best practice of any business. If entrepreneurs do not pay themselves a salary, accessing their profits allows the hardworking, risk-taking entrepreneur to purchase food, buy school uniforms, pay for water installation in their home, be used for emergencies, or used for future investments such as higher education for their children. Money provides opportunity and options for the business and the entrepreneur. Yet, not all entrepreneurs can and do allocate profits the same. It is a mistake to think that all entrepreneurs follow textbook business-school best practice. In general, entrepreneurs in developed and developing countries, regardless of belonging to a collective or individual society, can view profits, savings, and reinvestment very differently (Beck *et al.*, 2014; Karlan *et al.*, 2014). Cox and Fafchamps (2008) undertook a comprehensive study of the behaviors of transfers in family and kinship networks and concluded that there is great value in looking at evolutionary biology, and the evolution of a whole variety of mutually beneficial and harmful relationships that influence behaviors, often hidden under the visible attributes of a "cultural iceberg." More on this in a moment.

Governments and development sponsors are often puzzled by the lack of "sticking" of the best practices that have been tried to be grafted onto learners in an entrepreneurial development program. How could good educational programs fail? Much of the blame can be attributed to the failure to understand, up front, how profits are allocated in some cultures and other social units. According to some local customs and cultures, profits from businesses are often allocated foremost for (extended) family obligations, thus leaving little or no savings from the entrepreneurs to reinvest or use for their own needs. And all too often, it is not profits that are redistributed but revenues, leaving costs (employees, suppliers, and bank loans) unpaid. In some countries and cultures, extracting wealth (revenues or profits, livestock, root crops, etc.) is a customary behavior or right. It is this extraction, without regard to the long-term business health of the entrepreneur, that kills the golden goose of entrepreneurialism, at least in a number of Pacific countries (author's observation). Why would a new entrepreneur farmer, or ice cream seller, or umbrella maker take the risks and work hard only to have their profits extracted? In this case, there would be little or less return on investment of entrepreneurial capacity and hard work. Thus, for the entrepreneur, there is less of a motivation to continue working hard and taking normal business risks if the benefits of doing so, in various monetary and non-monetary forms, are less than the costs.

7.4 The Money Mechanics of a Culture

So, why are these very important lynchpin signals missed when development programs are conceived or when instructors are deployed to educate in developing countries or areas of conflict? Part of the problem comes from the fact that those from developed countries are looking at the things they can see—for example, that people have similar physical characteristics, or that there appears to be a particular need—and these features are somehow signaled to catalyze a development effort. Innovators and early adopters appear to be asking for knowledge and skill upgrades. This all makes sense—these are common "demand" signals. So, what is being missed? As a number of people looked at Hall's (1976) work and others, they observed that there was an additional dimension that tied into the collective/individual and high/low-context discussion (Hofstede, 1980; Hoft, 1996). This was the visible/invisible individual behavior that people display and act out (directly or indirectly). Some people call it the "cultural iceberg" (Fig. 7.1). These attributes are related to the collective/individual lists, but the fact that they are "below the waterline" makes it hard to appreciate that it is necessary to ask about them. How would a western-trained business instructor even imagine that an entrepreneur's profits might be in danger of being unapologetically redistributed by forces beyond the entrepreneur's control? The concept is formulated in many different ways but the unifying factor is that we do not know what we do not know. Nor do we know how to ask about what is not visible. The lack of true and in-depth data, DNA-like profiles, on a culture and potential individual behaviors can lead to programmatic failures.

Business-like attributes of this iceberg are discussed further below. Nevertheless, conducting more thorough pre-project work-ups could increase the likelihood of real, lasting impacts because the cultural factors that can be positively engaged, rather than aggravated, would be better known. Cultural blueprints or decision trees that help project planners actually see the generic flow of decisions can aid in identifying the "sweet spots" for mutually desirable education, evolution, and impact. These sweet spots may be small, or may be non-existent in some cases, but in other cases, the clarity might lead to significant outcomes. One such area that needs culture-by-culture, or perhaps even at the resolution of the family, clarity would be about exactly how profits are retained, or not, by the entrepreneur before they engage in a western-style business skill-building program.

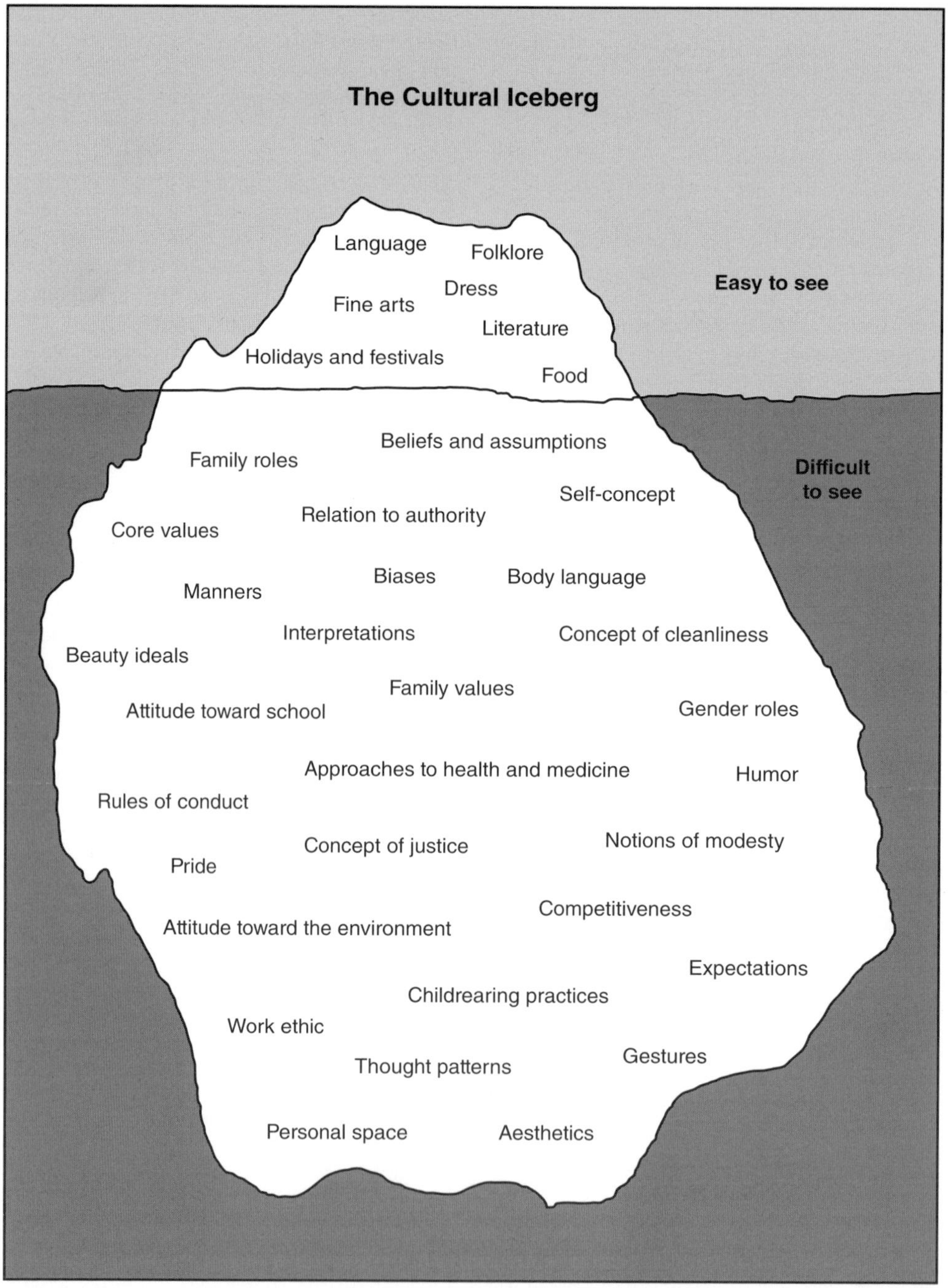

Fig. 7.1. The Cultural Iceberg, demonstrating the majority of challenges unseen that pose barriers to successful entrepreneurial training (reproduced with permission from Jean Slaga, http://www.janinesmusicroom.com/the-rest-of-the-iceberg.html).

7.5 Profit and Revenue Sharing with Others

Entrepreneurs generate revenues, and hopefully profits. Workers generate income. Whether it is profit or income, having money is important for life's necessities. Certainly, there are incomes from investments in property, stock market, and other financial instruments but this chapter focuses on self-generated profits from entrepreneurial activities. Businesses can be built and established in the home country or in a new country. Entrepreneurs in both cases, out of choice and/or cultural obligation, often share the money they make beyond the boundaries of their immediate family to parents, grandparents, uncles, aunts, and other family members. The entrepreneur may even share their gross revenues before paying business costs, thus putting their business and themselves in debt. In addition to their immediate family, an entrepreneur may be required to share their wealth with their extended family, their village, their place of worship, and with others (Connell and Brown, 2005). The practice of sharing wealth from a distance to family, friends, and others is called *remittance*. Remittances can be an important income source for receivers, especially in the Pacific Islands, many developing countries, and areas of conflict (Connell and Brown, 2005; Brown, 1997). Remittances are often from salary and wages, rather than from entrepreneurial activities, but entrepreneurial activities are a potential source of funds. Such draws upon wages or profits can strain a new business as it works to establish itself and build security and stability (Connell and Brown, 2005). Research by Brown *et al*. (2012) indicated that many factors influence giving from a distance, including how long someone has been away, how educated the sender is, and the density of people exerting social pressure on the individual to send money back home. Lubambu (2014) listed reasons for giving, including altruism, insuring the receivers from economic shocks and accumulating future "bargaining power." Fleischer (2007) indicated that the pressure to migrate to a location with better economic opportunities to support the family remaining home can be intense, as in the case of Cameroon migrants. With remittances, a spatial barrier exists between the generator and the remittance receiver that provides some relief from everyday watchful eyes.

In some cultures there can be immense pressure to give to a wide range of receivers when the revenue/profit-generating individual never leaves the home, and their farm, store, or service is visible for all to see and is readily accessible. There is often little relief from the "asks" in this case. "Social claimants" can be a significant extractive force for more successful individuals (Karlan *et al*., 2014). Thus, knowing and discussing the "what-ifs" scenario before exciting a group of (young) entrepreneurs about the possibility of starting their own business is critical to help them plan to be part of the family and communal system, but not at the cost of their entrepreneurial activities. These entrepreneurs need a sound strategy upfront to protect their business assets at some level or, at the very least, secure something of value in return so that they are not less well off in terms of the time and entrepreneurial capacity invested in a business. Therefore, it is critical that a more evolved understanding by national and regional governments, churches, village leaders, and families of what it will take to nurture new business start-ups in their area and country be the subject of international support conversations. Entrepreneurial training will have little impact and new businesses will have short lives if social claimants are too great of an extractive force. A country cannot have it both ways—a strong, thriving small-business sector and a high rate of family and community profit extraction that extinguishes the flame of entrepreneurial passion.

7.6 Using Monetary Assets Until They Are All Gone

Research (Hollyer, unpublished) in the Marshall Islands, the island of Kosrae, and on the Pacific Islands of Yap, Chuuk, Pohnpei, and Palau highlighted family/cultural obligations draining some on-island wage earners of some, most, or all of their disposable income—almost on a weekly basis. One of the Marshall Islands' largest employers noted that most of the employees in his company spent all their income each week and/or were in debt to the limit that the employer would allow (Hollyer, 2014). This lack of savings, and the relative economic freedom forgone because of the heavy income extraction by family obligations, can be very difficult for a wage earner. The wage earner is almost a slave to close and extended family obligations. Some evidence suggests that individuals may make inefficient choices on saving and investment allocations in order to prevent leakage to expectant social networks (Karlan *et al.*, 2014). Hollyer (unpublished) had a conversation with a Kosraean woman in 2013 who had a good job at a hotel and made a fairly good wage. When asked, "Do you save a lot of your income?" the individual said, "No, I eat it and drink it so that I don't have to give it to relatives." While certainly this would not be the case with everyone, the behavior appears to be somewhat pervasive, at least on some Pacific Island nations. Duncan and Nakagawa (2006) looked at six Pacific Island economies (Cook Islands, Fiji, Federated States of Micronesia, Kiribati, Samoa, Vanuatu) and concluded:

> The sharing of wealth within communities also creates problems for the development of enterprise within these communities. What seem to the outsider to be unreasonable demands by relations and others within the community for the income and assets of a budding entrepreneur have ruined many developing enterprises. For example, the advent of a wedding or funeral can lead to requests for the donation of a farmer's draught animal for the celebration. Success in business has often meant finding ways to quarantine the business from such demands.
>
> (Duncan and Nakagawa, 2006)

While these instances are probably rare, income-generating people rebel in their own way against social claimants. It is within this context of extraction that no new entrepreneur can stay in business very long without paying back loans, saving some of the profits to reinvest in the business, or having some profits to freely enjoy of the fruits of their labor and entrepreneurial capacity.

By contrast, in the typical European-American business, the "rugged" entrepreneur retains (quarantines) some-to-much of their profits, but this is not the case in countries with some collective cultures where a "local" business appears to be more of a group asset (immediate family, extended family, and their cultural obligations to an even larger group) than an individual asset. Cultural influences on locally owned businesses can be large and powerful in the non-European-American businesses that were studied by Danes *et al.* (2008). Thus, it is critical when discussing an entrepreneurial development strategy in a highly extractive culture that workable, i.e. mutually beneficial, solutions exist before rolling out the program. Ultimately, without some serious realignment of local perspectives on a local entrepreneur's right to allocate their profits as they see fit, at least to a larger degree, and for the long-term health of the business, there do not currently appear to be many good ways to "quarantine" profits and still have the entrepreneur considered to be a "good" member of the family or the community (Danes *et al.*, 2008). Perhaps it is a matter of showing entrepreneurs the value of savings and reinvesting to the long-term success of their business and also teaching the entrepreneur ways to quarantine some profit. Some savings "mechanisms" do impact savings rates, but tools on the one hand and intrinsic behaviors toward spending and extraction by social claimants on the other may not be enough to create a tipping point for a culture or country to support a Euro-American-style entrepreneur class (Karlan *et al.*, 2014).

7.7 Sex and Relative Power Matter to Savings

A number of studies have attempted to identify "social chromosomes" that impact savings decisions—this time, real ones: *XX* or *XY*, female or male (Beck *et al.*, 2014; Karlan *et al.*, 2014). Beck *et al.* (2014) working in Tanzania concluded that "female entrepreneurs seem more likely to save in formal institutions—perhaps to escape from redistributive pressures." Ganapathy and Mayilsamy (2013) observed that "compared to developed countries, women in developing countries have lower savings. With a low income, it is difficult to save money and hence, the probability of becoming an entrepreneur diminishes as well." Scott (2014) created a matrix of attributes that illustrate the support of female entrepreneurs in a few countries. This matrix contains a specific measurement that covers "entrepreneurial support" for women. It would be helpful to have a list that covers smaller developing nations that are looking to support new entrepreneurs.

7.8 Entrepreneurial Development and Success in Conflict Zones

One of the newest areas of development is work within active conflict zones—growing businesses while dodging bullets and other forms of violence. If life in developing cultures were not difficult enough for the average person, someone who wants to start and grow a business has an even more difficult time succeeding when life conditions can be life threatening. In a review of the post-Cyprus conflict, Katsos and Forrer (2014) stated the obvious:

> First, businesses promote economic development. This is the most basic form of violence reduction that a business can engage in, yet also one of the most powerful. Economic development in this context is simply business doing what it does best: providing jobs (Fort and Schipani 2004), technology transfers (Spencer 2008), and investment (Buckley and Ghauri 2004, Oetzel *et al.* 2007). By providing these basic inputs for economic development in conflict-sensitive regions, businesses help reduce violent conflict.
>
> (Katsos and Forrer, 2014)

Clearly, most businesses are good, but as the UN (2010) *Guidance on Responsible Business in Conflict-Affected and High-Risk Areas: a Resource for Companies and Investors* points out, some businesses create problems for the communities in which they operate. This guide provides suggestions to very large, mostly international businesses. It is suggested that these businesses should be kind, be communicative, have high standards, and be supportive. One suggestion that might be of value to large businesses could be to contribute to new entrepreneur development programs. Not everyone wants to be an employee.

When it comes to the cultivation of micro and small businesses in a conflict zone, the literature is scarce on best practices. These businesses, however, are often the difference between starvation and the start of a highly energized business idea by a young person interested in a better life. While the results of work in conflict zones has yet to be reported, a study on Nigerian business failures by Adisa *et al.* (2014) sums up many of the "normal" reasons for business failure in developing countries. Their list of business failure reasons includes:

- lack of adequate funding;
- poor record keeping and information management;
- inability to distinguish business capital from personal money;
- lack of crucial infrastructural facilities; and
- lack of proper business and management skills/knowledge.

It would be appropriate for development efforts to create a checklist to assess the conditions of an economy and culture and then to offer potential responses to identified less-than-ideal conditions. Alexander (2012) provided a summary of the destructive impacts and lost opportunities in conflict zones. Most things economic simply do not function well when there is massive social and political disruption. However, there are entrepreneurs who persist and succeed under extraordinarily harsh conditions in areas of conflict, though their number is not as great as in peaceful areas. Fortunately, development work led by institutions such as Mercy Corps, Bpeace, Cherie Blair Foundation, The Small Enterprise Impact Investment Fund, and Business Edge, has had a positive impact in conflict zones.

7.9 The Critical Value of Savings to Business Success

Regardless of the sex, age, or marital status of an entrepreneur, if they cannot pay off input costs or purchase new inputs for the next cycle of sales, the business does not last very long. In areas of conflict and subsequent economic instability, private savings that are kept as insurance (Loayza *et al.*, 2000) can be a source of capital to undertake an entrepreneurial activity (Lingelbach *et al.*, 2005). Successful entrepreneurs in areas of conflict are more likely to access private savings from family and social networks to start their business than entrepreneurs in more stable areas (Lingelbach *et al.*, 2005). Furthermore, in areas lacking easy access to credit (either informal or formal), personal and entrepreneurial savings can be important for business liquidity needs (Beck *et al.*, 2014).

A variety of savings options exist for the entrepreneur. Personal and entrepreneurial savings may be deposited into formal savings accounts with established financial institutions, placed in informal arrangements such as Rotating Savings and Credit Associations, held in a household by relatives, or held individually (such as under the mattress or in a jar). In many rural areas of most developing countries, as well as in areas of conflict, formal financial institutions may be absent, leaving the entrepreneur no choice other than informal types of savings. Often the informal savings options are within the entrepreneurs' social network. The savings method does not directly impact the likelihood of entrepreneurial re-investment (Beck *et al.*, 2014); consequently it is the act of saving that is important. Savings in areas of conflict may be especially challenging for entrepreneurs. Access to financial institutions, credit associations, and other financial institutions can be limited. Limited savings options and difficulty and risks associated in reaching formal savings institutions may act as a disincentive for saving by entrepreneurs (Mason, 2007).

One area that is popular for business development is the agricultural sector. The characteristics often used to describe the typical entrepreneur, for example the attributes of a farmer entrepreneur (Fig. 7.2), do not include saving. Nowhere is the entrepreneur asked "Do you have the ability to save some profits (and not to have to give remaining profits to family and others)?" or "Do you have the willpower to save some profits on a regular basis?" The business of business is missing from teaching the business of agriculture!

The ability to control profits is a large factor in the motivation to get up each morning and endure the ups and downs of being an entrepreneur; or at least it is in Euro-American cultures. This control over profits, savings and wealth distribution can be different in other cultures. The DNA of a culture may contain natural behaviors that can be tapped to create more successful businesses. A culture that has been without the need or ability to save anything, even food, during its existence may lack the notion of saving liquid assets, like money. Savings is simply not part of the "social genome" of the culture. Perhaps collective cultures

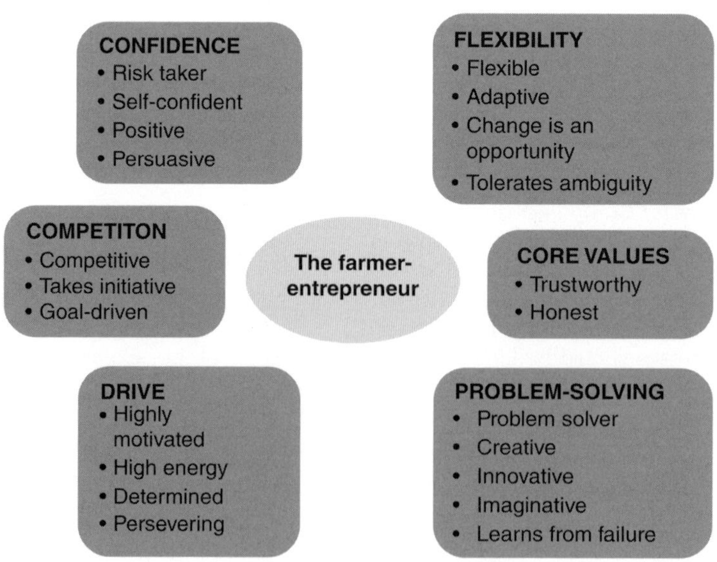

Fig. 7.2. Characteristics of an entrepreneur, from FAO's *Entrepreneurship in Farming* (Kahan, 2013).

evolved because single individuals, or small groups, could not exist and prosper, at some level, without the cooperation of others and so sharing of resources for the common good led to a personal zero-sum situation for individuals. Collective cultures, and even communally minded individual ones, do benefit from risk-sharing (Karlan *et al.*, 2014). Sharing may have an upper boundary where it no longer creates positive benefits, however. Recent studies by Karlan *et al.* (2014) documented that many people in Kenya, Nepal, Philippines, and other places do not save as much as people in Europe and that it is difficult to save in these areas. Scientists, however, have not asked, "Why, what is it about the culture or the social and physical conditions in the culture, that makes it so difficult for people to want to, and actually, save?" People classified as low savers may never have had enough of the right type of "wealth" that was savable. These low savers lived day-to-day and they failed to develop a longer-range perspective like the Swiss, who are one of the world's best net saving nations at a projected rate of 13.4% in 2015 (OECD, 2014). In many cultures, saving is part of the social DNA, whereas in other countries, development projects are trying to graft savings behaviors onto people's existing thought processes.

It is unclear how important having control over profits is in other cultures, especially collective cultures that are highly extractive. Robert Bishop (unrecorded personal communication, 2014) pointed out that the island country of Palau is mostly a reciprocal society. Benefits are derived from distributing (some) profits or excess goods to family and friends. For example, the entrepreneur may get something in return for money of lesser, greater, or equal value such as root crops, fish, or help from family members. The entrepreneur might even gain in stature in the eyes of the family, village, or church. An entrepreneur giver might also be rewarded many years into the future for contributions to the family and others—someone is always keeping score. This is the "social capital," another form of wealth, in some collective societies (Tamasese *et al.*, 2010). In many ways, social capital is as liquid as cash, if it can

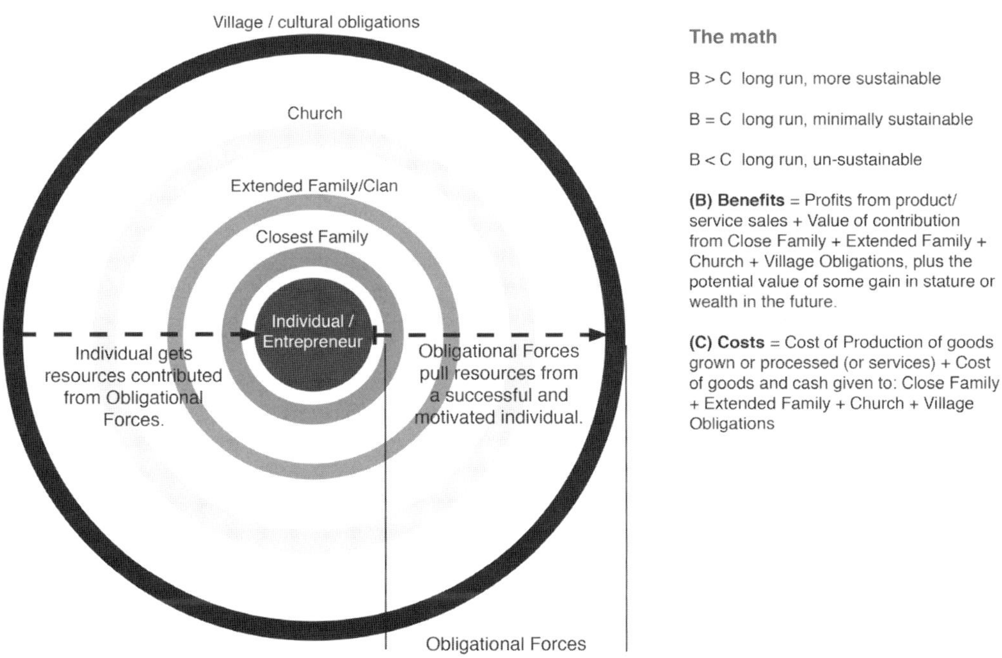

Fig. 7.3. The potential obligational forces acting upon an entrepreneur and the math associated with long-term sustainability of the give and receive relationships.

be employed strategically and as needed by an entrepreneur. The entrepreneur must balance their social capital with monetary capital when establishing a profit-making farm, shop, or food-processing business. A profit-making farm, for example, is different than a subsistence farm where barter with others may help balance the give-and-take of community life. The difference is that social claimants who do not make the money (business revenues or profits) may desire some of the money (or goods) generated by the farm. Figure 7.3 represents the potential pulls-against and contributions-to the motivated entrepreneur. There are rings of extractors and contributors around every entrepreneur. It is the sum of the monetary and non-monetary benefits emanating from the entrepreneur's business, minus the normal business costs, including the contributions to all obligations, that describes the potential sustainability of the business (assuming the entrepreneur has all the needed skills and that there is a market for the product or service where consumers have the ability and willingness to buy).

In general, benefits (B in Fig. 7.3) must outweigh costs (C in Fig. 7.3) for a business to be sustainable. Breaking even, where summed benefits less summed costs is about equal, may work if the entrepreneur either has no choice or little need for saving profits. Little need for savings may exist for unforeseen needs when there is a guarantee that there would be a "delayed benefit," like a balloon payment, sometime in the future if needed that would come from others who recognize the value or contributions of the entrepreneur. However, in a world where there is more and more of a need for cash to buy goods and services, having no profits to spare may eventually extinguish the fire of entrepreneurialism. This is why business development programs must talk about and understand the entrepreneur's savings potential. Mechanisms need to be put in place to help local entrepreneurs quarantine some of their profits to be used at their own discretion. In turn, those profits need to be allocated methodically if the business is to continue for the long run (Fig. 7.4). This type of savings acumen does not exist for many entrepreneurs in developing countries, but is required for long-term financial success. Dupas and Robinson (2013) noted, however, that there is growing interest in saving by individuals and families. Furthermore, additional technical mechanisms are available today than in the past to manage savings. Thus savings are increasing. A Gates Foundation, World Bank and Gallup World Poll study in 2012 documented savings and savings methodologies in 148 countries (Demirguc-Kunt and Klapper, 2012). Entrepreneurs can and do save if the mechanisms to save are present.

For example, in a project evaluation conducted on the island of Yap, it was discovered that a new feed mill for chicken feed was approved and funded based upon a plan to engage *x* number of people growing chickens (Hollyer, 1989). While a seemingly reasonable

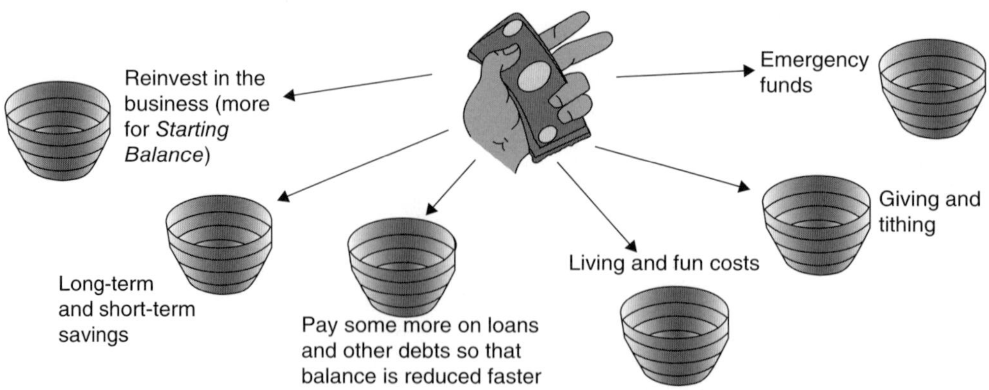

Fig. 7.4. The types of saving that needs to take place if an entrepreneur is to be successful in the short and long run (from Hollyer *et al.*, 2009).

plan with benefits for many—i.e. fresh eggs, chickens, and profits for the growers—there was a catastrophic flaw. The flaw would have been discovered had the planners had the cultural decision-making blueprint for the island's people. Those people to be engaged in growing chickens were subject to having their chickens, eggs, and profits extracted by social claimants of higher rank with little or no direct compensation. The project was doomed from the start as there was no incentive for the lower-rank members to raise chickens that they would most likely not benefit from. The many thousands of dollars of aid money was wasted as the mill stood idle until it fell into disrepair.

Social claimants have also had negative impacts on local entrepreneurs on the Pacific islands of Samoa and American Samoa. In these closely related countries, there is a cultural tradition called *fa'alavelave*. In the *fa'alavelave* custom, family and friends are expected to contribute money and other customary gifts, such as fine mats and food, for a wedding, funeral or on other significant occasions. While this might seem reasonable because goods and money are somewhat redistributed among family members, the growing demands by some chiefs have created great and long-lasting financial burdens on individuals and families. Historically, when this custom was about the exchange of pigs, fine woven mats, and other non-cash items, the burdens and benefits were less on the average person and a form of cultural sustainability was fostered. Now, however, the potential burden of *fa'alavelave* on a small Samoan business could be enough to cause the entrepreneur to close up shop (Feagaimaalii-Luamanu, 2013).

Similarly with young entrepreneurs participating in the USAID-funded Philippines project, Linking Out-of-School Youth to Agri-Entrepreneurship Development (UPLOAD), a number of young people noted that their family took much or most of their profits (personal observation, 2015). With no profits, it was difficult to purchase for the next round of products that would make additional profits (basket "Reinvest ...", Fig. 7.4). While it is understandable that people living with their parents may be required to contribute to family costs, if young entrepreneurs do not have some autonomy over the profits that derive from their entrepreneurial capacity, there is little incentive to stay motivated in the business.

While it may be hard to say "No" to requests for cash or free goods or services from family, friends, the church, or other people who desire the fruits of an entrepreneur's labor, this is something every entrepreneur needs to be empowered to say. Entrepreneurs need to understand the consequences and effects of not saving to their business if they want their business to be around for the long run. This is not to say that sharing and fulfilling cultural obligations is bad in any way, only to say that making moderate and reasonable requests, and reciprocating in some fashion to the giver, will increase the chances that a motivated entrepreneur, in a profitable business, will stay in business. This paradigm shift must be part of any discussion of business development.

7.10 Actionable Ideas for Improvement of the Entrepreneur Class

There are hundreds of studies pointing to what is happening, or not happening, in the support of entrepreneurs. Most studies point to the fact that cultural systems in developing economies are simply not as supportive of entrepreneurs as compared with Euro-American cultures. However, it is the Euro-American culture that supplies a great number of entrepreneurial educators to developing countries to teach people "the right way" to be successful in a business. Preparation and implementation of entrepreneurial trainings will be more impactful if researchers and aid agencies put effort into the following suggestions.

1. Create a physical "DNA social map" of the workings of cultures so that development agencies and trainers might find the spot where entrepreneurs can take business coaching and put it into practice with some success.
2. Design a software-based cultural mapping tool that would improve decision making by showing possible outcomes of decisions.
3. Identify how some collective cultures are able to have economic growth, while others stifle locally owned business with their extractive practices. Look for ways that more successful business behaviors can be borrowed from similar cultures and adapted by another.

4. Lead discussions on sustainability and resilience of small business which will require, perhaps for the first time, that entrepreneurs are able to make more direct decisions on what they must do with profits (and revenues).
5. Create a global goal of 15% savings for entrepreneurs after they pay off debts and taxes, and before contributing to social claimants.
6. Support a pledge of equivalent type of business support that men get, but for youth, women, and minorities.

7.11 Summary

Culture and how wealth is shared within a culture can have substantial impacts on entrepreneurs and the success of their businesses. Sharing wealth is common in most families and to some extent beyond the family. Sharing binds families (both immediate and extended), contributes to cultural obligations, supports religious beliefs, and establishes reciprocal relationships of mutual benefit. Giving and receiving is good and best when balanced to protect the entrepreneur and their business. In order for new entrepreneurs—those interested in generating cash from the sales of goods and services—to be successful, they must be allowed to allocate their profits with a significant level of autonomy. This aspect of business development is missing in many entrepreneurial trainings. If an entrepreneur's profits (or even revenues before all costs are paid) will be extracted by family members and others, without reasonable reciprocation, there is little incentive for a person to enter into or remain in business. This situation must be addressed up front in any development work. While solutions may be limited, it is important to have this tough conversation before skill or other training is completed. An entrepreneur's golden goose must be allowed to spread its wings and keep some of its eggs for personal use. It is clear in many developing economies that the dreams of many new entrepreneurs, and the entrepreneurs themselves, are caught under a thick and wide cultural web similar to the unacknowledged "glass ceiling" experienced by women and minorities who strive to be equals with typically white males in the workplace. These social webs can be repressive and inhibit budding entrepreneurs for being successful. The entrepreneur change-makers grow an economy and they must be supported. Rather like tithing, entrepreneurs should pledge at least 15% of their net profits to savings. A goal like this will help to create the tipping point that energizes a successful, global entrepreneur class.

References

Adisa, T.A., Abdulraheem, I. and Mordi, C. (2014) The characteristics and challenges of small businesses in Africa: an exploratory study of Nigerian small business owners. *Petroleum-Gas University of Ploiesti Bulletin, Technical Series* 66(4), 1–14.

Alexander, D.C. (2012) Facilitating economic activity in conflict zones. *Security Strategies Journal* 8(16), 1–20.

April, W. (2009) Fostering entrepreneurship in developing nations: tapping into the 'hidden potential' of the Namibian Rural Youth. *Journal of Asia Entrepreneurship and Sustainability* 5(1), 78–105.

Axtell, R.E. (1993) *Do's and Taboos Around the World*. Parker Pen Co., White Plains, New York.

Axtell, R.E. (1995) *Do's and Taboos of Using English Around the World*. John Wiley & Sons, Hoboken, New Jersey.

Beck, T., Pamuk, H. and Uras, B. (2014) *Entrepreneurial Saving Practices and Reinvestment: Theory and Evidence from Tanzanian MSEs*. Economics Series Working Papers, WPS/2014-15. Centre for the Study of African Economies, University of Oxford, Oxford, UK. Available at: www.csae.ox.ac.uk/workingpapers/pdfs/csae-wps-2014-15.pdf (accessed 14 December 2016).

Brown, R. (1997) Estimating remittance functions for Pacific Island migrants. *World Development* 25(4), 613–626.

Brown, R., Leeves, G. and Prayaga, P. (2012) *An analysis of recent survey data on the remittances of Pacific island migrants in Australia*. Discussion paper 457. School of Economics, University of Queensland, Brisbane, Australia. 42 pages.

Buckley, P.J. and Ghauri, P.M. (2004) Globalization, economic geography and the strategy of multinational enterprises. *Journal of International Business Studies* 35(2), 81–98.
Connell, J. and Brown, R. (2005) *Remittances in the Pacific, an Overview*. Asian Development Bank, Metro Manila, Philippines.
Cox, D. and Fafchamps, M. (2008) Extended family and kinship networks: economic insights and evolutionary directions. In: Schultz, T.P. and Evenson, R.E. (eds) *Handbook of Development Economics*, Vol. 4. Elsevier, New York, pp. 3710–3786.
Danes, S., Lee, J., Stafford, K. and Heck, R.K.Z. (2008) The effects of ethnicity, families and culture on entrepreneurial experience: an extension of sustainable family business theory. *Journal of Developmental Entrepreneurship* 13(3), 229–268.
Demirguc-Kunt, A. and Klapper, L. (2012) *Measuring Financial Inclusion: the Global Findex Database*. World Bank Policy Research Paper 6025. World Bank, Washington, DC.
Duncan, R. and Nakagawa, H. (2006) *Obstacles to Economic Growth in Six Pacific Island Countries*. World Bank, Washington, DC.
Dupas, P. and Robinson, J. (2013) Why don't the poor save more? Evidence from health savings experiments. *American Economic Review* 103(4), 1138–1171. Available at: http://dx.doi.org/10.1257/aer.103.4.1138 (accessed 9 December 2016).
Feagaimaalii-Luamanu, J. (2013) Burden of fa'alavelave is concern to Lolo admin. *Samoa News*. Available at: www.samoanews.com/node/76103 (accessed 27 July 2015).
Fleischer, A. (2007) Family, obligations, and migration: the role of kinship in Cameroon. *Demographic Research* 16, 413–440. Available at: http://www.demographic-research.org/volumes/vol16/13/16-13.pdf (accessed 8 December 2016).
Fort, T.L. and Schipani, C.A. (2004) *The Role of Business in Fostering Peaceful Societies*. Cambridge University Press, Cambridge, UK.
Ganapathy, S. and Mayilsamy, C. (2013) Women entrepreneurship: a global perspective. *Indian Journal of Applied Research* 3(1), 1–4.
Hall, E.T. (1976) *Beyond Culture*. Anchor Press/Doubleday, Garden City, New York.
Hall, E. and Hall, M. (1990) *Understanding Cultural Differences: Germans, French and Americans*. Intercultural Press, Yarmouth, Maine.
Hofstede, G. (1980) *Cultures Consequences: International Differences in Work-Related Values*. Sage, Beverly Hills, California.
Hoft, N. (1996) *Developing a Cultural Model. International users interface*. John Wiley & Sons, New York, pp. 41–73.
Hollyer, J. (1989) *Economic Assessment of the Yap State Feed Mill*. The Pacific Business Center Program, University of Hawai'i, Honolulu, Hawaii.
Hollyer, J. (2014) *Making positive and permanent steps toward reducing the prevalence of non-communicable diseases in the Republic of the Marshall Islands through increased production and consumption of fresh fruits and vegetables*. US Embassy, Marshall Islands.
Hollyer, J., Beniamina, M., Tokari, N., Avaki, J., Enoka, L., Raui, T., Manuela, T. and Bammann, H. (2009) *Cook Islands: Young Entrepreneurs Program. Business Development Manual*. Cook Islands Ministry of Agriculture, Rarotonga, Cook Islands.
Kaasa, A, and Vadi, M. (2008) *How does culture contribute to innovation? Evidence from European countries*. Working Paper Series, University of Tartu, Faculty of Economics and Business Administration. Tartu University Press, Tartu, Estonia.
Kaasa, A. and Vadi, M. (2010) How does culture contribute to innovation? Evidence from European countries. *Economics of Innovation and New Technology* 19(7), 583–604.
Kahan, D. (2013) *Entrepreneurship in Farming*. Food and Agriculture Organization, Rome.
Karlan, D., Ratan, A. and Zinman, J. (2014) Savings by and for the poor: a research review and agenda. *Review of Income and Wealth* 60(1), 36–78.
Katsos, J.E. and Forrer, J. (2014) Business practices and peace in post-conflict zones: lessons from Cyprus. *Business Ethics: A European Review* 23(2), 154–168.
Leake, D. and Black, R. (2005) *Cultural and Linguistic Diversity: Implications for Transition Personnel*. National Center on Secondary Education and Transition, Essential Tools series, October 2005. *NCSET, University of Minnesota*, Minneapolis, Minnesota.
Lingelbach, D., de la Viña, L. and Asel, P. (2005) What's distinctive about growth-oriented entrepreneurship in developing countries? Center for Global Entrepreneurship Working Paper 1. UTSA College of Business, San Antonio, Texas.

Loayza, N., Schmidt-Hebbel, K. and Servén, L. (2000) What drives private saving across the world? *Review of Economics and Statistics* 82, 165–181.

Lubambu, K.M.K. (2014) *The Impacts of Remittances on Developing Countries*. EXPO/B/DEVE/2013/34. Directorate-General for External Policies of the Union, European Parliament, Brussels.

Mason, A. (2007) Demographic dividends: the past, the present and the future. In: Mason, A. and Yamaguchi, M. (eds) *Population Change, Labor Markets and Sustainable Growth: Towards a New Economic Paradigm*. Elsevier, Amsterdam, pp. 75–98.

Meyer, E. (2014) *The Culture Map: Breaking Through the Invisible Boundaries of Global Business*. PublicAffairs, New York.

Morris, M., Mok, A. and Mor, S. (2011) Cultural identity threat: the role of cultural identifications in moderating closure responses to foreign cultural inflow. *Journal of Social Issues* 67(4), 760–773.

Nishimura, S., Nevgi, A. and Tella, S. (2008) Communication style and cultural features in high/low context communication cultures: a case study of Finland, Japan and India. In: *Proceedings of a subject-didactic symposium in Helsinki* 01/2008, pp. 683–796.

OECD (2014) Household saving rates – forecasts – percentage of disposable household income. Organisation for Economic Co-operation and Development, Paris. Available at: http://www.oecd-ilibrary.org/economics/household-saving-rates-forecasts_2074384x-table7 (accessed 8 December 2016).

Oetzel, J., Getz, K.A. and Ladek, S. (2007) The role of multinational enterprises in responding to violent conflict: a conceptual model and framework for research. *American Business Law Journal* 44(2), 331–358.

Peters, T. and Waterman, R.H. (2006) *In Search of Excellence – Lessons from America's Best-Run Companies*. Harper Business, New York.

Scott, L.M. (2014) Thinking critically about women's entrepreneurship in developing countries. Unpublished paper. Said Business School, Oxford, UK.

Spencer, J. (2008) The impact of multinational enterprise strategy on indigenous enterprises: horizontal spillovers and crowding out in developing countries. *Academy of Management Review* 33(2), 341–361.

Tamasese, T., Parsons, T., Sullivan, G. and Waldegrave, C. (2010) *A Qualitative Study into Pacific Perspectives on Cultural Obligations and Volunteering.* Pacific Section and The Family Centre Social Policy Research Unit, Lower Hutt, New Zealand.

UN (2010) *Guidance on Responsible Business in Conflict-affected and High-risk Areas: a Resource for Companies and Investors*. United Nations, New York.

Warner-Søderholm, G. (2013) Beyond a literature review of Hall's context dimension: scale development, validation & empirical findings within a Norwegian study. *International Journal of Business and Management* 8(10), 27–40.

8 Assessing Gender Gaps in Information Delivery for Better Farming Decisions: The Case of Albania

Edvin Zhllima[1]* and Klodjan Rama[2]

[1]*Agriculture University of Tirana, Albania;* [2]*European University of Tirana, formerly Leibniz Institute of Agriculture Development in Transition Economies (IAMO) and Martin Luther University, Halle-Wittenberg, Germany*

8.1 Introduction

Public sector agricultural advisory services remain one of the most crucial and critical mechanisms to enhance farmers' efficiency and profitability in both developed and developing countries. These services are means for disseminating and changing the modality of supply and adoption of innovative technologies. In transitioning economies, including those exiting conflict or the transitional ones, where structural and institutional changes happen rapidly during the transition, well functioning agricultural advisory services are critical for delivering effective new practices and knowledge to farmers. Thus, it is extremely challenging yet important for advisory or extension services to function well in transitional economies.

Public extension service is a key element for the development of agriculture due to the need for more technological skills and vocational education in rural areas. The extension service plays a key role in providing support and technical assistance to on-farm and off-farm subsectors. The access to extension service in terms of region, sector, and gender is crucial for increasing both fairness and efficiency of the service and boosting agricultural productivity (Lockheed *et al.*, 1980; Ragasa *et al.*, 2013). Not surprisingly, where women have had access and rights to agricultural services, there has been an increase in farm productivity, efficiency, and profit (Saito and Weidemann, 1990; Jiggins, 1994; Ragasa *et al.*, 2013). If gender inequalities in access to land, capital, agricultural extension or inputs were eliminated, agricultural production could increase between 10% and 20% (Sevi *et al.*, 2010). Therefore it is very important to analyze women's access to agricultural extension services as a vulnerable group in terms of access during rapid institutional and structural changes. Albania is an illustrative case on how extension service is being impacted by the transition.

Extension service delivery in Albania experienced a tremendous change as part of the accompanying changes related with land reform and privatization of the agricultural sector. There is a consensus of institutional theories, though, between different approaches that some institutions are easier to change than others where formal rules can change faster than the informal ones such as social norms and values (Williamson, 1998; Roland, 2004).

*ezhllima@ubt.edu.al

In Albania it seems that, despite the formal institutional changes that transferred the land from the state to the household very quickly, the process of informal institutional changes happened very slowly. The increasing role of men in bringing cash income, remittances and the revival of customary behaviors with the demise of state institutional structures in the rural areas may have further weakened women's social and economic influence in the decision-making processes within the household and the local community.

Women, who before 1990 were guaranteed employment under communism, were adversely impacted after the reform. In the aftermath of the closure of state farms and agricultural cooperatives, women struggled with a lack of wage-earning employment opportunities, while men migrated to urban areas seeking jobs (Kodderitzsch, 1999; INSTAT, 2014). By default, women began playing a more active role in subsistence farming to support their households. Despite the important role that women play, they often do not have equal rights to own land and do not enjoy the same accessibility to credit and related services as men.

In addition to the aforementioned challenges, rural women have limited access to reliable scientific and technological information through extension services (Daman, 1997). Studies in the past have reported that male farmers have more access to agricultural extension services than women due to various reasons such as lack of gender sensitivity on providing extension services, women-accessible extension locations, scheduling of extension service activities appropriate for women participants, and lack of female trained staff and extension agents (Obinne, 1995; Daman, 1997; Osuman, 1997). Lack of equal access to extension services may limit women farmers' information and knowledge and reduce their contribution to the rural economy. The negative consequences of limited access to extension services may be more severe in those countries where civil conflicts exist as well as in those areas experiencing male migration resulting in increased incidence of female-headed rural households (FAO, 2011b).

This chapter aims to describe the key issues influencing access by rural women to extension and advisory services in Albania by providing various illustrative figures from supply structures of a typical post-socialist country like Albania. The methodological approach used in this chapter is described in Section 8.2. Section 8.3 summarizes the main programs and strategic documents describing the situation of rural women in Albania. A detailed description of the extension service in Albania and the access women have to this service is provided in Section 8.4. The chapter concludes with a summary and some recommendations with a focus on transition economies.

8.2 Approach

This chapter examines supply structures for agricultural extension services and the gender imbalances in rural areas in terms of access to advisory services as a key element for increasing their economic opportunities. Primary data were collected using structured questionnaires with extension service workers and focus groups with women in rural areas (primary data), in addition to the existing reports/studies that served as secondary sources of data. In order to achieve a high level of participation in the survey, a step-wise process of identification of interest groups and direct involvement of stakeholders operating at central (national institutions such as ministries and national agencies), regional (council regions and regional agencies) and local governance levels (municipalities and local non-profit institutions) was pursued. The identification of the main stakeholders and opinion leaders was done in cooperation with the Ministry of Agriculture, Rural Development and Water Administration (MARDWA) via a snowball method.

The findings from the interviews were analyzed in conjunction with the results of focus groups with women farmers and structured interviews with extension specialists. Two focus group exercises with women farmers were carried out in Berat and Shkoder to identify the main problems hampering women's access to agricultural extension services. Structured interviews were conducted with 128 public extension specialists with the aim of understanding their perception on the involvement of women and men in daily and monthly agricultural and household activities. The respondents were targeted based on a list of extension advisory services provided by MARDWA for each region. Since the goal of this activity was to reach as many extension specialists as possible, the sample was not random but based on the extension officers' availability. Respondents were contacted by telephone while the questionnaires were sent via mail. The questionnaire was based on an extensive literature review, and on consultations with several agro-economist scholars and practitioners. The descriptive analysis of the data was conducted using SPSS.

8.3 Main Constraints Facing Rural Women Empowerment in Albania

The perceived role of women in Albania's rural society currently differs from their real contribution in terms of their workload in agricultural and other economic activities. Although women's role in agriculture is not studied in depth, secondary data suggest that women make up more than half of the agricultural labor force in Albania (INSTAT, 2015). Despite their important role within the agricultural labor force, women remain economically disadvantaged. The position of females in paid activities is inferior to males. According to reports from Albania's Institute of Statistics, women contribute more to the unpaid and informal labor force than men (INSTAT, 2015). Also, women have a higher rate of unemployment (14.3% female, 12.4% male, respectively) and earn less than men, by about 17.4% (INSTAT, 2011).

The generally small farm size (average 1.26 ha) and extreme land fragmentation in Albania have increased farmers' predisposition to look for employment opportunities outside the farm. The extent of off-farm activities make up on average 30% of farmers' working days, and often depend on the availability and types of employment opportunities arising in other sectors of the economy, such as construction, services and transport. These activities are generally considered a domain of men in rural areas, mainly due to distance and the male-dominant structure of these professions. For instance, in the Albanian farms of Puka, Lushnja and Vlora, generally women work more as full-time farmers compared with their spouses (50% to 44%) (USAID/Albania, 2008). Traditionally, off-farm work is performed by the males of the family, while women are relegated to domestic household chores.

Women also face disadvantages in access and ownership of natural resources such as land, forests and pastures. The current legislative framework does not stipulate legal co-ownership of land, and as a result, the majority of land titles do not name the female partner (Mandro-Balili, 2012). This gender-based inequality in access to land and other natural resources is strengthened and even increased by the social relations that are prevalent in most of Albania's rural areas. Zhllima *et al.* (2010) found that females, despite being aware of their rights, did not claim their ownership of land in order to preserve the dignity of the household head. Most of the on-going informal arrangements on property relations among household members, relatives and friends are still based on customary rules (Lemel, 2000; Wheeler and Waite, 2003; Guri, 2007).

Another implication of such customary arrangements on ownership relations relates to women's access to financial resources. A report of the Convention for Elimination of all forms of Discrimination Against Women (CEDAW) found that, due to lack of property rights, women are not able to access credit and other services from financial institutions (UN, 2010). To be granted a loan from financial institutions in Albania, women must document and provide official titles of land ownership, which are required to meet the collateral criteria set by the banks. These land titles are usually signed under the name of the male head of household (the husband or the father-in-law). Although the Albanian legislation does not differentiate between male and female borrowers in the banking credit system, the real-life situation in rural areas is very different.

Improving access to land, capital, information, and especially advisory services (public and private) is important not only for the empowerment of women in rural areas, but also for increasing the overall productivity of farms (FAO, 2011b). Indeed, extension played an important role in the past and continues to do so in the present for disseminating technical and market information to farmers (Anderson and Feder, 2007). In Albania, studies have shown that females in rural areas—especially those born in the past three decades and without access to agricultural vocational education—were particularly more vulnerable in accessing knowledge compared with men in rural areas (Bezhani, 2012). Additionally, rural women suffer from limited direct access to public and private programs and services to support them in their productive role in Albania (IDM, 2010). Although the prominent contribution of women in agriculture is widely acknowledged worldwide (FAO, 2011b), in Albania the focus on gender issues related to rural and agricultural development is still insufficient (World Bank, 2007; FAO, 2011a; Mandro-Balili, 2012).

8.4 Access to Extension and Farm Advisory Services by Gender

To review the accessibility of women to extension and farm advisory services, this section will first describe the role of farming functions by gender, followed by structural characteristics of the extension services pre and post transition, and end with proposing what is needed for the future. In Albania, the division of tasks and responsibilities within the household during the post-socialist transition period has gone hand-in-hand with changes in household structure and transformation of rural institutions. Rural women have, in many cases, become the responsible actor for household food security (Snyder, 1990). The share of family farms officially headed by women is approximately 7% (UN Women, 2015). Although these figures point to a relatively small share of female-headed farms, the actual role of rural women in farm activities is nevertheless far more prominent. Many farms officially headed by males are de facto run by women as a result of high prevalence of male migration (USAID/Albania, 2008). In these cases, women perform most of the farm tasks, including some of the ones performed by men.

Zhllima (2013), based on a focus group survey, found that females generally were responsible for feeding animals, milking, and selling milk (Table 8.1). Activities such as planting, harvesting, land preparation, and application

Table 8.1. Major responsibilities of farm activities by gender in Albania (from Zhllima, 2013).

Tasks	Females	Males	Females and males equally split
A. Livestock			
1. Grazing		√	
2. Milking	√		
3. Processing	√		
4. Trading of meat / milk / dairy products	√		
5.Transporting of meat/milk/milk products			√
B. Crops			
1. Purchasing seeds and fertilizer		√	
2. Irrigating		√	
3. Plowing			√
4. Hoeing			√
5. Harvesting and post-harvest management	√		
6. Transporting		√	
7. Selling fruits and vegetables		√	
8. Renting and leasing land		√	
9. Deciding what crops to plant			√
C. Other activities			
1. Cash management			√
2. Deciding on monetary gifting	√		
3. Purchasing of food	√		
4. Purchasing of durable goods			√
5. Attending meeting in the commune or village		√	
6. Receiving training		√	

of fertilizers were jointly shared by both female and male household members. Males, on the other hand, were generally responsible for irrigation, taking agricultural products to the market (except milk) and for maintaining the relations with the local government in terms of paying taxes, participating in meetings or applying for certain services. The latter tasks were often taken over by women when the male head of the household had migrated away from the farm. Meanwhile, animal grazing was frequently a duty of children.

Purchasing of farm inputs was also mostly carried out by men. Interviews with agrochemical salespeople pointed out that women made up more than 25% of the customers. Women's involvement in these practices was more frequent in areas where males seasonally migrated abroad or were intensively engaged in off-farm employment in urban centers. The roles were found to be less divided in the lowlands and more developed areas of the country as compared with mountainous areas, where land is more marginal.

8.4.1 Assessment of past and current extension service

Transition economies experience more abrupt political and economic reforms due to the need to transform an economy quickly. The rapidness of the transitions has varying outcomes in terms of cost/benefit, returns to investments, and equity rebalance (Alex *et al.*, 2004). Relevant examples that best illustrate the processes of institutional change are the transformative processes that accompanied the collapse of communist regimes in Albania and other countries of Central Eastern Europe (CEE). As a result, the institutional reforms and outcomes were far from being uniform, and the emergence of new institutional arrangements depended on economic dimensions, as well as political and social factors (Swinnen, 1999; Theesfeld, 2005; Stahl, 2010). The same factors also impact the development of advisory and extension services during the transition.

Albania is an illustrative case on how extension service is being impacted by a country's transition. In the early 1990s, Albania was confronted with high inflation (more than 300%) and reduced annual gross domestic product (GDP) growth (–25%), and ended up with a huge budget deficit (50% of GDP) when the country undertook radical market reforms. The aim was to deregulate, decentralize and liberalize institutions in order to have the markets reallocate natural resources (World Bank, 2003). The reforms aimed at controlling and stabilizing the macroeconomic situation with the dismantling of major industries, privatization of agricultural enterprises, retail trading, and small- and medium-sized enterprises, and at the same time were accompanied with a huge flow of foreign aid (Segrè *et al.*, 1999). The Albanian government's determination to establish a market-oriented economy led to the transfer of land and other state properties to private entities. Land redistribution and reform were endorsed by the government in 1991 resulting in transferring free-of-charge land ownership titles to former members of agricultural cooperatives and state farms (Zhllima *et al.*, 2010). About 500 state farms and agricultural cooperatives that operated until 1991 were replaced by about 400,000 small family farms (Zhllima *et al.*, 2010). As a result, a reconstituted centrally based delivery service was required due to the emergence of a very fragmented agricultural sector. This required provisions of new types of activity for extension services and partnership with private actors, as well as transformation in extension service governance.

Similar to the land reform, the extension service delivery change was driven by the government and international donors who played a key role in filling voids in the delivery of agricultural extension services during the transition from a centrally planned economy to a market-oriented economy. Donors often funded and facilitated the creation of the reformed extension service support particularly in some regions of Albania. For example, the Dutch supported the semi-private Regional Agricultural Advisory Centers (RAACs) that provided (for 2 years) advisory services to commercially oriented farmers of the coastal plains (Hotland and Tarelli, 2000).

The provision of extension advisory services changed radically during the first two decades of the Albanian transition, especially in the way farmers accessed information and advisory services. Before 1990, a public agricultural extension service did not exist as a separate organization but was under the state research institutions and state agricultural units of production (the former state farms and agricultural cooperatives) (Daku, 1997). The extension service under that arrangement was logical, given the small number and the size of state farms and cooperatives.

The transition from a centrally planned economy to privately-owned family farming has made farmers' access to advisory services more challenging, particularly for women farmers. Due to the large number of private family farms that emerged with the implementation of land reform in 1991, extension specialists face several difficulties in reaching all farmers to provide advisory services. In cooperatives and state farms, an agricultural extension specialist followed a rigid advisory schedule plan and then provided the advisory and technical assistance plans and schedules to the chief of brigade in state farms as to when and what services were provided. The chief of brigade then executed the plan monthly. To execute the plan, every morning the chief of brigade updated farm workers with the weekly and daily schedule of extension activities. Using this model, women in state farms and cooperatives gained their farm skills and experiences. Due to the routine rotation of tasks between and within brigades (shifts from one sector to another and from one plot to another) most women became knowledgeable on the different production techniques and a wide variety of agricultural production systems. In the aftermath of the 1991 land reform, a different approach was used to carry out extension services. Extension centers were no longer part of centralized agricultural production units, but were attached to and located in the local government premises. Various donor projects (EU-PHARE, Dutch Bilateral Aid, etc.) in 18 districts in Albania (EBRD and FAO, 2007; World Bank, 2007) supported this decentralized approach of service delivery. Each district also had several extension information centers staffed with one or more extension service specialists.

Similar to other Western Balkan countries, the current level of extension service in Albania from both private and state organizations remains rather poor due to the small number of staff who lack up-to-date information and training (World Bank, 2007; FAO, 2011a). The extension service's failure to recruit motivated and innovative staff with new attitudes and behaviors is also a great concern. Few efforts have been undertaken to prepare the next generation of extension staff, a crucial activity as most of the existing staff are close to retirement age (Fig. 8.1). The average age of extension officers is approximately 50 years. Their overall work experience in the agricultural sector is quite long (on average more than 20 years) but their experience in extension service is only a decade. This shorter duration of extension experience might be attributed to the frequent changes in and transferring of staff between different government agencies, such as agricultural directories, agricultural technology

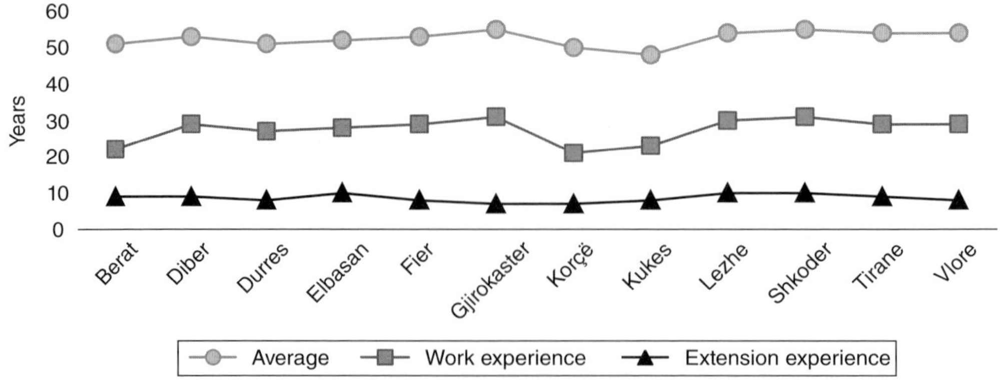

Fig. 8.1. Albania extension staff age and working experience by region for year 2013 (expert information from MARDWA, 2013).

transfer centers, and extension service directories. As shown in Fig. 8.1, for instance, Korçë and Berat regions, which are endowed with rich agricultural resources, have extension staff who on average have lower work experience and in the extension service as compared with other regions. This might have happened as a result of new recruitment and the concentration of agricultural information centers in these regions.

Equipping extension specialists with different technical backgrounds is also an important issue that needs to be addressed. Currently, the majority of extension specialists have degrees in agronomy and animal science (including veterinary sciences, zootechnical sciences, aquaculture, etc.), whereas fields like economics, marketing, and finance are underrepresented (Fig. 8.2).

8.4.2 Capacity-building needs for extension specialists and implications on gender

Despite efforts to enhance the capacities of extension specialists in a broad range of topics, trainings on issues related to finance, marketing, budgets, gender implications, natural resource management, and community development are scarcely covered.

In a study on evaluation of competencies and skills of extension specialists, Zhllima and Kromidha (2012) found that respondents perceived themselves to have skills in new farming technology but a lack of capacity in regards to gender issues, marketing, natural resource management and land-use planning and maintenance (Table 8.2). Extension specialists were more confident in new technology including food safety, cooperation and Instrument for Pre-accession Assistance in Rural Development (IPARD) procedure. Agriculture Technology Transfer Center (ATTC) personnel perceived themselves to have good skills and to be knowledgeable in areas of food safety and post-harvest technology. Non-government organizations (NGOs) perceived themselves to have sufficient skills in new farming technology, post-harvest technology, IPARD compliance, gender issues, and cooperation. Farmers felt competent in new farming technology, farm management, post-harvest techniques and cooperation.

Referring to the perceptions of extension specialists with regards to the majority of tasks covered by women in farming activities, various gaps were identified. The extension specialists perceived that they have competencies on issues less likely to be women's main activities, such as new farming technology and management. The majority of farmers adopting new technologies in the past decade have been mostly male migrant workers returning from EU countries such as Greece and Italy. Also,

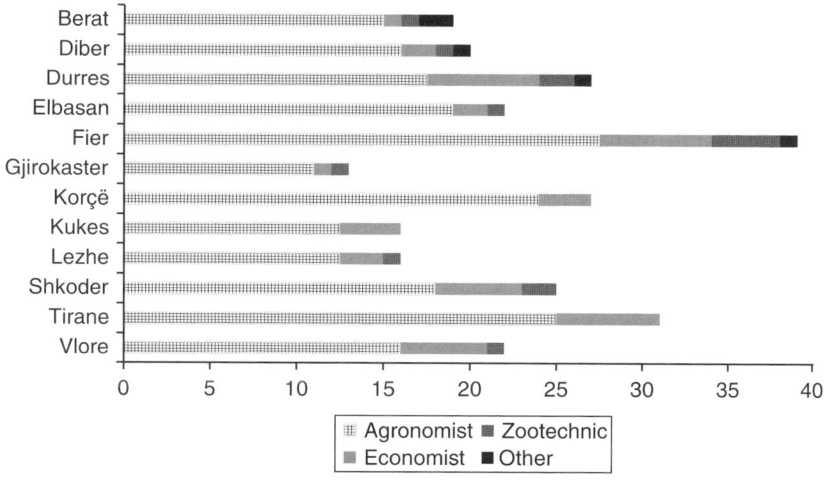

Fig. 8.2. Extension staff profile by specialization by region for year 2013.

Table 8.2. Perception by various stakeholders regarding their competence in various skills (1 = not competent at all; 5 = very competent) (from Zhllima and Kromidha, 2012).

	Advisory stakeholders groups				
Types of skills	Extension service	Input suppliers	ATTC[a]	NGOs	Farmer groups
New technology	5	5	5	5	4
Marketing	1	1	1	2	4
Farm management	4	1	1	1	4
Business plan	2	1	1	2	1
Natural resource management	1	1	4	1	1
Food safety	1	1	3	1	1
Post harvesting	2	2	3	5	3
IPARD[b] compliance	2	1	2	2	1
Gender issues	1	1	1	5	1
Cooperation	3	1	1	5	5
Land consolidation and maintenance of agricultural infrastructure	3	1	1	1	1

[a]Agriculture Technology Transfer Centers, where extension and research experts are housed.
[b]Instrument for Pre-accession Assistance in Rural Development.

the provision of advisory services by the extension specialists is biased towards male activities such as farm management and new farming technology. Moreover, on issues where women are more involved, such as post-harvest technology and food safety, extension specialists appeared even less knowledgeable as compared with other advisory services.

In terms of training needs required to improve their technical skills, extension specialists ranked strengthening of knowledge on business plans, farm management, post-harvest technologies, and food safety standards as the areas with the highest training needs. Another training need identified is knowledge relating to minimal environmental standards. However, none of the advisory service experts mentioned the need for training on gender or related issues such as impact of new value-added small-scale food processing, or land tenure. This highlights the low awareness of the importance of women's contribution to agriculture by extension service specialists.

The level of services provided by the extension staff is based on funding and motivation, level of farm fragmentation and other geographical factors. The budget share of public extension service in the overall expenditures of the MARDWA has been shrinking over the years, accompanied by a reduction in human resources. According to a survey carried out in 2005, only 34% of rural households in Albania had direct access to extension (World Bank, 2007). In less developed (typically mountainous) regions of Albania, there is less access to extension services (both private and public). The Albanian Household Budget Survey of 2006 showed that the services were less available to poorer farm households (28%) in the northern mountainous areas than to the more prosperous farm households in the coastal areas (41%) (World Bank, 2007). Currently, on average each extension specialist in Albania is designated to serve more than 2000 farms but on a regular basis can cover only 20–25% of them, which tend to be the highly commercialized farms. According to previous estimates by the Ministry of Agriculture, only 10% of the farmers contacted by extension staff have been females. In addition to direct communication, women farmers have in many cases been approached indirectly, through contact farmers. This is a service provision approach where the extension specialist meets with an active member of a farmer community who is responsible for diffusing information and training materials to other farmers. In this case, the individual contact has been complemented with a group contact.

This bias in favor of male farmers is illustrated by the farmer training participation and meeting/training location. According to responses gathered by a mail survey with extension specialists during 2012–2013 (Zhllima, 2013), extension activities have scarcely covered women's involvement, although 67% of the extension agents contacted more than 200 farmers per year. More than 60% of the respondents confirmed that less than 10% of the beneficiaries of their service assistance were women.

No women farmers' groups are represented in the planning of the extension program. Furthermore, gender balance is not a requirement in the design of targeted farmers' groups. The target of the extension service in Albania is the mid-size farm with the potential to produce for the market. No criteria exist for defining mid-size farms. Rather, it depends on the interest of the farmers (this excludes farms that can purchase services) and their potential to produce for the market (this excludes small farms producing only for household consumption). No monitoring indicators that divide beneficiaries of the advisory services by gender are in place either. Lack of women farmers as an important group in the activities of extension comes about as a result of the limited representation of women in extension service program planning and in the farmers' contact groups. Approximately 65% of extension staff state that women are not present when and where extension meetings take place. For them, the primary reasons for women's absence in extension activities are "women's hesitation to participate" and "lack of adequate transportation to reach women farmers" as well as their inability to access women farmers (Table 8.3). Women's lack of involvement comes due to lack of time and lack of incentives but also low awareness on the importance of the advisory services to the success of their operations.

A special concern related to women's access to agricultural advisory services is the low involvement of female extension specialists. The share of female extension specialists to total number of extension specialists is 22% (data from interviews with Extension Service Department in 2014) (Fig. 8.3). In some regions the percentage of women extension specialists is extremely low. The recruitment of women compared with men, especially in some mountainous regions, is low. Moreover, few women are included in field advisory services, as the majority of female staff are given office duties such as data entry, reporting, and survey activities. Furthermore, few women are heads of extension service departments in the regional offices.

The education track at the university or training to become an extension agent is also highly saturated with males. A very low number of female students are currently enrolled in agronomy, animal science, and forestry degree programs. Female students avoid these fields since they prefer other disciplines or majors. The government plays no role in supporting the enrollment of female students to study agriculture in disciplines such as plant protection, veterinary sciences, food safety, and

Table 8.3. Problems and constraints of Albanian agricultural extension services in reaching out to women farmers, as perceived by extension personnel (numbers represent percentage of responses in a category) (from Zhllima, 2013).

Rank of importance	No answer	Extremely important	Very important	Somewhat important	Total
Communication skills	2.3	6.3	26.6	64.8	100
Technology	3.1	8.6	47.7	40.6	100
Transportation	1.6	36.7	32.8	28.9	100
Teaching and communication equipment	3.1	17.2	38.3	41.4	100
Too much official work	3.1	6.3	41.4	49.2	100
Women's hesitation	0.8	63.3	28.9	7	100
Coverage of too many target groups	3.9	28.1	41.4	26.6	100
Subject-matter specialists	5.5	18.8	31.3	44.5	100

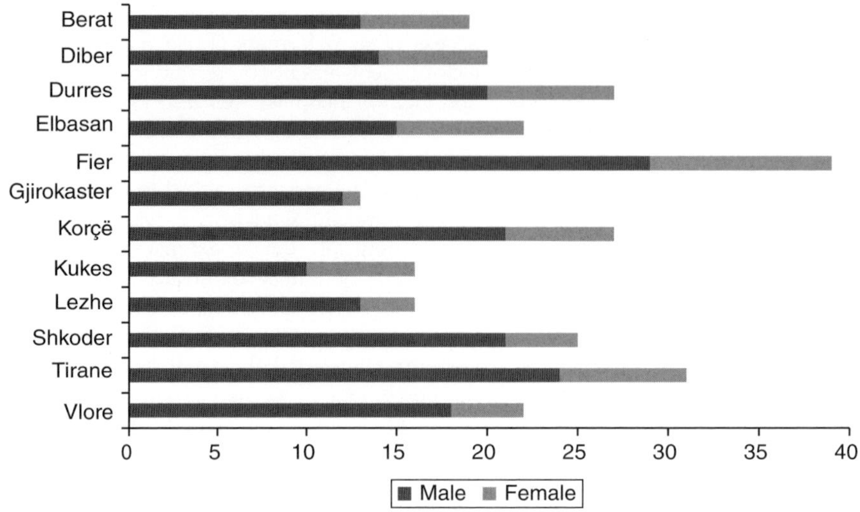

Fig. 8.3. Gender composition by region for year 2012 (expert information from MARDWA, 2013).

environmental sciences. Therefore gender stereotypes are witnessed in these areas of study which also influences the low enrollment of female extension service staff in the future.

According to the opinion of extension specialists, the most important factor in reduced access to extension services for women is the low number of female extension staff (Zhllima, 2013). Female extension specialists claim to establish more contacts with women farmers than do male extension specialists. Another highly ranked problem is coverage of too many target groups by one agent, such as market-oriented farms, cooperatives, subsistence farms and other types of farmers. A less important problem was "lack of subject-matter specialists to serve women's specific problems" such as home economics and services based on some types of tasks and operations that are traditionally conducted by women.

Home economic activities, such as cooking, food preservation, food safety, and handicrafts, are not included in extension (Zhllima, 2013). Very few agricultural extension topics are directed towards women farmers. Instead, more extension training is directed on the integration of women farmers in livestock and annual crop production (mainly vegetable cultivation). Such program contents reinforce gender stereotypes in the extension service and in the community.

8.5 Main Findings and Recommendations

One of the most important factors for progress in agriculture and rural development is advisory services. The agricultural research and extension services in Albania have undergone a radical restructuring since transitioning from a command economy to a market economy in the 1990s. Donor organizations have played a key role in this context. However, despite improvements in some private and public services, most farms in Albania are poorly served. Women are an especially vulnerable group in terms of access to services. The access of women to extension service falls far short of what is needed considering that the public sector investments in agricultural research and extension are under pressure and when approximately 58% of women are employed in the agricultural sector.

Interviews with agricultural experts indicate that, during the past two decades, no systematic efforts to include women farmers in either extension planning or gender-specific agricultural program development have taken place. No budget has been allocated specifically for extension program activities for women farmers. Agricultural extension services in Albania still do not attach much importance to reaching women farmers or women on the farm.

No official working instructions have been given by the extension service to field agents to target women in their regular extension program. Moreover, no extension strategy exists to work with women farmer groups. The data on the limited number of women farmers being reached by agricultural extension activities shown by monthly or quarterly reports, meetings, discussions and visits exemplify gender inequity and could in part be remedied by a directive from the service agency to rebalance the service to women. The extension staff addressing the situation for women are already burdened by multiple tasks, heavy workloads, and the directive to contact many target groups, leaving them no time to focus on the extension advisory needs of women farmers.

Some of the factors contributing to gender inequity can be alleviated through provision of adequate training and capacity building for advisory service staff. As a primary source of information, public agricultural extension services play an important role in agricultural development and can contribute to improving the welfare of farmers and other people living in rural areas. Therefore, the realignment towards greater gender inclusion in supplying extension services in Albania is imperative for balanced and sustainable development. The Albanian government should aim to end gender inequality in order not to limit the growth of agricultural output. Gender inequities in accessing key agricultural information, advice, assets and inputs will limit rural development.

Various steps should be taken in order to improve the access of women farmers to extension services. In a survey of the views of extension staff as to ways to improve the effectiveness of agricultural extension services to women farmers, more than 67% of respondents found it extremely important to have a dedicated budget item for carrying out extension projects for women. More than half of the respondents suggested increasing the budget for demonstration materials/equipment for the implementation of gender-specific projects. Moreover, half of the respondents indicated that they would require adequate transportation budgets in order to expand their reach to women farmers. A smaller proportion of respondents suggested increasing the number of female contact farmers. The following are seven recommendations that would close the gender gaps in advisory services offered to women farmers.

8.5.1 Policy

To develop a policy on gender inclusiveness in delivery of advisory services, an action plan should be designed, implemented and monitored from an inter-ministerial board, along with a multi-stakeholder advisory group. The policy agenda should be adjusted by establishing actions and monitoring tools for increasing the access of women to advisory services. Gender-based budgeting should be allocated to this action plan. The annual program plan of work of the extension service should be studied through a gender lens to ensure that progress is made and goals are achieved.

8.5.2 Stakeholder engagement

Participation by stakeholders such as civil society and private operators to share ideas and knowledge is important, as well as to provide opportunities, resources, and connections to other people through networking. As donor contributions to agricultural development decrease in a transitional economy, the role of civil society and the private sector must increase. Civil society must play an important role in promoting gender equity in advisory services in local areas. Businesses can participate in mentoring activities or in field training. Business partners can introduce skill sets that are not always prevalent in the non-profit sector. For example, during 2013, MARDWA included private entities for carrying extension services in topics related to environmental protection and organic agriculture.

8.5.3 Partnership

Public and private partnerships are equally important in improving gender equity in rural areas. Interviewed respondents regarding this arrangement reflected that training activities went better when local government was assisting the training through sharing offices and conference rooms. Such cooperation allows the implementation of trainings and professional courses as well as assistance on credit applications and development of business plans without obliging women to travel far from home.

8.5.4 Employment

Recruitment of more women is vital to delivering extension services specifically to women farmers. There should be hiring protocols that highlight equal employment opportunities to establish a robust team of women field extension agents trained to cover women-specific training topics. The entire extension staff should be trained to be gender sensitive. Training programs for extension staff to recognize and identify gender differences and use gender analysis tools should be planned and organized.

8.5.5 Programming

New or updated training programs should be introduced that are traditionally tasked by women farmers, such as post-harvest technology, milk storage, animal and poultry husbandry, formation of women groups/cooperatives, food processing and inherent women activities (handicrafts, family planning, child development and cooking).

8.5.6 Mode of delivery

Delivery approaches of extension activities should be evaluated to bring the best effort forward. Traditional extension activities involve training seminars, workshops and field demonstrations conducted in small groups as well as through media dissemination of documentaries and via farmer journals such as *Agrobiznesi* and leaflets (Zhllima and Kromidha, 2012). However, other approaches such as individual mentoring and coaching are needed and may be more effective in addressing gender equity. Mentoring and coaching approaches are used by many donor projects, especially those focusing on the development of small and medium-size enterprises (SME).

8.5.7 Donor aid projects

These recommendations apply also to donor aid projects. Agriculture is a source of self-employment and entrepreneurship; therefore decision and policy makers should promote activities that would give the best value for donors contributing to agriculture and related off-farm entrepreneurial activities in rural areas by targeting women farm operators and women's groups and increasing the participation of women in extension activities.

References

Alex, G., Byerlee, D., Helene-Collion, M. and Rivera, W. (2004) *Extension and Rural Development: Converging Views on Institutional Approaches?*Agriculture and Rural Development discussion paper no. 4. Extension. World Bank, Washington, DC.

Anderson, J. and Feder, G. (2007) Agricultural extension. In: Evenson, R. and Pingali, P. (eds) *Handbook of Agricultural Economics. Agricultural Development: Farmers, Farm Production and Farm Markets* (Vol. 3), Elsevier, Amsterdam, pp. 2348–2378.

Bezhani, M. (2012) *Stocktaking of Potential Training Providers in North-eastern Region of Albania.* EU Policy Brief No. 7. Report for the Swiss–Danish Formulation Mission, Tirana, Albania.

Daku, L. (1997) Investing in Agricultural Extension: The Case of Albania. Research project submitted to the Virginia Polytechnic Institute and State University for the fulfillment of the requirements for the degree of Master of Science in Agricultural and Applied Economics. Virginia, USA.

Daman, P. (1997) *Cooperatives and Poverty Reduction: Enhancing Social and Economic Imperatives.* ICA ROAP, New Delhi, India.

EBRD and FAO (2007) *A Systematic Analysis of the Agribusiness Sector in Transition Economies: The Albanian Olive Oil Sector*. Commissioned by the European Bank for Reconstruction and Development (London) and the Investment Centre of the Food and Agriculture Organization of the United Nations (FAO), Rome, Italy. Available at: http://www.eastagri.org/files/ALB_Oliveoil_VC_final.pdf (accessed 8 December 2016).

FAO (2011a) *Assessment of the Human Capacity Development Needs for, and Gaps in, the Agricultural Advisory Services in Western Balkans*. Final Report, FAO technical cooperation programme facility report. FAO, Rome. Available at: http://www.fao.org/docrep/015/i2119e/i2119e00.pdf (accessed 8 December 2016).

FAO (2011b) *Women in Agriculture: Closing the Gender Gap for Development*. The State of Food and Agriculture 2010–2011. FAO, Rome, Italy. Available at: http://www.fao.org/docrep/013/i2050e/i2050e.pdf (accessed 8 December 2016).

Guri, F. (2007) Study of the Land Market in Albanian Coastal Area (Durres-Kavaje Area): The Importance of Tourism and Land Management Policies. PhD thesis, Agricultural University of Tirana, Albania.

Hotland, G. and Tarelli, I. (2000) *The Albanian AKIS in Transition. Institutional Development and Organizational Changes in the Albanian Extension Service in the 1990s*. Paper presented at International Conference on Eastern European Agricultural Extension. Gogollo, Hungary: 13–18 May.

IDM (2010) *Empowering Albanian Rural Women: Economic and Social Perspectives*. Gender Alliance for Development Centre. Institute for Democracy and Mediation, Tirana, Albania.

INSTAT (2011) *Albania Time Use Survey 2010–2011*. Instituti i Statiskes (INSTAT), Tirana, Albania. Available at: http://www.instat.gov.al/media/171100/albanian_time_use_survey_2010-2011.pdf (accessed 8 December 2016).

INSTAT (2014) *Migration in Albania*. Instituti i Statiskes (INSTAT),Tirana, Albania. Available at: http://www.instat.gov.al/media/242045/migration_in_albania.pdf (accessed 8 December 2016).

INSTAT (2015) *Labor Force Survey 2014*. Instituti i Statiskes (INSTAT),Tirana, Albania.

Jiggins, J. (1994) *Changing the Boundaries: Women-centred Perspectives on Population and the Environment*. Island Press, Washington, DC.

Kodderitzsch, S. (1999) *Reforms in Albanian Agriculture: Assessing a Sector in Transition*. Technical Paper. World Bank, Washington, DC.

Lemel, H. (2000) Tenure security, land use and investment. In: Lemel, H. (ed.) *Rural Property and Economy in Post-communist Albania*. Berghahn Books, New York and Oxford, pp. 50–73.

Lockheed, M., Jamison, D. and Lau, L. (1980) Farmer education and farm efficiency. *Economic Development and Cultural Change* 29, 37–76.

Mandro-Balili, A. (2012) *The Property Rights of Women in Albania: Analysis of Legal Standards and its Implementation in Practice*. UN Women, Tirana, Albania.

MARDWA (2013) Situation of extension service covering main facts. Reports for internal use. Department of Statistics, Ministry of Agriculture, Rural Development and Water Administration, Tirana, Albania.

Obinne, C. (1995) Rural women in food chain activities: A case of Ibo speaking ethnic community of Delta State, Nigeria. *Agrosearch* 1(1), 35–40.

Osuman, G. (1997) *The Education of Women in Developing Countries*. Osuman & Co., Makurdi, Nigeria.

Ragasa, C., Berhane, G., Tadesse, F. and Taffesse, A.S. (2013) Gender differences in access to extension services and agricultural productivity. *The Journal of Agricultural Education and Extension* 19(5), 437–468.

Roland, G. (2004) Understanding institutional change: Fast-moving and slow-moving institutions. *Studies in Comparative International Development* 38, 109–131.

Saito, K.A. and Weidemann, C.J. (1990) *Agricultural extension for women farmers in Africa*. Policy, Research, and External Affairs working papers, no. WPS 398. World Bank, Washington, DC.

Segrè, A., Bogucanin, H., Brandani, A., Canali, M., Giorgi, S., Gomez y Paloma, S. and Zappacosta, M. (1999) *Agricultural and environmental issues for sustainable development in Albania : an agrarian system analysis-diagnosis in Divjaka region*. BESA, Nardò, Italy.

Sevi, S., Manuel, C. and Blackden, M. (2010) *Gender Dimensions of Investment Climate Reform*. The International Bank for Reconstruction and Development / The World Bank, Washington, DC. Available at: http://www.ifc.org/wps/wcm/connect/ec3a5f004885538eaf94ff6a6515bb18/Gender_IC_reform_book.pdf?MOD=AJPERES (accessed 8 December 2016).

Snyder, M. (1990) Women: The Key to Ending Hunger. *The Hunger Project Papers* 8. The Global Hunger Project, New York.

Stahl, J. (2010) *Rent from the Land. A Political Ecology of Postsocialist Rural Transformation*. Anthem Press, New York and London.

Swinnen, J.F.M. (1999) The political economy of land reform choices in Central and Eastern Europe. *The Economics of Transition* 7(3), 637–664.

Theesfeld, I. (2005) *A Common Pool Resource in Transition: Determinants of Institutional Change for Bulgaria's Postsocialist Irrigation Sector*. Institutional Change in Agriculture & Natural Resources Series. Shaker Verlag, Aachen, Germany.

UN (2010) *Concluding Observations of the Committee on the Elimination of Discrimination against Women*. UN Committee on the Elimination of Discrimination against Women (CEDAW), Geneva, Switzerland. Available at: http://www2.ohchr.org/english/bodies/cedaw/docs/co/CEDAW-C-EGY-CO-7.pdf (accessed 8 December 2016).

UN Women (2015) UN Women Action Plan, Albania. Strategic document prepared for UNWOMEN Albania, Tirana, Albania. Available at: http://www.ekonomia.gov.al/files/userfiles/Plani_Veprimit_Ang.pdf (accessed 8 December 2016).

USAID/Albania (2008) *Albanian Women in Agriculture: Case Studies*. Produced for USAID by Kara Nichols Barrett, Development & Training Services, Inc., under the Greater Access to Trade Expansion (GATE) Project, WID Office IQC contract number GEW-I-00-02-00018-00, Task Order No. 02. USAID, Washington, DC.

Wheeler, R.S. and Waite, M. (2003) *Albania Country Brief – Property Rights and Land Markets*. Land Tenure Center, University of Wisconsin, Madison, Wisconsin. Available at: https://www.nelson.wisc.edu/ltc/docs/albaniabrief.pdf (accessed 8 December 2016).

Williamson, O.E. (1998) Transaction cost economics: how it works; where it is headed. *De Economist* 146(1), 23–58.

World Bank (2003) *Rural Development Strategy*. World Bank Office, Tirana, Albania.

World Bank (2007) *Albania Strategic Policies for a More Competitive Agriculture Sector*. World Bank, Washington, DC.

Zhllima, E. (2013) *Gender Analyses of Extension Services*. Report prepared for the Ministry of Agriculture, Food and Consumer Protection in Albania. Tirana, Albania.

Zhllima, E. and Kromidha, G. (2012) *Agriculture Extension and Research in the Context of Albanian Rural Development*. Report prepared for FAO 'Capacity Development of Agricultural Technology Transfer Centers and Extension Services for Rural Diversification Projects'. Tirana, Albania.

Zhllima, E., Viaggi, D. and Müller, D. (2010) Property rights of land and its perception in rural part of Central Albania. *New Medit* 3, 56–64.

9 Is Marketing Intelligence Necessary in Conflict and Transitional Region Markets?

Pauline Sullivan,[1]* Michelle Ortez[2] and Lusille Mission[2]
[1]*Tennessee State University, Nashville, Tennessee;* [2]*College of Business, Southern Christian College, Midsayap, Philippines*

9.1 Introduction

Customers expect to receive a product or service they want, when they want it, how they want it, and in the form they want it. In order to meet customer demand, businesses require information. Marketing intelligence is a support system that provides businesses with the information necessary to respond to market conditions (Feinberg *et al.*, 2012). It gathers existing or provides new information about the marketplace, consumer segments, and government policy that could affect entrepreneurship (Suttle, 2015a). This information is important when considering long-term trends and planning business activities. Marketing intelligence monitors consumer demand and product or service production (Jones and Rowley, 2011). Information obtained from marketing intelligence helps entrepreneurs evaluate long-term trends and plan business activities.

The chapter discusses marketing intelligence within the context of micro, small, and medium-size entrepreneurs (MSMEs), firm size and agri-entrepreneurs, relative to conducting business in transitional and developing countries. In addition, this chapter covers agri-entrepreneurship in conflict zones, as well as challenges in gathering information and disseminating it.

9.2 Importance of Marketing Intelligence to Entrepreneurs

Marketing intelligence is important in identifying potential target customers and business strategies for all size businesses (Jones and Rowley, 2011). This includes identifying what the customer wants, why they want something, and reasons why a product or service fulfills their needs (Czinkota and Ronkainen, 2012). Marketing intelligence is particularly important for MSMEs when making business decisions. In fact, market orientation contributes to firm knowledge, which positively influences company performance (Micheels and Gow, 2012). Components of MSME firm knowledge include understanding that learning helps businesses succeed, is an investment, is necessary for firm survival, and requires continually challenging perceptions about their customer, as well as the market place, and adapting to change. In addition, marketing intelligence provides knowledge about what the customer wants, why they want something, and reasons why a product or service fulfils their needs (Czinkota and Ronkainen, 2012). Marketing intelligence is important in identifying potential target customers and business strategies for all size businesses (Jones and Rowley, 2011). However, MSMEs have

*psullivan10605@gmail.com

fewer available resources than large firms and many are more likely to be self-financed (Cole *et al.*, 2004; Kelliher and Reinl, 2009). Thus, MSMEs need to use available marketing intelligence strategically because they cannot easily absorb the costs of business failure (Longenecker *et al.*, 2013).

Since MSMEs do not have money readily available to gather marketing intelligence, they rely on available programs. These programs provide information and teach MSMEs how to use marketing intelligence (Vega Rodríguez and Rojas Berrio, 2011; Zamberi Ahmad, 2012). Other available sources of marketing intelligence include MSME employees, business systems, social media, sales records, and clients (Ross *et al.*, 2012). Also, MSMEs have external sources for marketing intelligence, such as industry networks, professional magazines, exposure in social media and traditional communication channels, and competitors' websites.

9.3 Marketing Intelligence: Generic Entrepreneurs and Agri-entrepreneurs

In general, marketing intelligence gives entrepreneurs objective information needed for evaluating potential opportunities, current customers, competitive position, prospective customers and company goals (Arline, 2015). Marketing intelligence helps identify best practices and reduce business costs and risks, which in turn contribute to an increase in sales and profit (Gagalyuk, 2012; Tzemach Lemmon and Harden, 2012). Also, MSMEs develop marketing intelligence networks around their business and relationships (Hall, 2007; Jones and Rowley, 2011; Donnelly *et al.*, 2012). These networks can include other supply-chain members (Simatupang *et al.*, 2002). Marketing intelligence shared among members in the supply chain helps entrepreneurs match product assortments and inventory levels to demand, which leads to an increase in profits (Jones and Rowley, 2011). MSMEs gain from the breadth of information a variety of marketing intelligence sources provide. Comparative advantages of having marketing intelligence are efficiency and profit (Simatupang *et al.*, 2002). Marketing intelligence in the agricultural sector began as a relatively unsophisticated system, using telephone calls, fax, in-person communication and informal reports (Lowe, 2003). When marketing intelligence is collected through official organizations or government entities generally it is analyzed, synthesized, and disseminated to agribusinesses, primarily MSMEs. Also, efficiency and profit are achieved when agri-entrepreneurs use available marketing intelligence to identify a profitable niche that gives the business a competitive advantage (Simatupang *et al.*, 2002).

Research conducted by agricultural extension or public organizations, like the United Nations, is freely available online. Marketing intelligence needs to be communicated to the appropriate decision maker so that business can focus on selling consumer-driven products (Mishra and Sarkar, 2014). Marketing intelligence is valuable when mentoring new agri-entrepreneurs (Tzemach Lemmon and Harden, 2012). Overall, marketing intelligence serves as a tool that aids the agricultural sector and agribusiness in responding quickly to challenges resulting from dynamic market conditions and unpredictable situations.

9.4 Developed versus Developing Countries

Arguably marketing intelligence is of value to agri-entrepreneurs, regardless whether or not they operate in conflict zones, transitional economies, or developed countries (DCs). In DCs, marketing intelligence is a relatively sophisticated support system that gathers information from MSMEs, supply-chain members, and agricultural extension, as well as state and federal government, so that agri-entrepreneurs can make informed decisions (Renko *et al.*, 2009). Marketing intelligence helps agri-entrepreneurs in DCs respond to the growth of niche markets, demand for specialty foods, and desire for socially responsible food (Diamond *et al.*, 2014). In one example, marketing intelligence found that consumers prefer environmentally friendly goods (Khoi and Thuy, 2013). Agri-entrepreneurs in DCs benefit from information disseminated through available, low-cost technology. For example, almost all agri-entrepreneurs have access to the internet, which makes marketing intelligence easy to access.

On the other hand, learning how to use marketing intelligence is a challenge for agri-entrepreneurs in DCs. However, marketing intelligence is particularly important in DCs for agri-entrepreneurs, because they may be the only source of income for people living in areas where there is high unemployment and high numbers of out-of-school youth (OSY).

There are few opportunities for entrepreneurial training in less developed countries (LDCs). LDCs have low gross per capita income, possess limited human assets (i.e. quality of life), and are vulnerable to economic shocks (Alonso *et al.*, 2015). The population of LDCs is almost a billion people. Some of the LDCs have conflict zones or a post-conflict situation. In LDCs, some businesses consider marketing intelligence as moderately important to their company operations (Damaskopoulos and Evgeniou, 2003). Marketing intelligence in transitional economies helps agri-entrepreneurs increase their attention to consumer demand and respond to situational changes (Kumar, 2014). The information enables entrepreneurs to exploit market opportunities proactively. Marketing intelligence provides concrete information that agri-entrepreneurs can use when evaluating potential opportunities, current customers, competitive position, potential customers, and company goals (Arline, 2015). Also, agri-entrepreneurs can use marketing intelligence to answer the following questions:

1. What types of activities should resources support?
2. Which markets will provide the next business opportunity?
3. What do loyal customers purchase and how often?
4. What additional products could sell to existing customers?
5. What additional demographics segments will find products desirable?

Cost-effective marketing intelligence can be obtained from internal and external information resources (Table 9.1).

Entrepreneurs in the Caribbean believe marketing intelligence decreases uncertainties in agribusiness (Williams and Smith, 2008). A pilot project in Uganda, Kenya and Tanzania demonstrated how marketing intelligence contributes to rural entrepreneurs' profits (Nyende, 2011).

Multinational corporations (MNCs) also use marketing intelligence in decision making and in strategy development (Jamali and Mirshak, 2010). Marketing intelligence contributes to gains received by MNCs when entering war zones, exiting conflict zone markets, or proactively supporting efforts to improve security in the area. In addition, marketing intelligence is valuable to MNCs in evaluating conflict zone industry sectors, investment structure, and political, security, reputation, legal, and economic risks. MNCs rely on marketing intelligence to ensure that their target consumers view the company as moral and supportive of human rights. Marketing intelligence is essential to the ability of MNCs to protect stock prices and profits.

The following are two examples of how marketing intelligence influences business decisions related to consumer behavior. Both examples pertain to consumers' concerns about human rights issues in conflict zones and MNC businesses' location decisions. Firstly, in response to consumers' concerns, Orange SA (a French telecommunications company) renegotiated its contract and license with Partner in order to exit the Israeli market 10 years before the original agreement ended (Pine, 2015). Similarly, in the second example, consumers' concerns about human rights affected the British–Dutch multinational Unilever's decision to move its snack-food plant operating outside the Israeli border in the conflict zone between Israel and Palestine. Unilever set up a snack-manufacturing operation in Israel.

Also, marketing intelligence is important to local agri-entrepreneurs in LDCs as it helps them target retailers who will sell their products. However, the supply chain for local entrepreneurs is different from multinationals. For local agri-entrepreneurs who manufacture a small quantity of their products, it is most cost effective for them to sell directly to retail outlets. Depending upon the level and length of conflict, agri-entrepreneurs producing small quantities of goods can establish a horizontally integrated supply chain among local firms producing the same product, which allows them, collectively, to produce a quantity of products necessary to contract with large or foreign companies (Khoi and Thuy, 2013). Cooperation among individual

Table 9.1. Selected examples of marketing intelligence available to agri-entrepreneurs in transitional economies.

Country	Article Content	Reference
Czech Republic	A survey of 135 Czech farmers provided data about their use of information and communication technology (ICT). Almost 40% of farmers own computers or smart phones for business purposes. A majority of the farmers (96%) employ nine or fewer people, have one plant (90%), and engage in animal production (52%). Yet over half of the farmers use 100 or more hectares in farm operations. Electronic information resources are available. Those most used are e-mail, eAGRI website, and land registry, registry of plants, protection product, and portals for subsidies. While marketing intelligence information is available to Czech farmers, there is little knowledge about how to apply data in decision making.	Ulman *et al.* (2015)
India	Farmers and traders receive marketing information through different communication outlets. Small farmers obtain word-of-mouth information from friends, neighbors and relatives, which may be due to illiteracy. Large farmers obtain information from television magazines and internet. Most farmers are unaware of crop production marketing information. Radio, newspapers, and commission agents are the most common sources of information. Traders receive information from farmers, fellow traders, contacts in other markets, and mass communications, i.e. radio. Few traders receive information from the Agricultural Produce Market Committee.	Vadivelu and Kiran (2013)
Kenya	Marketing intelligence aids entrepreneurs' transition into deregulated agricultural trade. Barriers to transition include limited extension services, little use of modern agricultural technology, and low farm productivity. A study of 375 farmers in three areas of Kenya finds adoption of information and communication technology (ICT) varies by province. The difference is attributed to awareness of available ICT. ICT has a significant positive influence on farmers' use of better seeds and fertilizers which improve farm productivity and significant negative effect on the use of hired farm workers and family labor.	Ogutu *et al.* (2014)
Mexico	A case study of knowledge systems in the Yaqui Valley, Sonora, Mexico finds a unique combination of organizations and individuals provide place- and situation-specific information that addresses their agricultural needs. The developed knowledge systems respond quickly to changes in the environment. When public support of agriculture decreased, credit unions found ways to address farmers' needs. Also, the Yaqui River Irrigation District approached scientists to resolve the knowledge gap accentuated by drought. Researchers worked together with community members, which created reciprocal communication.	McCullough and Matson (2011)
Rwanda	Product assortment influences consumers out-shopping. Variables that influence this decision including lower prices (especially for those with lower incomes), broad product assortment, lack of fresh foods available in town, merchants' honesty, quick checkouts, and an opportunity to socialize and enjoy time with friends. In-shopping was influenced by convenience and opportunity to receive credit. Younger consumers, rather than older consumers, are more likely to out-shop.	Musonera *et al.* (2009)
South Africa	A survey was sent to 40 South African Competitive Intelligence (CI) experts in different industry sectors, including agriculture, hunting, and forestry and fishing industries. The response rate was 60%. Information sources used by CI experts include secondary resources (information from regulatory boards) and primary data sources (feedback from customers, employees and staff reporting information gathered at conferences and seminars). Information about the firm's CI was obtained through industry analysis (50%), SWOT analysis (40%), and benchmarking and gap analysis (35%). This information was disseminated by the business through briefings (55%), company reports (45%), presentations (41%), email (27%) and personal communication (23%).	Sewdass and Du Toit (2014)

agri-entrepreneurs manufacturing small quantities of goods is beneficial when selling value-added products.

Agri-entrepreneurs have the option of establishing a vertically integrated supply chain among local businesses at different stages in the product-to-consumer distribution systems (Khoi and Thuy, 2013). The vertically integrated supply chain helps local agri-entrepreneurs establish a reputation for specialty products, increase productivity, and monitor product quality. These activities contribute to the success of local agri-entrepreneurs in export markets. Use of this type of distribution system by agri-entrepreneurs in conflict zones is dictated by firm size, level of conflict, and transportation systems.

In contrast, MNCs, such as Nabisco, use marketing intelligence differently because they have sophisticated distribution systems and they manufacture products in different locations. MNCs also license product production to companies in foreign markets as it is an efficient way to reach customers and decreases risk associated with entering foreign countries, especially those with conflict. Marketing intelligence informs MNCs where consumers will find their products desirable, retail formats that reach consumers and foreign markets, and the benefits that appeal to these consumers. MNCs also work with export agents who have contacts that facilitate the distribution of their products in different retail formats, ranging from a grocery store to the local 7-Eleven or sari-sari store.

9.5 Marketing Intelligence Networks

In India, agriculture employs 67% of the population. However, small-size farms often are not profitable because information about new technologies and practices fail to reach farmers (Behera *et al.*, 2015). Information and communications technology (ICT) enables a marketing intelligence network of researchers, scientists and administrators to freely discuss current practices and disseminate information about new technology through online agricultural courses. ICT helps inform agri-entrepreneurs about efficient supply chains that can move products to retailers for distribution.

Collaborative networks use a system of activities to create value for all partners (Zott and Amit, 2010). For instance, collaborative networks can link producers with buyers to provide opportunities for increased sales. There are different types of networks: personal contact networks (PCNs), networks between entrepreneurs and their customers, and industry networks (Jones *et al.*, 2013). Networks between businesses are valuable resources for marketing intelligence. Successful entrepreneurs are more protective of their personal network, while less successful ones frequently increase their personal networks (Batjargal, 2006). A study of female entrepreneurs indicates that they use social networks as resource for business activities and to decrease potential danger from militants (Anugwom, 2011). Networks create social capital that aids agri-entrepreneurs in business development (Zimmer, 1986).

During the first stage of business development, entrepreneurs rely on the small network of their closest friends and family. Foley and Klyver (2012) examined the influence of culture on entrepreneurship networks in different countries. In Denmark, the average entrepreneur considered a business-related social network as a resource, while in Hawaii these networks were culturally accepted. As such, Danish entrepreneurs participated in a business-related social network as a result of rational choice, while Hawaiian entrepreneurs participated in a business-related social network as a demonstration of respect for others. Hawaiian marketing intelligence networks overlapped with their social networks. In contrast, Danish involvement in the marketing intelligence network was contingent on the business lifecycle stage.

Similarly, agri-entrepreneurs in LDCs use their business-related networks according to their stage in the business lifecycle (Zimmer, 1986). In the start-up stage, agri-entrepreneurs in LDCs or conflict zones rely on a small network of their closest friends and family. In the second stage, agri-entrepreneurs establish growth plans and increase the size of their network, as well as the type of support received from it. In the third stage of the business lifecycle, entrepreneurs' networks decrease to a core group of actors who are most supportive of the firm and the agri-entrepreneur.

The following is an example of how a marketing intelligence network helps Ugandan farmers earn fair prices and producers purchase at competitive price. The Ugandan network manager is a member of the Rural African Ventures Investment, which is a registered Transaction Security Service (TSS) (Dennig *et al.*, 2011). Rural African Ventures Investment has 23 agents who operate throughout East Africa. Agents collect crop availability and process information from farmers, buyers and other agri-entrepreneurs. The network works as follows. A buyer calls the Ugandan network manager regarding a potential purchase. They discuss purchase information, such as quality and price, and product. The network manager quotes an estimated price for the buyer and calls TSS agents, who then collect information about local prices from farmers. TSS agents post market prices, create alerts, and marketing intelligence on the village AgriNet Information Boards to find interested farmers. Once the agents collect enough information about price and crop availability, they calculate whether or not the deal is viable. The TSS agents' costs and a fair price for farmers' products are part of the cost of goods sold. Agents provide information to the Ugandan network manager, who contacts the purchaser with the details. Once business transactions can be agreed upon, the buyer sends the Ugandan network manager a Local Purchase Order, which confirms the deal. The Ugandan network manager earns a commission from the deals brokered.

Marketing intelligence networks also provide information to increase coping skills necessary for agri-entrepreneurs when rules of law and government fail (Woolcock and Narayan, 2000). These problems are relatively common in LDCs and rural areas.

9.6 Lack of Marketing Intelligence in Conflict Regions

Entrepreneurs who start businesses in conflict zones have a steep learning curve to obtain the knowledge necessary to manage daily operations under conditions of uncertainly (Farquharson, 2011). Conflict-zone start-ups are especially challenging for young entrepreneurs with little experience. Countless hurdles, such as political tensions and bureaucratic policies, are barriers to success as they impede the business start-up process for agri-entrepreneurs. Established agribusinesses have different barriers to success. These barriers include limited mobility, check points that delay delivery of goods to market, and poor communication technology that slows delivery of marketing intelligence information (Farquharson, 2011; Vlassenroot and Perro, 2012).

Telephone communication was a problem for one entrepreneur with a business in the West Bank during the 2011 Israeli–Palestinian conflict (Farquharson, 2011). The telephone service in the West Bank was politically divided, with one type of technology used in Israel and another in the West Bank. The systems could not connect. In order to manage daily business transactions, the West Bank entrepreneur used two telephones to circumvent the politically divided telephone operating system.

Also, the amount of marketing intelligence available to conflict-zone entrepreneurs is limited (Table 9.2). Table 9.2 also provides an overview of the content and academic research pertaining to entrepreneurship in conflict zones. Issues affecting entrepreneurs in conflict zones range from conflict-resistant crops to brand awareness-related or brand equity. Other research not included in the table examines innovation and agricultural production products (Demont *et al.*, 2012). While this research exists, conflict zone entrepreneurs cannot readily access it.

Conditions of uncertainty plague conflict-zone entrepreneurs (Skorobogatykh *et al.*, 2015). This increases the need for marketing intelligence. Some marketing intelligence results from research collaborations between universities and communities, but availability of this information often is limited due to communication infrastructure, mobility, or active conflict. Commercial research is also available. Euromonitor International publishes Consumer Lifestyles in Nigeria for US$2100. Costs of these reports or information are often prohibitive for conflict-zone entrepreneurs. Thus, there is an apparent disconnect between available information and the conflict-zone entrepreneurs who would benefit from this information.

Marketing intelligence helps MSMEs understand the market, their roles, and interrelationships among their business and the value chain (Lusby and Panlibuton, 2007). Yet, MSMEs

Table 9.2. Selected examples of marketing intelligence available to agri-entrepreneurs in conflict zones with the least peace and stability.

Country	Article content	Reference
Iraq	This study examines how price awareness influences consumers brand equity. Questions were asked in the context of a Syrian sportswear brand. Data was collected from 319 university students. Data analysis indicated that price awareness had a positive significant relationship with brand awareness and image. This example here is how marketing intelligence provides information about developing brand and grade.	Alhaddad (2014)
	A study of 200 consumers in Basra, Iraq indicated that social, physical and marketing mix elements influence choice of consumer electronics.	Furaiji *et al.* (2012)
Somalia	In North-Western Somalia (Somaliland) analysis revealed a need to develop a system that established quality standards and processing practices in slaughter houses to help local producers to compete against imports. Transportation to market decreased the ability of the slaughterhouse to deliver products on time. Additionally, drought and export markets challenged the slaughterhouses operations and led to its closure.	Massimo *et al.* (2013)
Democratic Republic of the Congo	A violence-resistant crop was needed. Quinquina remains unaffected when farmers are displaced from their land. The bark of this tree contains quinine. The trees are easily transformed into cash but farmers cannot add value in its raw form. Once processed, quinquina is used in a variety of products from Vermouth to hair-care products. Development of quinquina is discussed within the context of successful entrepreneurship.	Kibriya *et al.* (2014)
Pakistan	Factors such as education, occupation and income level are known as social class. Social class influences store selection and product purchases. People who cannot read are unlikely to shop at large self-service stores, as they cannot read signs or product information. People with monthly fixed incomes tend to purchase their groceries in bulk and to last a month or longer. Consumers who are paid throughout the month frequently visit grocery stores. Income and household size affect store selection. Men pay for most purchases in Pakistan, so gender does not influence store selection.	Iqbal *et al.* (2013)
Russia	Russian consumer preferences for retail-type product result from market segmentation by customer loyalty and demographics. Regardless of market segment, a consumer's desire to purchase alcohol influences the selection of type of retail outlet to patronize.	Zhigalova (2014)
Nigeria	This study of crop diversification in North Central Nigeria collected data from 150 farmers. Farm size, labor, planning materials agrochemical, fertilizer, and capital had a significant positive effect on small-size farm output. The determinants of crop diversification were farm size, farmers' experience, and land ownership in contact with extension agents.	Ojo *et al.* (2014)

often have limited knowledge regarding how information can be used by their business (Vega Rodríguez and Rojas Berrio, 2011; Zamberi Ahmad, 2012). The need for marketing intelligence is crucial to the success of conflict-zone entrepreneurs but it is difficult for them to obtain this information. Moreover, entrepreneurship education is nonexistent or limited in conflict zones and thus marketing intelligence can only help if conflict-zone entrepreneurs learn how to use the information (St-Jean and Audet, 2012; Collier and Decron, 2013).

9.7 Case Study

Marketing intelligence is not a stand-alone solution to joblessness and poverty in conflict zones. A combination of marketing intelligence, extension and entrepreneurial training, product research and ability to pass the information on to potential agri-entrepreneurs is an integrative approach to creating new businesses. Marketing intelligence enables MSMEs to respond to changing consumer preferences and a dynamic market place, which enhances product delivery (Micheels and Gow, 2012). Three handcrafted soap products are examined below as a case study with recommendations.

This case study describes how a partnership between university faculty at Southern Christian College on the island of Mindanao, in the Philippines, and the Department of Natural Resources and Environmental Management at the University of Hawai'i, Honolulu collaborated with community members to develop marketing intelligence that agri-entrepreneurs could use for business development and by MSMEs. The island of Mindanao is a conflict zone with poverty, high unemployment, a high number of unemployed youth between the ages of 16 and 24, and few opportunities to create new jobs. Mindanao's 2011 unemployment rate was 26%. Entrepreneurial activities provide a means of generating income for families and creating jobs for the unemployed.

The partnership between all stakeholders, consistent personal interactions, direct communication, and application of market research and business planning contributed to the success of this project. Communication between stakeholders provided the support and information needed for an entrepreneur to develop handcrafted soap from postharvest crop waste. Postharvest crop waste exists because farmers in rural conflict zones earn too little money to offset the cost of harvesting crops (Parfitt *et al.*, 2010). Poor communities lack the money or the technology to deal with postharvest crop waste.

The entrepreneur's creativity and inclination to innovate helped identify an opportunity that existed in problematic areas: unemployment and postharvest crop waste. Wasted tomato and calamansi crops could be used as ingredients in handcrafted soap. The entrepreneur created several product prototypes but faced several dilemmas. The first was which ingredient, tomato or calamansi, would be preferred by consumers. A second concern was what attributes consumers seek in handcrafted soap and how much was desired. Also the entrepreneur needed to know the price consumers were willing to pay for the handcrafted soap. This information was important in determining whether or not the business would be viable. A market study could answer the entrepreneur's questions.

The collaborative partnership between university faculty at Southern Christian College and the Department of Natural Resources and Environmental Management at the University of Hawai'i, Honolulu provided the entrepreneur with access to faculty familiar with conducting this type of research. Together, they determined that survey data could be used to analyze consumer preferences for value-added handcrafted soap made from postharvest calamansi and tomato fruit. Faculty had the knowledge and software available to conduct a Conjoint Choice Experiment (CCE) to determine consumer preferences for each of the attributes and calculate willingness-to-pay for calamansi and tomato soaps. The first step was to collect data from focus group discussions, carry out a literature review, and gather local market information. This information revealed the following.

- Types of soap preferred were: (i) tomato; (ii) calamansi; and (iii) regular soap with nothing added.
- The forms of soap desired were: (i) glycerin (clear); or (ii) oil-based (opaque).
- Consumers wanted: (i) three small-size soap bars (40 g each); or (ii) one large-size bar (120 g).
- Consumers desired: (i) international; (ii) national; and (iii) local artisan soap.
- The preferred price points were: (i) 65 PHP; (ii) 80 PHP; and (iii) 100 PHP.

Sample choice options generated by Sawtooth Software, Inc. for evaluation in the consumer survey based on the focus-group information were as follows.

- *Option 1:* Calamansi, oil-based (not clear), international brand, 1 large bar (120 g), 80 PHP.
- *Option 2:* Regular (nothing added), oil-based (not clear), local artisan, 1 large bar (120 g), 65 PHP.

- *Option 3:* Tomato, glycerin (clear), national brand, 3 small bars (40 g each), 100 PHP.

Data analysis identified four market segments:

- *Brand seeking* (6.9%)
  Strong preference for tomato and strong preference against calamansi and regular types.
  Strong preference for international brands, rather than national brands or local artisan products.
  No preference for form, size, or price of soap.
- *Price sensitive* (12.7%)
  Strong preference for oil-based soap and preference against glycerin soap.
  Significant preference against international brands.
  Preferred one large bar rather than three small bars.
  Higher prices decreased demand.
- *Type-seeking cluster* (56.3%)
  Strong preference for calamansi and regular soap, but did not desire soap.
  Price was also negative and significant in their purchase decisions.
  Form, brand, and size were not important for these consumers.
- *Size-seeking cluster* (24.1%)
  Strong preference for calamansi soap.
  Did not prefer tomato and regular soaps.
  Strongly preferred one large bar and strongly disliked three small bars.

Also, data analysis established the price points that would result in a purchase of value-added handcrafted calamansi soap. The different prices were used in a break-even analysis to identify the optimal price that would cover costs and provide the most profit for the entrepreneur. Break-even analysis indicated net sales of 715.4 PHP when 500 bars of value-added beauty soap are sold at 24.45 PHP. Break-even occurs when 29 bars of handcrafted soap are sold. Results suggested that targeting the market segment *Type-seeking cluster*, with 56% of the sample population in this cluster preference, would optimize net profits. Results from this study could be shared with other soap makers in the conflict region.

9.8 Role of Strong Institutions: Extension and Government Policies

Strong institutional support is needed for entrepreneurship in transitional countries and conflict zones because policy makers view entrepreneurship as essential to building resilient communities and economic growth (Birchall and Ketilson, 2009; UNDP, 2013). Iraqi data indicated that a 10% increase in entrepreneurs' training had resulted in an additional 10% in sales of goods and services (Iyengar *et al.*, 2011). As a consequence, violence decreased sales by 10%. Also reconstruction programs that created a 10% increase in income helped increase sales and reduce periods of violence.

The partnership between the entrepreneur and university faculty creates value in entrepreneurship (Jones *et al.*, 2013). The Mindanao marketing intelligence case study about soap products shows how the entrepreneur/faculty collaboration provides quality information to use in business planning (Atuahene-Gima and Ko, 2001) and help small-sized firms grow (Larson, 1991). It also illustrates the value of entrepreneurship education in executing a business plan.

This case study demonstrates an alternative method of creating marketing intelligence for conflict-zone entrepreneurs who do not have access to ICT. The partnership between the entrepreneur and faculty combined with information sharing among the network of soap makers reinforces the need for interpersonal contact to communicate marketing intelligence in conflict zones (Woolcock and Narayan, 2000). Lastly, this case study provides evidence of the importance of knowledgeable mentors in increasing entrepreneurs' skill sets and social capital.

Agricultural extension agents in LDCs are an important marketing intelligence resource. They are able to share information with poor farmers living in rural areas (Aker, 2011). Farmers receive information about product prices, weather, transportation, new technology, or other relevant information from agricultural extension agents. Traditionally, agricultural extension agents use radio broadcasts or direct contact with farmers to provide agri-entrepreneurs with marketing intelligence. They also provide entrepreneurship education.

Social entrepreneurs, not associated with government programs, have goals that benefit society rather than focus on profit. Their goals may include creating institutions that give individuals in conflict zones increased confidence in the political and economic system (McEachran, 2013). Another organization, Nahdet El Mahrousa network, trains new social entrepreneurs and helps them turn ideas in to actionable programs (McEachran, 2014). For example, Impulse Social Enterprises and Impulse NGO Network offer alternative career opportunities for areas and victims of human trafficking in North-east India (Vijayann, 2014). This social enterprise developed the "Meghalaya Model" (or Impulse Model) to help more than 20,000 local female artisans. Social enterprises create new opportunities for the poor to participate in and benefit from economic growth (Pillsbury, 2015). About 39% of social enterprises in India are started by females and a program objective is to enable these females to return to their communities and teach entrepreneurship skills to others, aiding them in adapting to situational change.

Marketing intelligence is important as it presents objective information that can be used in business decisions. However, there needs to be strong institutional support to effectively use entrepreneurship to alleviate poverty and provide jobs for OSY. Marketing intelligence is more readily available in transitional economies. In most conflict zones, it is unlikely that MSMEs have access to the internet and related technologies. Therefore networks, such as Rural African Ventures Investment, are essential in providing MSMEs with information about market conditions.

References

Aker, J.C. (2011) Dial 'A' for agriculture: A review of information and communication technologies for agricultural extension in developing countries. *Agricultural Economics* 42(6), 631–647.

Alhaddad, A. (2014) Does price awareness matter to brand equity? *European Journal of Economics, Finance and Administrative Sciences* 65, 21–27.

Alonso, J.A., Cortez, A.L. and Klasen, S. (2015) LDC and other country groupings: how useful are current approaches to classify countries in a more heterogeneous developing world? Chapter 10 in: Alonso, J.A. and Ocampo, J.A. (eds) *Global Governance and Rules for the Post-2015 Era: Addressing Emerging Issues in the Global Environment.* Bloomsbury Academic, New York.

Anugwom, E.E. (2011). 'Wetin We for Do?' Women entrepreneurs and the Niger Delta conflict. *Journal of Small Business & Entrepreneurship* 24(2), 243–252.

Arline, K. (2015) What is Market Intelligence? *Business News Daily*, 18 February 2015. Available at: http://www.businessnewsdaily.com/4697-market-intelligence.html (accessed 28 June 2015).

Atuahene-Gima, K. and Ko, A. (2001) An empirical investigation of the effect of market orientation and entrepreneurship orientation alignment on product innovation. *Organization Science* 12(1), 54–74.

Batjargal, B. (2006) The dynamics of entrepreneurs' networks in a transitioning economy: the case of Russia. *Entrepreneurship and Regional Development* 18(4), 305–320.

Behera, B.S., Panda, B., Behera, R.A., Nayak, N., Behera, A.C. and Jena, S. (2015) Information communication technology promoting retail marketing in agriculture sector in India as a study. *Procedia Computer Science* 48, 652–659.

Birchall, J. and Ketilson, L.H. (2009) *Resilience of the cooperative business model in times of crisis.* Sustainable Enterprise Programme, International Labour Organization (ILO), Geneva.

Cole, R.A., Goldberg, L.G. and White, L.J. (2004) Cookie cutter vs. character: The micro structure of small business lending by large and small banks. *Journal of Financial and Quantitative Analysis* 39(02), 227–251.

Collier, P. and Dercon, S. (2013) *African agriculture in 50 years. Smallholders in a rapidly changing world?* Stanford Symposium Series on Global Food Policy and Food Security in the 21st Century. Center on Food Security and the Environment (FSE), Oxford University.

Czinkota, M. and Ronkainen, I. (2012) *International Marketing.* Cengage Learning, Mason, Ohio.

Damaskopoulos, P. and Evgeniou, T. (2003) Adoption of new economy practices by SMEs in Eastern Europe. *European Management Journal* 21(2), 133–145.

Demont, M., Zossou, E., Rutsaert, P., Ndour, M., Van Mele, P. and Verbeke, W. (2012) Consumer valuation of improved rice parboiling technologies in Benin. *Food Quality and Preference* 23(1), 63–70.
Dennig A., Scheuermeier, J. and Lightfoot, C. (2011) Entrepreneurs in agricultural marketing. *New Agriculturalist*, Available at: http://www.new-ag.info/en/focus/focusItem.php?a=1989 (accessed 27 June 2015).
Dentoni, D., Hospes, O. and Ross, R.B. (2012) Managing wicked problems in agribusiness: the role of multi-stakeholder engagements in value creation. *International Food and Agribusiness Management Review* 15, Special Issue B.
Diamond, A., Tropp, D., Barham, J., Frain, M., Kiraly, S. and Cantrell, P. (2014) Food value chains: creating shared value to enhance marketing success. Agricultural Marketing Service, US Department of Agriculture, Washington, DC. Available at: http://ageconsearch.umn.edu//handle/176263 (accessed 22 November 2015).
Donnelly, C., Simmons, G., Armstrong, G. and Fearne, A. (2012) Marketing planning and digital customer loyalty data in small business. *Marketing Intelligence & Planning* 30(5), 515–534.
Farquharson, V. (2011) Doing business abroad: West Bank draws Canadian entrepreneurs *The Globe and Mail*, 9 September 2011. Available at: http://www.theglobeandmail.com/report-on-business/small-business/sb-growth/west-bank-draws-canadian-entrepreneurs/article593866 (accessed 29 June 2015).
Feinberg, F., Kinnear, T. and Taylor, J. (2012) *Modern Marketing Research: Concepts, Methods, and Cases*. Cengage Learning, Mason, Ohio.
Foley, D. and Klyver, K. (2012) Networking and culture in entrepreneurship. *Entrepreneurship and Regional Development* 24(7–8), 1–28.
Furaiji, F., Łatuszyńska, M. and Wawrzyniak, A. (2012) An empirical study of the factors influencing consumer behaviour in the electric appliances market. *Contemporary Economics* 6(3), 76–86.
Gagalyuk, T. (2012) *Goal achievement in supply chain networks: a study of the Ukrainian agri-food business*. No. 64, Studies on the agricultural and food sector in Central and Eastern Europe. Available at: http://www.econstor.eu/bitstream/10419/58998/1/690233604.pdf (accessed 21 September 2015).
Hall, A. (2007) *Challenges to Strengthening Agricultural Innovation Systems: Where Do We Go from Here?* UNU-MERIT Working Paper 2007-038, United Nations University, Maastricht, Netherlands.
Institute for Economics and Peace (2015) *Global Peace Index*. Available at: http://www.visionofhumanity.org/ #/page/our-gpi-findings(accessed 24 June 2015).
Iqbal, H.K., Ghafoor, M. and Shahbaz, S. (2013) Impact of demographic factors on store selection: an insight in Pakistani society. *Journal of Marketing Management* 1(1), 34–45.
Iyengar, R., Monten, J. and Hanson, M. (2011) *Impact of aid on the labor market for insurgency*. Working Paper 17297. National Bureau of Economic Research (NBER), Cambridge, Massachusetts. Available at: http://www.nber.org/papers/w17297 (accessed 27 June 2015).
Jamali, D. and Mirshak, R. (2010) Business-conflict linkages: revisiting MNCs, CSR, and conflict. *Journal of Business Ethics* 93(3), 443–464.
Jones, R. and Rowley, J. (2011) Entrepreneurial marketing in small businesses: a conceptual exploration. *International Small Business Journal* 29(1), 25–36.
Jones, R., Suoranta, M. and Rowley, J. (2013) Strategic network marketing in technology SMEs. *Journal of Marketing Management* 29(5–6), 671–697.
Kelliher, F. and Reinl, L. (2009) A resource-based view of micro-firm management practice. *Journal of Small Business and Enterprise Development* 16(3), 521–532.
Khoi, N.V. and Thuy, L.T.T. (2013) Exporting Vietnam's agricultural and fishery products to European Union market under green trade barriers. *International Journal of Diplomacy and Economy* 1(3–4), 309–328.
Kibriya, S., Partida, V., King, J. and Price, E. (2014) Conflict resistant agribusiness in Democratic Republic of Congo. *International Food and Agribusiness Management Review* 17(B), 75–80.
Kumar, A. (2014) Need of business intelligence by Indian MSME. *Social Science and Management* II(8), 27–29.
Larson, A. (1991) Partner networks: leveraging external ties to improve entrepreneurial performance. *Journal of Business Venturing* 6(3), 173–188.
Longenecker, J., Petty, J., Palich, L. and Hoy, F. (2013) *Small business management*. Cengage Learning, Mason, Ohio.
Lowe, J.C. (2003) Evolution of international market information services for the Canadian agriculture and agri-food sector. *Quarterly Bulletin of the International Association of Agricultural Information Specialists* 48(3/4), 173–177.
Lusby, F. and Panlibuton, H. (2007) *Value chain program design: promoting market-based solutions for MSME and industry competitiveness*. Action for Enterprise (AFE), Arlington, Virginia.

Massimo, C., Kinyanjui, W., Njue, S., Joseph, G., Nyorangi, O.J. and William, O. (2013) Opportunities and challenges of setting up HACCP system in slaughterhouses in countries emerging from conflict: an appraisal of North-Western Somalia. *Food and Nutritional Sciences* 4, 126–130.

McCullough, E.B. and Matson, P. (2011) Evolution of the knowledge system for agricultural development in the Yaqui Valley, Sonoro, Mexico. *Proceedings of the National Academy of Sciences of the USA* 113(17), 4609–4614.

McEachran, R. (2013) The difficulty of starting up a social enterprise in a warzone: with a little help, social enterprises can help rebuild countries overrun by conflict. *The Guardian, sustainable business: Social enterprise blog.* Available at: http://www.theguardian.com/social-enterprise-network/2013/aug/16/social-enterprises-in-warzone (accessed 21 September 2015).

Micheels, E.T. and Gow, H.R. (2012) The value of a positional advantage for agricultural SMEs. *Small Enterprise Research* 19(2), 54–73.

Mishra, P. and Sarkar, S. (2014) Customer-Based Corporate Brand Equity (CBCBE) in business-to-business firms: an emerging market perspective. In: *Brand Management in Emerging Markets: Theories and Practices*. IGI Global, Hershey, Pennsylvania, 73–90.

Musonera, E., Nyamulinda, I.B. and Ganesan, P. (2009) Determinants of rural out-shopping behavior in developing countries: a case of grocery shopping in the Republic of Rwanda. *The Journal of International Business Research and Practice* 3, 51–63.

Nyende, P. (2011) *Building networks for market access: lessons learned from the rural knowledge network (KN) pilot project for East Africa (Uganda, Kenya and Tanzania).* Research and Extension Branch, Food and Agriculture Organization of the United Nations, Rome.

Ogutu, S.O., Okello, J.J. and Otieno, D.J. (2014) Impact of information and communication technology-based market information services on smallholder farm input use and productivity: the case of Kenya. *World Development* 64, 311–321.

Ojo, M.A., Ojo, A.O., Odine, A.I. and Ogaji, A. (2014) Determinants of crop diversification among small–scale food crop farmers in North Central, Nigeria. *PAT* 10(2), 1–11.

Parfitt, J., Barthel, M. and Macnaughton, S. (2010) Food waste within food supply chains: quantification and potential for change to 2050. *Philosophical Transactions of the Royal Society of London B: Biological Sciences* 365(1554), 3065–3081.

Pillsbury, A. (2015) Empowering women social entrepreneurs in India. *The Guardian*. Available at: https://womennewsnetwork.net/2013/11/08/empowering-women-social-entrepreneurs-india/ (accessed 8 December 2016).

Pine, J. (2015) BDS in the boardroom: campus protests are loud, but corporate pressure is more effective. *Jewish News Weekly*, San Francisco, California. Available at: http://www.jweekly.com/article/full/75384/bds-in-the-boardroom-campus-protests-are-loud-but-corporate-pressure-is-mor/ (accessed 8 December 2016).

Renko, M., Carsrud, A. and Brännback, M. (2009) The effect of a market orientation, entrepreneurial orientation, and technological capability on innovativeness: A study of young biotechnology ventures in the United States and in Scandinavia. *Journal of Small Business Management* 47(3), 331–369.

Ross, P., McGowan, C.G. and Styger, L.E. (2012) A comparison of theory and practice in market intelligence gathering for Australian microbusinesses and SMEs. *19th International Business Research Conference: Research for Re-thinking*, pp. 1–17. Social Science Research Network, University of Wollongong, Australia.

Sewdass, N. and Du Toit, A. (2014) Current state of competitive intelligence in South Africa. *International Journal of Information Management* 34(2), 185–190.

Simatupang, T.M., Wright, A.C. and Sridharan, R. (2002) The knowledge of coordination for supply chain integration. *Business Process Management Journal* 8(3), 289–308.

Skorobogatykh, I., Tarasenko, E. and Shirochenskaya, I. (2015) Possible scenarios of Russian luxury market development under economic uncertainty and turbulence. *Review of European Studies* 7(9), 131.

St-Jean, E. and Audet, J. (2012) The role of mentoring in the learning development of the novice entrepreneur. *International Entrepreneurship and Management Journal* 8(1), 119–140.

Suttle, R. (2015a) Difference between marketing intelligence and marketing research. *Houston Chronicle*. Available at: http://smallbusiness.chron.com/difference-between-marketing-research-marketing-intelligence-25562.html (accessed 9 December 2016).

Tzemach Lemmon, G. and Harden, E. (2012) Doing business in post-conflict zones: why SMEs are key. *Next Billion Development through Enterprise*, 28 May 2012. Available at: http://nextbillion.net/blogpost.aspx?blogid=2815 (accessed 29 June 2015).

Ulman, M., Tyrychtr, J., Vostrovský, V. and Jarolímek, J. (2015) Information and knowledge engineering in agricultural enterprises in the Czech Republic. *Future Communication, Information and Computer Science*, 139.

UNDP (2013) The rise of the South: human progress in a diverse world. Human Development Report 2013. United Nations Development Programme, New York.

Vadivelu, A. and Kiran, B.R. (2013) Problems and prospects of agricultural marketing in India: an overview. *International Journal of Agricultural and Food Science* 3(3), 108–118.

Vega Rodríguez, R.A. and Rojas Berrio, S.P. (2011) Uso del mercadeo estratégico en las Mipymes de Bogotá. *Poliantea* 7(13), 97–111. Available at: http://journal.poligran.edu.co/index.php/poliantea/article/view/142 (accessed 9 December 2016).

Vijayann, M. (2014) Staying safe in conflict zones: advice from female social entrepreneurs. *The Guardian, Guardian sustainable business: Social enterprise blog*. Available at: https://www.theguardian.com/social-enterprise-network/2014/jun/18/women-social-entrepreneurs-conflict-zones (accessed 8 December 2016).

Vlassenroot, K. and Perrot, S. (2012) Ugandan military entrepreneurialism on the Congo border. In: Utas, M. (ed.) *African Conflicts and Informal Power: Big Men and Networks*. Zed Books, London, pp. 35–59.

Williams, T.O. and Smith, R. (2008) *Rethinking agricultural development: the Caribbean challenge*. Paper presented at the XL (40th) Annual Monetary Studies Conference, 11–14 November 2008, Basseterre, St Kitts.

Woolcock, M. and Narayan, D. (2000) Social capital: implications for development theory, research, and policy. *The World Bank Research Observer* 15(2), 225–249.

Zamberi Ahmad, S. (2012) Micro, small and medium-sized enterprises development in the Kingdom of Saudi Arabia: problems and constraints. *World Journal of Entrepreneurship, Management and Sustainable Development* 8(4), 217–232.

Zhigalova, A. (2014) Building consumer-oriented strategy for a specialized regional store chain needs. *Regional Formation and Development Studies* 8(3), 275–286.

Zimmer, C. (1986) Entrepreneurship through social networks. In: *The Art and Science of Entrepreneurship*. Ballinger, Cambridge, Massachusetts, pp. 3–23.

Zott, C. and Amit, R. (2010) Business model design: an activity system perspective. *Long Range Planning* 43(2), 216–226.

10 Urban Consumer Preferences for Food in Post-conflict Economies: The Case of Kosovo

Maurizio Canavari,[1] Drini Imami,[2]* Muje Gjonbalaj,[3] Ekrem Gjokaj[4] and Anera Alishani[5]

[1]University of Bologna, Bologna, Italy; [2]Agricultural University of Tirana, Tirana, Albania; [3]University of Priština, Priština, Kosovo; [4]Ministry of Agriculture, Forestry and Rural Development (MAFRD), Priština, Kosovo; [5]University of Prizren, Prizren, Kosovo

10.1 Introduction

Kosovo is located in the Western Balkans, with a land area of 10,908 km^2, a population of 1.8 million and a density of 177 people/km^2 in 2013. More than half of the population lives in rural areas. The largest city is Priština, which is also the capital. Kosovo is considered a post-conflict transition country because it used to be a centrally planned economy under Yugoslavia till the late 1980s, while it underwent a notorious conflict in the 1990s and emerged as an independent country in the following decade. According to the Kosovo Agency of Statistics (KAS), despite economic growth in the last few years, Kosovo remains one of the poorest European transition countries with a per capita income of €3100 in 2014 (KAS, 2015).

The agri-food sector is a major contributor to gross domestic product (GDP) and has always been a key sector for Kosovo's economy despite the recession prior to and during the post-war period (1990–1999). According to the Ministry of Agriculture, Forestry and Rural Development (MAFRD) for the Republic of Kosovo, agriculture's contribution to GDP was about 30% in 1995, but it has decreased since the post-war period (MAFRD, 2002). However, the agricultural sector is still one of the main employing sectors as it accounts for about 12% of the GDP (MAFRD, 2014b). This contribution would have been higher if it were not for the damage caused in the last armed conflict in the late 1990s. During the conflict, 50% of agricultural assets were reported as either damaged or lost, with an estimated monetary value of over US$700 million (World Bank, 2000).

About 60% of the total population lives in rural areas, most of whom are engaged in the agricultural sector (MAFRD, 2014b). Even before the transition process in 1991 the rural population was high, namely 63% (KAS, 2001). It was expected that Kosovo would have followed the experience of other Balkan countries, but the migration to urban areas has been slow. Albeit this phenomenon was pronounced after the war, the outmigration to the city was minimal, thus the current proportion of the rural population remains high. In addition, this rural population is characterized by a large

*dimami@ubt.edu.al

proportion (about 30%) of young people between 15 and 29 years old (KAS, 2006).

Data on total consumption for the post-conflict period are available from the year 2003 and are shown in Table 10.1. Total consumption value has almost doubled from €1493 million in 2003 to €2292 million in 2012 but with an irregular growth pattern. The main consumption categories are food and non-alcoholic beverages that on average comprise 40% of total consumption, followed by housing at approximately 30% of total consumption (KAS, 2014). Consumption of food and non-alcoholic beverages shows an increasing trend in total value, following the same trend of total consumption expenditure (Fig. 10.1). A large part of the increasing food consumption expenditure trend is due to a higher demand for local horticulture and livestock products since the post-conflict period, as income has been increasing. In particular, during the last decade, the demand for high-value horticultural products has shown an increase more than for any other food category (MAFRD, 2013). In the post-conflict period the agri-food sector is still an important sector, complemented with many diverse choices. Knowledge of consumer preference is essential to capture any shifts in food demand.

Table 10.1. Total consumption expenditures in Kosovo for the period 2003–2012 (from KAS, 2014).

	Total consumption expenditure		Consumption per household		Consumption per capita		Food and non-alcoholic beverages share
Year	million €	% change	€	% change	€	% change	%
2003	1493		5400		840		48
2004	1529	2.4	5600	3.7	910	8.3	43
2005	1549	1.3	5900	5.4	950	4.4	40
2006	1414	–8.7	5700	–3.4	980	3.2	39
2007	1403	–0.8	5700	0.0	980	0.0	40
2008	1798	28.2	6707	17.7	1156	18.0	38
2009	1911	6.3	6847	2.1	1161	0.4	36
2010	1937	1.4	7110	3.8	1226	5.6	35
2011	1928	–0.5	7010	–1.4	1210	–1.3	38
2012	2292	18.9	7657	9.2	1380	14.0	45

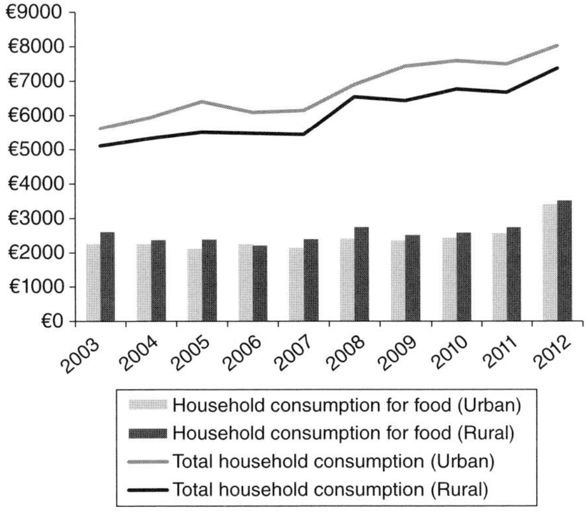

Fig. 10.1. Household consumption in Kosovo by the main consumption group (food) and by location (urban versus rural) (from KAS, 2014).

Despite a recent increase in local food production to meet higher and various local demands, the country still suffers a negative trade balance that remains problematic in the agricultural sector and to the Kosovo economy (Fig. 10.2). As such, since the end of the conflict, Kosovo has been a net importer of most of the demanded agricultural products. However, this was not the case before the war when Kosovo had a trade deficit mainly for cereals, while for many fruits and vegetables it was a net exporter (Fischer, 2004).

The negative trade balance is due to many challenges that domestic producers face. There is a lack of access to market intelligence and consumers' increasing demand and there are preferences for high quality imports. As a country in transition, the common problem identified was a lack of a proper and tighter linking of domestic supply and demand for economic development. In a centrally planned economy as in the past, the government would set production targets. With market reform, market intelligence becomes the main driver of production; however, either the demand preference information is missing and/or the link to the market is a challenge. Establishing a strong link to the market is imperative and is a priority for making the agricultural industry competitive, efficient, and able to introduce innovation for quality and food safety that benefits consumers and the actors in the supply chain. The dearth of information on the food industry and consumption trends in the country is variable and sometimes not systematic, as demand preferences change due to two drivers: income growth and the flood of cheaper imports. Cheaper imports are becoming more common due to regional free trade agreements and the expansion of supermarket chains. The government's role in data collection has been dismal, while various agricultural support schemes have not been based on in-depth understanding of consumer expectations and preferences for local food products. This has caused a disconnection between what is being produced and what is being demanded.

In this context, this chapter aims to provide an insight on demands and preferences of the urban food consumers' market in Kosovo, highlighting the most relevant issues in this post-conflict transition country and discussing potential options and policies to support the development of the food industry in a post-conflict situation. The chapter is structured as follows. Section 10.2 provides an overview of the country's agri-food sector and socio-economic background. Section 10.3 explains the approach and methods that were deployed for the study.

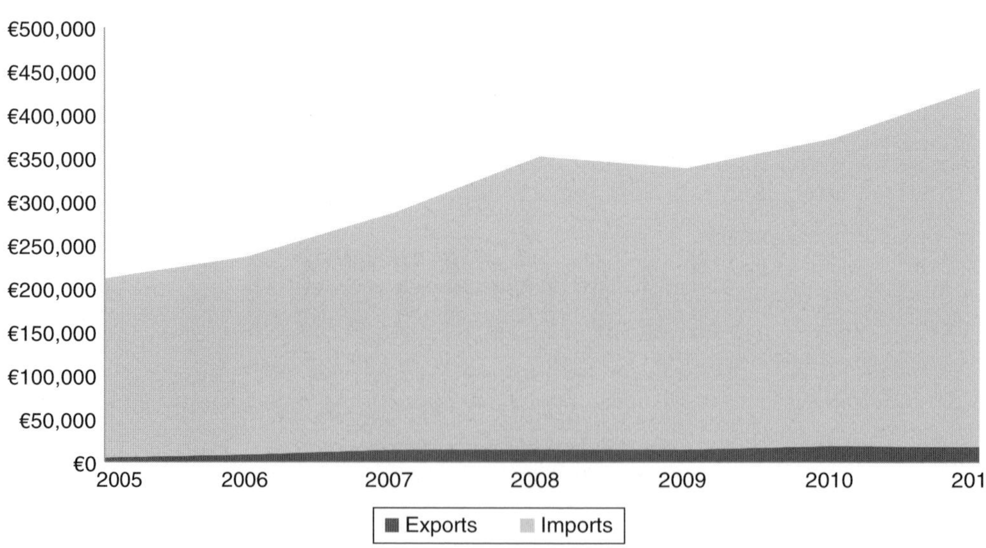

Fig. 10.2. Kosovo food and live-animals trade balance (in 000 €), 2005–2011 (from KAS, 2011).

Section 10.4 analyzes key findings from the consumer surveys. This is followed by conclusions and recommendations.

10.2 Kosovo Background and Economy

Kosovo is Europe's youngest state, having unilaterally declared its independence from Serbia in February 2008. It is part of the so-called Western Balkan region together with other relatively small-sized economies, namely Albania, Bosnia and Herzegovina (BiH), Croatia, the Former Yugoslav Republic of Macedonia (FYRoM), Montenegro, and Serbia (EC, 2002). Western Balkans is defined as a region of south-east Europe that includes countries aiming for integration and harmonization with the European Union. Apart from Albania, all the other states included in the definition of this region emerged from the former Yugoslavia, which started splitting apart in early 1990s due to a series of armed conflicts. Slovenia is the only former Yugoslavian state that is not included in this region. Croatia became a member state of the European Union in 2014. Although Serbia still considers Kosovo an autonomous province, since 2008 Kosovo has been recognized as an independent state by 23 out of the 28 member states of the European Union, and by 108 out of 193 member states of the United Nations. Common economic characteristics of the Western Balkans are small populations, low GDP per capita, high unemployment, and low foreign direct investments (FDI) (Montanari, 2005).

Kosovo's population is roughly 1.8 million people with a median age of 27 years (SDC, 2013). Kosovo is one of the poorest countries in Western Balkans and Europe in general, with 34.5% of its population living below the poverty line, i.e. on less than €1.55/day, with about 12% living in extreme poverty, i.e. on less than €1/day (SDC, 2013). Unemployment is at 45%, and 75% of young people under the age of 25 are considered unemployed (SDC, 2013). High poverty and unemployment rates make migration an attractive option, in particular for young people.

On the other hand, Kosovo's economy is rather dynamic and it is evolving as a free-market system, but the transition process that Kosovo has been facing since the 1990s along with the armed conflict of 1999 are some of the reasons that Kosovo still faces many economic, political, and social challenges. Kosovo has not yet established a strong economic and social record since its transition to a market-oriented economy. Compared with the other Western Balkans countries, Kosovo is behind with respect to many macroeconomic indicators.

The GDP growth after the war averaged between 2% and 4% and with strong international donor support and remittances (from diaspora) it reached its highest level of 8.30% in 2007 (Fig. 10.3). However, due to adverse external developments as a result of the global financial crisis of 2007, most of the countries of the Western Balkans, including Kosovo, performed poorly subsequently. Compared with neighboring countries, though, GDP growth in Kosovo was proceeding at a higher rate than in the EU-27 since during the period between 2008 and 2011 Kosovo's average GDP growth has been fluctuating around 4% while for EU-27 it was just above 3% (EC, 2010). During 2012 and 2014 the GDP growth has been averaging about 3%, which is attributable to both domestic and external demand dropping (EBRD, 2014).

The GDP per capita experienced continuous growth, indicating that the economic well-being of the citizens of Kosovo is improving (Fig. 10.4). However, high poverty and unemployment levels remain challenging.

According to Bartlett (2008), since the GDP of Western Balkans is only 0.4% of the overall GDP in the EU, international trade and opening to the global economy are considered fundamental elements for progress. Regarding international trade, both merchandise exports and imports experienced growth from 2001 onwards. However, in nominal terms imports greatly outpaced exports, which resulted in persistent and large negative trade balances (Fig. 10.5). The negative trade balance has deepened each year, which makes it a major problem for the country's sustained economic growth.

The negative trade balance indicates weak exports and strong imports, which contribute to a growing imbalance. This imbalance deepened greatly in 2008 prior to independence and recovered a little in 2009 due to the decrease

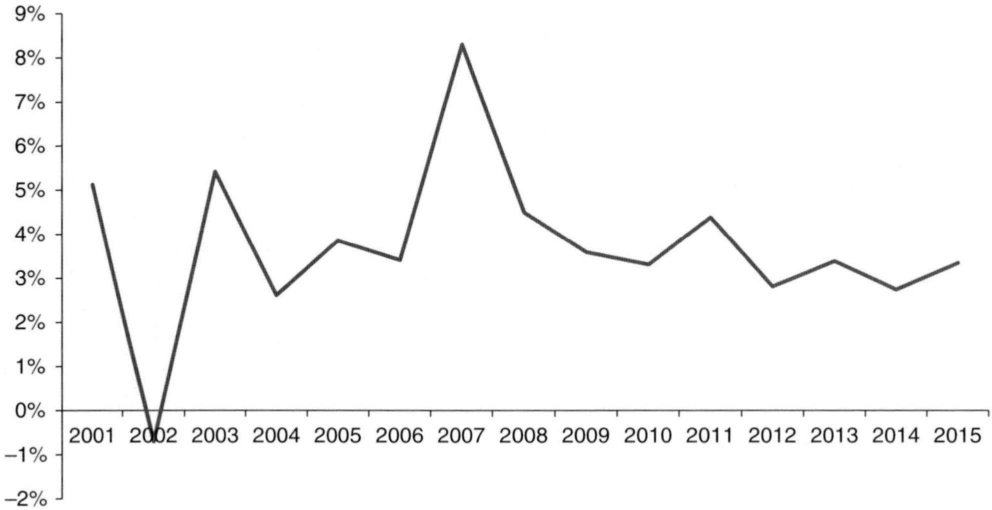

Fig. 10.3. GDP growth (%) for the period 2001–2014 (from International Monetary Fund, World Economic Outlook Database, April 2015).

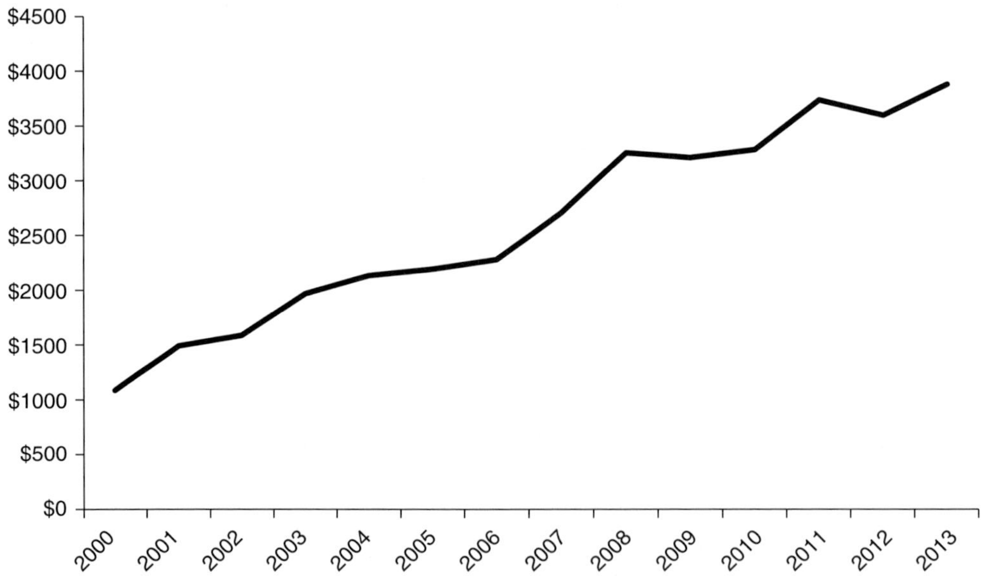

Fig. 10.4. GDP per capita (in US$), 2000–2013 (from World Bank, World Development Indicators, 2014).

in trade as a result of global financial crises. In 2010, due to increased demand for goods at the global level, exports started to increase in many states in south-east Europe, including Kosovo, which resulted in the continued reduction of deficit, especially in 2011 (Central Bank of Kosovo, 2010). For the government of Kosovo, the goal continues to be increasing competition within the economy and at the same time increasing the export capacity to reduce Kosovo's trade deficit.

In terms of international investment, Kosovo, in general, attracts a low amount of FDI but between 2004 and 2007, due to privatization in strategic sectors such as telecommunication, banking, and oil refining (Bartlett, 2009), the total amount of FDI increased dramatically in nominal value (Fig. 10.6). The level of FDI is an important element in integrating Kosovo into international markets (Bartlett, 2009).

Remittances represent another factor that plays a significant role in the economic performance of Kosovo. Kosovo, as many other states in the region, is known for its high dependency on remittances. Remittances have been an important source of income for household consumption as well as for private sector investment, including farming.

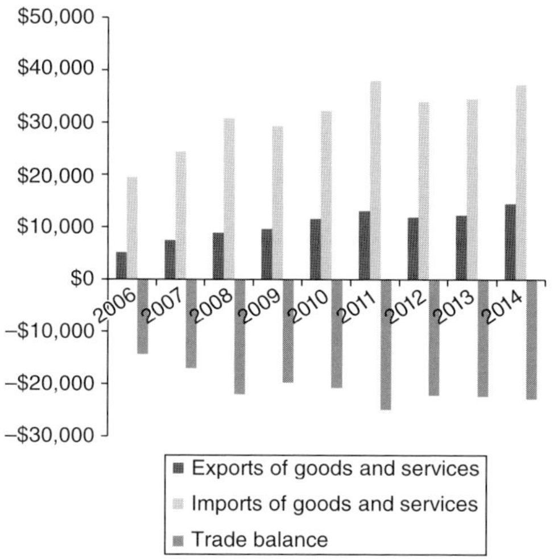

Fig. 10.5. Trade of goods and services (in million US$), 2006–2014 (from World Bank, World Development Indicators, 2014).

Despite the GDP growth, the labor demand markets remain a serious problem in Kosovo (Koro, 2007). Unemployment remains very high, with approximately 75% of young people under 25 considered unemployed. Moreover, 25,000 young adults enter the labor market each year (SDC, 2013). The high unemployment rate and poverty in Kosovo requires high economic growth, and particularly stronger support for the agricultural sector, given its importance in terms of employment and income generation for rural households.

Agriculture in Kosovo is considered a key sector for economic development since more than half of the population lives in rural areas. Most of those living in rural areas are employed in the agricultural sector and agriculture still represents a large share of the economy (14.1% of GDP and 16% of exports) (MAFRD, 2010, 2014a).

To support the development of the agricultural sector, a better access to markets and market intelligence are the key issues. Due to

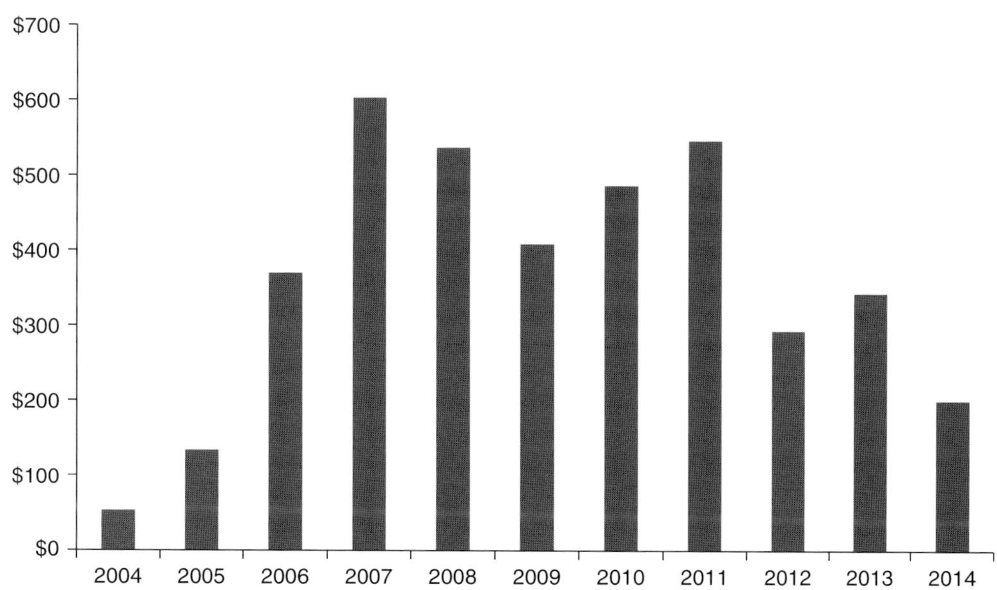

Fig. 10.6. Foreign direct investment (in million US$), 2004–2014 (from World Bank, World Development Indicators, 2014).

its post-conflict condition and its transition from a centrally planned to a market-oriented economy, agriculture and the food industry are still behind modern standards and this affects the quality of local agricultural produce relative to imports and its ability to produce what the consumer wants.

10.3 Objective and Approach

The needs for and expectations of food by consumers and practitioners in the domestic market are evolving and changing, focusing not only on availability and price but also on quality and safety issues. As mentioned in the Introduction, the link between domestic supply and demand is missing and provides a major challenge that may require specific policy measures to fix. The study reported in this chapter is therefore aimed at addressing this challenge of a post-conflict country, i.e. the scarcity of available information about the food industry and specifically post-conflict consumer expectations and consumption habits and attitudes. The study's goal is to highlight the importance of such information.

To collect information regarding food supply and demand in the post-conflict era, questionnaires were developed and the data analyzed. To begin, a series of key industries within the agricultural sector in Kosovo and cross-cutting issues were identified by a committee of experts. The issues were identified based on the most relevant problems that could contribute to increasing the welfare of consumers in the post-conflict era. To be comprehensive, the study therefore included market research activities that have been implemented targeting both consumers and food industry practitioners. The main aims of these surveys were to explore opinions and describe beliefs, attitudes and perceptions, purchasing behavior, and preferences of consumers for products in the following industries: (i) dairy; (ii) meat; and (iii) fruit and vegetables. These industries were chosen as they are the most relevant in the agri-food sector and they can be targeted for investments resulting from the delays due to the slow post-conflict developments. The results can help domestic producers to become more competitive and able to compete with foreign producers.

Livestock is the most important agri-food sector in terms of both production and consumption: in particular meat and dairy production. The livestock sector consists largely of cattle (dairy and meat), chickens (egg and meat) and, to a lesser extent, small ruminants (lamb and goat kid) and pork. Per capita consumption of dairy products in Kosovo is quite high compared with other countries and it covers a variety of products; and dairy production has increased significantly in recent years. After dairy, meat production contributes substantially to the family income of rural households in Kosovo (Bytyqi *et al.*, 2012).

Fruit consumption in Kosovo, although growing because of higher income, is very low compared with other countries in the region. Apples, bananas, grapes, and citrus are the most consumed fresh fruits, while peaches and strawberries are consumed as processed fruit (Giron, 2001; Prism Research, 2009). Vegetable consumption is mainly tomatoes and cucumbers. Apart from the tropical species, Kosovo has traditionally been producing a wide range of fruits and vegetables, mainly for the domestic market.

In addition to analyzing the performance of key industries, two main cross-cutting key issues were also analyzed: food safety and organic foods. These issues were studied because information on them has been considered crucial to the design of strategies and policies aimed at fostering the development of a competitive and thriving domestic production system. The cross cutting issues are:

- food safety—consumer awareness and perception of food safety including current practices, trust, and food safety labeling, and subsequent purchasing and consumption behavior; and
- organically grown products—consumer awareness and perception of organic products, a popular value-added niche food category in developed countries and facing a growing interest in emerging and developing countries.

Finally, since consumer preferences, attitudes and lifestyles are among the main drivers for the development of the domestic market, their analysis (Brunsø *et al.*, 1995; Brunsø and Grunert, 1995; Dimech *et al.*, 2011) has been undertaken, specifically the identification of consumer segments and description of consumer profiles, and identifying consumers who are more interested in higher-quality value-added products.

The research method consisted of the following activities: (i) literature and secondary data review; (ii) expert interviews; (iii) consumer focus groups; (iv) consumer survey; (v) round-table discussions; and (vi) workshop.

10.3.1 Literature and secondary data review

A thorough review of previous market and consumer studies on Kosovo was completed. When applicable, the research was extended beyond Kosovo in order to make comparisons with other countries. Literature review included previous scientific publications and project reports, and international database data collection and analysis.

10.3.2 Expert interviews

Two types of interviews with value-chain actors and experts were developed: open interviews and structured interviews. Over 15 open interviews with individual food-chain actors (e.g. wholesalers, retailers) and experts were conducted to obtain a more in-depth understanding of the issues related to consumers, producers, markets, and food-chain development. In addition, 30 food-chain actors (processors, traders) and 25 experts were interviewed using a structured questionnaire that focused on specific information about the selected sectors (fruit and vegetables, meat, and dairy) as well as the food chain in general.

10.3.3 Consumer focus groups

Altogether, four consumer focus groups were conducted. Each focus group had eight to nine participants representing different types of socio-economic status. The focus group was carried out based on a protocol/guideline that was prepared in advance. The objectives of the focus groups were:

- to obtain information and get a better understanding of the latest market development trends in Kosovo for the main agri-food products; and
- to explore consumer preferences and purchasing behavior for the main agri-food products that are produced in Kosovo, with the aim of providing useful information for the design of the following quantitative survey.

10.3.4 Consumer survey

Five separate structured surveys were conducted. Three focused on the dairy, meat, and fruit and vegetables sectors, respectively, and two focused on food safety and organic products, respectively. Each survey consisted of about 300 face-to-face interviews with randomly selected respondents. The food-related lifestyle section was included in all of the five surveys; therefore, the analysis of the consumer segments was based on a sample of about 1500 respondents. The surveys were administered between December 2013 and January 2014. The interviews were carried out by ten graduate students who were trained and supervised throughout the survey implementation by the authors of this chapter. The survey was administered in the three largest cities of Kosovo: Priština (capital city), Prizren, and Gjilan. Distribution of the sample was based on the non-probabilistic intercept sampling technique and the sample size was determined in proportion to the respective size of the population of the city. Questionnaires were designed based on literature review, focus groups, and expert interviews, and were pre-tested through a pilot study of direct interviews with consumers.

10.3.5 Round-table discussion

In May 2014, a panel of experts gathered to discuss preliminary findings of the above-mentioned studies. Results were reported, and comments and discussion outcomes were used for validation and interpretation purposes.

10.3.6 Workshop

During the workshop, there were presentations on the findings of the analysis of the Kosovo consumer and agri-food chain actors' perceptions, concerns, and preferences related to food products, which are reported in this chapter. Overall, the findings of the study were well received, while comments and questions that were made before, during and after the workshop were addressed in the final report.

10.4 Study Findings

10.4.1 Key findings from the value-chain actors surveys

Survey results of food industry practitioners and experts who expressed opinions on future developments are highlighted below.

- Food practitioners are confident that the food sector, particularly dairy, fruit, and vegetable sectors, will continue to grow.
- Imports will grow in comparison with domestic food supply and this will affect domestic production as well, both at the general level and when looking at the three specific industries analyzed in this chapter.
- Prices of the food products will grow (both national and international).
- Development of new market infrastructure is dependent on agri-food policy and market services (e.g. financial and monetary policy). They are deemed very important to industry growth and modernization.
- Growth in the number of local producers is seen very likely for the near future.
- Implementation of food safety standards is seen as the biggest potential bottleneck slowing down the industry's growth.
- Quality of human capital and management, as well as the quality of inputs, are key factors that may hamper growth in product value chains.
- Most of the people interviewed advocate a change in food quality standards.
- Experts are aware that the macroeconomic situation is crucial to industry development; therefore, policies should take this into account.

10.4.2 Consumer survey results on the main challenges and opportunities of the food industry

In this section we highlight the problems and opportunities identified by analyzing the consumer survey data obtained from the five surveys administered (three focused on fruit, vegetable, meat and dairy products; and two focused on cross-cutting issues, namely food safety and organic food, respectively). This section focuses on what might support the development of a more efficient and competitive agricultural sector. The analysis of the data from the five structured consumer surveys provided a rich variety of information that for the sake of brevity will not be reported in its entirety here.

Food safety perceptions can be briefly summarized as follows.

- A substantial degree of concern exists in terms of food safety, with about two-thirds of consumers interviewed saying they are somewhat concerned.
- About 50% of consumers interviewed agree or strongly agree that they do not have enough information to decide whether the food they purchase is safe.
- Consumers are especially afraid of microbial contamination and are worried about pesticides in food.
- Most consumers are not familiar with safety-related quality management/assurance schemes and certifications, such as Hazard Analysis and Critical Control Points (HACCP) and Global Good Agricultural Practices (GAP), but they have heard about International Organization for Standardization (ISO) quality standards.
- The main responsibility in assuring food safety is perceived to be associated with food manufacturers rather than retailers.
- Consumers are not confident they can assess food safety, even though they feel they are responsible for assuring food safety.
- Brand reputation and expiration date are important factors, which are ways to show evidence of food safety for most consumers.

Consumer preference for horticulture (fruit and vegetable) and livestock (dairy and meat) products are summarized below:

- Many consumers consider origin of their food important, specifically in the case of apples. Overall, there is a strong preference for locally produced food.
- Consumers trust domestic food more than imports except for when the imports come from the EU. Region of production within Kosovo is important too. An example is the case of cheese from Sharr, for which consumer surveys and focus groups data indicate a strong preference.
- Consumers are willing to pay a higher price for cheese not produced with powdered milk.
- "Certified stamp from the veterinarian inspector" is preferred to "knowing the butcher" and "knowing the origin" for lamb meat. Thus there is trust in the veterinary system, which is different from the consumer preference of neighboring Albania (Imami *et al.*, 2011; Zhllima *et al.*, 2015).
- Consumers prefer to purchase fresh fruit and vegetables at the large farmers/green markets versus supermarkets as they are perceived to be cheaper and fresher.
- For fruit and vegetables, consumers prefer to buy their produce uniform in size and shape, and of certain varieties.
- Buyers of fruit prefer basing their purchasing decisions on the seller's information and brand.
- The most important attributes when purchasing fresh fruit were freshness and cleanliness of the outlets.
- Consumers are concerned about pesticide residues on fruit.
- Consumers are concerned about soil contamination on their fresh produce.

The concern on the use of pesticides and soil contamination on buyer's fresh produce is related to the preference for organic food, which was a topic in a separate survey whose findings are summarized below:

- Many consumers are familiar with the word "bio" or "organic", but do not have a clear understanding of this label's meaning. However, most consumers are aware that producing through the organic method implies several limitations imposed on the agricultural practices, such as not using agro-chemicals and hormones.
- Most consumers have positive or strongly positive opinions of organic food in terms of environmental friendliness, safety, taste, and, for fruit and vegetables, freshness.
- The main reason why consumers prefer organic food products is that it is perceived to be safer for human consumption.

However, important barriers hinder consumers purchasing organic food:

- Many consumers cannot find them in the stores.
- Price is perceived as too high, although people basically stated they are willing to pay a premium, with the average amount of less than 20%, but it varies significantly by consumers and it is probably overstated.

10.4.3 Survey results on consumer food-related lifestyle

In this section the lifestyle findings for the data obtained in each of the surveys (1500 respondents) are summarized. Lifestyles are defined as patterns in which people live and spend time and money (Engel *et al.*, 1990), therefore its analysis may help in understanding what matters to people. The food-related lifestyles instrument used in this study (a condensed version of the original one developed by Brunsø *et al.*, 1995) was useful to describe consumers on the basis of a set of characteristics (called dimensions) that reflect their way of living or how they perceive food in their daily life.

Lifestyle data is a useful basis for consumer segmentation, a concept that applies to markets as well as to political audiences or target beneficiaries of policy interventions. Segmentation consists of splitting the overall study population into smaller and mutually exclusive groups (that is, one individual belongs to only one group). Each group consists of individuals with similar lifestyles that are different from those in the other groups. The benefit of segmentation for policy and strategic decision making are apparent: (i) it allows one to avoid making decisions on the basis of an unclear definition of an "average" customer or user, focusing instead on the strategy of more clearly defined "segments"; (ii) it focuses decisions by designing better targeted policies, and for businesses this is important for competitiveness and, ultimately, profitability; and (iii) it clearly defines segments, allowing for fine-tuning solutions and communication, thus leading to a higher satisfaction of the target consumers.

Summarizing the results of the analysis of Kosovo consumers using the "reduced" Food Related Lifestyle instrument (Dimech *et al.*, 2011), the following seven dimensions of lifestyles were extracted from the data:

- care and commitment in food choice;
- role of food in the individual's social life;
- fun associated with food and price awareness;
- curiosity and variety seeking;
- awareness and interest for the sensory characteristics of food;
- extent to which food purchase and preparation is perceived as a woman's task; and
- unplanned versus planned meals as a sign of degree of centrality of food in daily life.

Each consumer's attitudes in the sample have been measured according to these dimensions, which allowed us to aggregate them into groups. Individuals in the same group have similar attitudes, and those attitudes are clearly different from those belonging to the other segments.

Table 10.2 shows how consumer profiles are segmented, using relative rating scale (extremely high, high, medium high, neutral medium low, low, extremely low) relative to the average. When a rating is left blank, the group position on a specific dimension reflects the overall average of the sample. The size of each group in this case is more or less similar (ranging from 13% to 20% of the sample); therefore, it is not shown in the table.

These groupings show that consumers have different preferences and that probably different strategies should be designed by sellers to align their sales goals with their target consumer group. In order to match the different attitudes and needs of consumers, sellers can customize their product to appeal to different groups.

Decision makers can use these results as guidance and the results can be derived from looking simultaneously at the single dimensions (rows in the table) and by groups (columns in the table). For instance, segments 2,

Table 10.2. Segmentation of consumers into homogeneous groups according to the dimensions of their food-related lifestyles and socio-demographic variables (from survey data, 2014).

Lifestyle dimensions and socio-demographics	Consumer groups						
	1	2	3	4	5	6	7
Care in food choice	Extr. Low	–	Low	High	Very high	–	High
Social life	–	High	Low	Low	Low	–	–
Fun and price awareness	–	Extr. High	Med-Low	Med-High	–	Med-Low	Extr. Low
Curiosity and variety seeking	High	–	Med-Low	Med-High	Med-Low	Extr. Low	High
Sensory awareness	–	High	Low	High	Extr. Low	High	High
Woman's task	Low	Very high	High	Extr. Low	Med-Low	Med-Low	High
Unplanned meals	High	Med-Low	Very Low	Very Low	Med-High	High	Med-High
Age	Younger	–	–	–	–	–	
Gender	–	–	–	Mostly females	–	More males	More males
Income	–	–	–	–	–	Lower	–

4, 6 and 7 show a high sensory awareness and are supposed to be more responsive to a strategy involving the value-enhancement of food products with exceptional sensory features. Instead, segments 1 and 6 that score high in the dimension "unplanned meal" category may be the most suitable target for the development of new products aimed at increasing the level of convenience (e.g. ready meals). Furthermore, it can be argued that groups 4 and 7 could be categorized as potential consumers for quality food produce. All combinations provide useful insight for sellers.

These findings reflect post-conflict era consumer preference information, which is useful to producers and value-chain actors who are still used to making decisions as if they were in an economy based on central plans and not an economy that is consumer driven. An increase in similar studies will help the transition into thinking about consumer preferences.

10.5 Conclusions and Recommendations

Kosovo, emerging as an independent country from a planned economy after going through an armed conflict, has been facing the challenges of strengthening institutions, adapting to free market economy demands, and attracting investments. At the same time, income and lifestyles are changing fast with pending European Union integration, which has influence and implications on shifting consumer preferences and behavior. Growing domestic demand for food comes with opportunities and challenges. Increasing awareness for food quality and safety, for example, should be perceived as a strong signal for locally produced food companies to adapt. Otherwise imports will keep increasing, which will adversely affect the viability of the agricultural sector.

With the dearth of market intelligence information, this chapter serves to highlight recent consumer perceptions of their food supply and preferences. It presented survey results that highlight some post-conflict key developments: a large segment of consumers in Kosovo now prefer high-quality products, such as organic or of local origin, and the domestic origin placed on product labels is important. Therefore, the government should take an active role to enhance competitiveness of local producers and the food industry. For example, in areas that are already perceived as having a good reputation, the government should facilitate the introduction of location-specific labels in these areas, like in the case of cheese from Sharr. This activity would satisfy consumers who are seeking and are willing to pay more for high-quality cheese products. In turn, the cheesemakers in Sharr should be leaders and advocates of innovations that demonstrate how the delivery of higher-quality differentiated products, sold on the domestic market at a higher price, works.

The market intelligence information is especially necessary and essential to connect the farmers and their produce with the appropriate market segments in post-conflict and transition countries like Kosovo. Utilizing market information to make production decisions is novel to farmers and supply-chain actors emerging from a centrally planned economy. Obviously, privatized business in post-conflict countries must make use of market information to exploit relevant opportunities. However, it is not sufficient for success, and policy makers have the responsibility to create the conditions to make these farms and agribusinesses sustainable, thus allowing consumers and producers to converge on a win–win solution. It should also be noted that, in Kosovo and other Balkan countries, direct relationships of consumers with producers (or with local small processors/dealers, such as butchers) still exist due to the historical trust relationship built through generations. The emergence of modern retail stores and supermarkets will inevitably break these close relationships and therefore there will be a need for other mechanisms to create the trust of consumers who have migrated to or were raised in urban areas.

Following EU rural development policies, several initiatives to develop local markets and efficient food supply chains have already been put in place to support and benefit farmers, distributors or consumers. Examples of initiatives are Local Action Groups (LAGs), which are common in the EU and can be supported in Kosovo if it joins the EU in the future. Objectives of initiatives deserving support could include modernization of agricultural holdings, improvement and guarantee of the quality of agricultural production and products, and support for actions fostering business creation, youth entrepreneurship, networking and synergies among local actors, and collaborative new product/service development.

Kosovo has been the last country in the region to gain independence and to enter the post-conflict era; therefore, there is an urgency to move quickly and make a special effort to help farmers and producers catch up to innovations that were introduced long ago in the EU and in many neighboring countries. The changes need to be quick and the economy needs to be stable to ensure that investments are effective. Thus, promotional measures supporting the initiatives mentioned before need to be paired with appropriate interventions: specifically, sanitary norms and hygiene rules, as well as the implementation of important trust-related tools in agriculture, such as quality management schemes, collective marketing, and branding.

Introducing rural development policies and support of local producers are not to be misunderstood as protecting the status quo of the agricultural sector. On the contrary, it suggests the recognition of the needed changes in supply and accommodation of shifting consumer preferences to make the agricultural sector competitive. Market intelligence information is also a key connector between suppliers and consumers, allowing the former to become aware of the needs and wants of the consumer. Market research shows that a large segment of consumers prefer organic produce, they would welcome a lower use of pesticides in production, and they expect informative and reliable food labeling. This is essential information to help producers make better decisions, but it is also useful to policy makers, who should define the framework conditions allowing producers to exploit market opportunities. For instance, it is the responsibility of policy makers to create an environment enabling technologies (such as information and communications technology) and potential marketing/distribution management tools—such as barcodes, quick response (QR) codes, radio-frequency identification (RFID) tags—to be readily accessible and viable to potential users/adopters.

A very wide variety of initiatives and experiences (with successes and failures) already exist (ENRD, 2011; Moroney *et al.*, 2013) and may be taken as possible templates and best practices for the implementation of a country-specific strategy suitable to achieve the objectives of rural development policies. Because of this, the identification of the potential tools available would not be the main challenge. The main challenge in post-conflict countries is rather the availability of sufficiently developed human capital, which calls for investments in education and training, especially for the younger generations.

The key to developing a competitive and innovation-based agri-food sector for emerging economies capable of producing high quality products should be based on building human capacity that generates science-based knowledge on the food supply determinants. Government has a key role in funding institutions and providing public goods and services such as extension resources to help Kosovo and other Balkan countries increase their exports and at the same time compete with imports. Simultaneously, soft infrastructure such as quality management principles, approaches and procedures also need to be developed and widely disseminated and practiced.

The pivotal investment, therefore, should be made in human capital and organization skills, meaning that it is necessary to create amongst local food practitioners the expertise necessary to manage single businesses, networks, and whole supply chains using a quality-oriented continuous improvement approach. Assuming that an overarching policy aimed at achieving this goal is put in place, the most appealing policy measures are those that are coherent with a long-term strategy of full alignment with EU standards. In short, they could be identified as follows:

- Improvement of the regulatory framework on food safety, supported by accurate and up-to-date statistics about the incidences of food-borne diseases and food frauds and by a continued effort to increase the availability of information about food safety hazards and risk levels.
- Incentives to farmers aimed at fostering the adoption of international standards such as FAO GAP, or Global GAP, or implementation of a fully equivalent national standard. Given the specific structure and the current state of farming development in Kosovo, it

is advisable that certification is based on a collective approach to certification suitable to the local/domestic market, thus facilitating the implementation of internal control systems or participatory guarantee systems (Nelson *et al.*, 2009; Sacchi *et al.*, 2010, 2011; Pugliese *et al.*, 2013) together with the usual third-party certification. The use of the latter will probably be left to well structured and organized companies that are more export oriented.

- Adoption of a national legal framework for Geographical Indications compatible with EU regulations, with the goal of enabling the registration of local food specialties in the EU.
- Implementation of organic agriculture regulations, fully harmonized with the EU and equivalent to other main international standards such as USDA, Japanese Agricultural Standards (JAS), etc.

This is obviously a demanding and ambitious set of goals, but it is necessary to add that competitiveness and innovation are strongly linked with research capacity, which in turn is dependent on the quality and accessibility of higher education. Typically, there is a vicious cycle of correlations between research for development (R4D) and economic development in the case of transition or post-conflict economies; in other words, limited resources for R4D constrained by poor economic performance which imply less innovation and slower pace of economic development and vice versa. Therefore, an adequate and consistent effort in R4D should also be a prerequisite to accompany these measures.

In a report published a few years ago (Demaj *et al.*, 2009), it was highlighted that the economy in Kosovo was too weak to support research and development, and that research activities were sporadic and marginal because of low government support (0.1 % of the GDP and funds mainly used to cover operational cost) and because higher education institutions mainly focused on teaching. Opportunities available to support advanced research from international funding sources, such as the 7th Framework Program (now Horizon 2020) of the European Commission, were not sought actively by Kosovo's higher education institutions (Demaj *et al.*, 2009). Under these conditions, it would be difficult to pursue the goals mentioned above; however, the Strategy for Development of Higher Education in Kosovo (MEST, 2004) lists improving the capacity for scientific research as one of its objectives and hopefully this objective will be pursued and achieved. Research is a crucial connection that could improve Kosovo's economy by connecting sellers with their consumers' preferences and thus increasing exports while simultaneously increasing locally desired supplies as well.

Acknowledgment

This chapter was prepared based on the studies developed in the context of the FAO Project "Policy assistance to Kosovo to identify support measures linking local agricultural production with the domestic market TCP/KOS/3401." The views expressed in this chapter are those of the authors and do not necessarily reflect the views of FAO or MAFRD.

References

Bartlett, W. (2008) *Europe's Troubled Region: Economic Development, Institutional Reform and European Integration*. Routledge, London and New York.

Bartlett, W. (2009) Economic development in the European super-periphery: evidence from the Western Balkans. *Economic Annuals* 54, 21–44.

Brunsø, K. and Grunert, K.G. (1995) Development and testing of a cross-culturally valid instrument: food-related life style. *Advances in Consumer Research* 22, 475–480.

Brunsø, K., Grunert, K.G. and Bredahl, L. (1995) The comparison of food-related lifestyles across countries. *Appetite* 24(3), 286–287.

Bytyqi, N., Verçuni, A., Pllana, M., Jahja, A. and Bytyqi, H. (2012) Analysis of consumer behavior in regard to the beef meat in Kosovo. *Food and Nutrition Sciences* 3(11), 1514–1521.

Central Bank of Kosovo (2010) *Bulletin of Payment Balance 2010*. Central Bank of Kosovo, Priština.
Demaj, A., Kaçiu, S. and Mehmeti, A. (2009) *National Background Report on Agro-Food Research for Kosovo (under UNSCR 1244)*. Priština, Kosovo. Available at: http://wbc-inco.net/object/event/6300/attach/AgroFoodReportKOSOVOFinal.pdf (accessed 9 December 2016).
Dimech, M., Caputo, V. and Canavari, M. (2011) Attitudes of Maltese consumers towards quality in fruit and vegetables in relation to their food-related lifestyles. *International Food and Agribusiness Management Review* 14, 21–36. Available at: http://ageconsearch.umn.edu/bitstream/117602/2/20110027_Formatted.pdf (accessed 8 December 2016).
EBRD (2014) Innovation in Transition. Transition Report, 2014. *European Bank for Reconstruction and Development, London*. Available at: http://2014.tr-ebrd.com/ (accessed 9 December 2016).
EC (2002) *Western Balkans in Transition Report*. European Commission, Brussels. Available at: http://www.csd.bg/fileadmin/user_upload/reporten.pdf (accessed 8 December 2016).
EC (2010) *Eurostat News release: Euro indicators 2010*. European Commission, Luxembourg. Available at: http://ec.europa.eu/eurostat (accessed 9 December 2016).
Engel, J.F., Blackwell, R.D. and Miniard, P.W. (1990) *Consumer Behavior (6th edn)*. Dryden Press, Oak Brook, Illinois.
ENRD (2011) The European Agricultural Fund for Rural Development: Examples of Food projects. ENRD, European Network for Rural Development, Brussels. Available at: http://enrd.ec.europa.eu/enrd-static/fms/pdf/18EC541F-CB32-ED81-55DF-AFB25B27E01E.pdf (accessed 9 December 2016).
Fischer, C. (2004) *Assessing Kosovo's Horticultural Potential – The market for fruit and vegetables on the Balkans. Institute of Agricultural Development in Central and Eastern Europe*, University of Bonn, Bonn, Germany. Available at: http://www.iamo.de/fileadmin/documents/dp67.pdf (accessed 8 December 2016).
Giron, L. (2001) Consumption Habits and Purchasing Behaviour of Fruits and Vegetables in Kosovo. Swiss Project for Horticulture Promotion – Kosovo (SPHP-K), Priština, Kosovo.
Imami, D., Chan-Halbrendt, C., Quanguo, Z. and Zhllima, E. (2011) Conjoint analysis of consumer preferences for lamb meat in central and southwest urban Albania. *International Food and Agribusiness Management Review* 14(3), 115–126.
IMF (2015) *World Economic Outlook Database, April 2015*. International Monetary Fund, Washington, DC. Available at: https://www.imf.org/external/pubs/ft/weo/2015/01/weodata/index.aspx (accessed 8 December 2016).
KAS (2001) *Statistics on Agriculture in Kosovo*. Kosovo Agency of Statistics, Priština, Kosovo. Available at: http://ask.rks-gov.net/en/kosovo-agency-of-statistics/agriculture (accessed 12 December 2016).
KAS (2006) *Agricultural Household Survey 2005*. Kosovo Agency of Statistics, Priština, Kosovo. Available at: http://www.esiweb.org/pdf/kosovo_Agricultural%20Household%20Survey%202005.pdf (accessed 8 December 2016).
KAS (2011) Export and import by SITC in value by trade, commodity and time 2005–2011. In: External Trade Statistics 2011 (Table 3, p. 10). Kosovo Agency of Statistics, Priština, Kosovo.
KAS (2014) *Statistical Yearbook of the Republic of Kosovo 2014*. Kosovo Agency of Statistics, Priština, Kosovo. Available at: http://ask.rks-gov.net/en/kosovo-agency-of-statistics/agriculture (accessed 12 December 2016).
KAS (2015) *Kosovo in Nutchsel Figures*. Kosovo Agency of Statistics, Priština, Kosovo. Available at: https://ask.rks-gov.net/eng/ (accessed November 2015).
Koro, L. (2007) *Economic Performance of the Western Balkan Economies for 2007*. UNDP Europe and CIS, New York and Istanbul. Available at: http://europeandcis.undp.org/home/show/41D6901C-F203-1EE9-BF623728E26C6241.
MAFRD (2002) Opportunities of investment in the subsector fruits and vegetables in Kosovo. Presentation summary of the Kosovo Investment Conference, Olten, Switzerland. Ministry of Agriculture, Forestry and Rural Development, Priština, Kosovo.
MAFRD (2010) *Agriculture and Rural Development Plan 2010–13*. Ministry of Agriculture, Forestry and Rural Development, Priština, Kosovo. Available at: http://app.seerural.org/wp-content/uploads/2013/08/Kosovo-ARDP-2010-2013-ENG.pdf (accessed 9 December 2016).
MAFRD (2013) *Green Report 2013*. MAFRD, Priština, Kosovo.
MAFRD (2014a) *Strategy for a new programme 2014 to 2020*. MAFRD, Priština, Kosovo. Available at: http://www.ardp-kosovo.eu/en/new-ardp-2014-2020/strategy-for-a-new-programme.html
MAFRD (2014b) *Green Report 2014*. MAFRD, Priština, Kosovo.

MEST (2004) Strategy for Development of Higher Education in Kosovo (2005–2015). Available at: www.esiweb.org/pdf/bridges/kosovo/6/22.pdf (accessed March 2016).

Montanari, M. (2005) EU trade with the Balkans large room for growth? *Eastern European Economics* 43, 59–81.

Moroney, A., O'Shaughnessy, M. and O'Reilly, S. (2013) *Facilitating and Encouraging Short Food Supply Chains*. Department of Agriculture, Fisheries, Food (DAFF), Dublin, Ireland.

Nelson, E., Gómez Tovar, L., Schwentesius Rindermann, R. and Gómez Cruz, M.Á. (2009) Participatory organic certification in Mexico: an alternative approach to maintaining the integrity of the organic label. *Agriculture and Human Values* 27, 227–237.

Prism Research (2009) *Horticulture Promotion in Kosovo: Consumption Habits and Purchasing Power of Fruits and Vegetables in Kosovo – Follow-up Survey*. Prism Research, Priština, Kosovo.

Pugliese, P., Zanasi, C., Atallah, O. and Cosimo, R. (2013) Investigating the interaction between organic and local foods in the Mediterranean: the Lebanese organic consumer's perspective. *Food Policy* 39, 1–12.

Sacchi, G., Zanasi, C. and Canavari, M. (2010) The theory of conventions as a framework to analyse participatory guarantee systems for organic food. In: V Research Workshop on Institutions and Organizations (V RWIO) October 3–5, 2010, Gonçalves – Minas Gerais, Brazil. *Faculdade de Economia, Administração e Contabilidade da Universidade de São Paulo, São Paulo, Brazil.*

Sacchi, G., Zanasi, C. and Canavari, M. (2011) Modelli alternativi di garanzia della qualità dei prodotti biologici alla luce della teoria delle convenzioni. *Economia Agro-Alimentare* 13, 57–80.

SDC (Swiss Agency for Development and Cooperation) (2013) Country Strategy – Kosovo, 2013–2015. SDC Priština, Priština, Kosovo. Available at: http://www.seco-cooperation.admin.ch/laender/05151/05164/index.html?lang=en&download=NHzLpZeg7t,lnp6I0NTU042l2Z6ln1ad1IZn4Z2qZpnO2Yuq2Z6gpJCDd39_gmym162epYbg2c_JjKbNoKOn6A-- (accessed 8 December 2016).

World Bank (2000) Kosovo: damage assessment in agriculture. ECSSD (Environmentally and Socially Sustainable Development) Working Paper #22. World Bank, Washington, DC. Available at: http://lnweb18.worldbank.org/ECA/ECSSD.nsf

Zhllima, E., Imami, D. and Canavari, M. (2015) Consumer perceptions of food safety risk: evidence from a segmentation study in Albania. *Journal of Integrative Agriculture* 14(6), 1142–1152.

11 Characterizing Farmer Innovation Behavior for Agricultural Technologies in Transitional Areas Facing Environmental Change

Jacqueline Halbrendt[1]* and Bikash Paudel[2]

[1]*Wageningen University, Netherlands;* [2]*Local Initiatives for Biodiversity, Research and Development (LI-BIRD), Pokhara, Nepal*

11.1 Introduction

Nepal has gone through tumultuous developments in the recent past and the country has a strong desire to bring about peace and political stability. During this transitional phase, the country faces a lot of challenges, among them a desire for economic development while facing poverty and natural disasters. In recent years, many development agencies have targeted Nepal as a country that requires aid assistance due to its poor humanitarian indicators. A popular mechanism is assisting small farmers with relevant information and technology to increase their productivity for enhanced nutrition and income, but not much is understood regarding farmers' behavior toward innovations. This chapter attempts to discover some of the salient characteristics of farmers who adopt innovations and where they get their information.

With rapid population growth and increasing food insecurity, a subsequent shift towards the cultivation of marginal land by smallholder farmers is taking place. In the case of Nepal, such marginal lands tend to be highly sloping, and prone to erosion and accelerated soil degradation (Shrestha *et al.*, 2004). This calls for improved land management strategies, such as the increased implementation of conservation agriculture (CA) practices, to mitigate the effects of degradation and improve the long-term productive capacity of the land.

CA applies the principles of minimum soil disturbance, permanent soil cover, and crop rotations to reduce land degradation, improve crop yields, and promote sustainable agricultural systems (Hobbs *et al.*, 2008; FAO, 2014). Due to its relatively low implementation costs and potential for improving food security, CA has been promoted by various donor organizations as an approach to improve the agricultural system and livelihoods of farmers in developing countries (Knowler and Bradshaw, 2007). Those introducing CA practices have faced challenges in establishing long-term adoption.

Globally, there are more than 500 million smallholder farms, generally defined as having less than 2 ha of land per household (IFAD, 2011). These households comprise the majority of the world's poor and are characterized as highly vulnerable to food insecurity and environmental change. With few technical and financial resources available, they often struggle

*jhalbrendt@msn.com

 Enabling Agri-entrepreneurship and Innovation: Empirical Evidence and Solutions for Conflict Regions and Transitioning Economies (eds C. Chan *et al.*)

to adapt to climate change and to maintain sufficient crop productivity (Shang, 2014). In Nepal, where 80% of the population comes from smallholder subsistence farms and 70% of farms are less than 1 ha in size, the plight of the rural poor is further exacerbated. The country's remote communities and rugged terrain limit access to markets and extension services, inhibiting economic development, while climate shifts in the form of drought and erratic rainfall contribute to reduce yields and create food scarcity (IFAD, 2014). These areas are prime candidates for interventions such as CA, requiring few input costs while decreasing degradation and improving yield. Yet historical evidence shows that introduced practices are often abandoned following the completion of development projects (Yadav, 1987; Bunch, 1999; Cochran, 2003). Thus, in order to develop effective conservation outputs, it is important to understand the dynamics of decision making by farmers for introduced CA technologies.

Implementing conservation practices can be an inherently difficult process, as the mitigating effects of reduced erosion and degradation are often not observable for many years and may show variability in terms of positive, negative, or neutral impacts during the initial years of CA integration. Furthermore, CA provides many indirect benefits via soil fertility and water conservation improvements. The delayed benefits of CA work in contrast to the relatively short time horizon of subsistence farmers, who often must make decisions based on the current season's yield (Giller *et al.*, 2009). The benefit delay of CA can result in lower adoption rates, as farmers' livelihoods and household food security are dependent on the capacity to produce viable and sufficient outputs from year to year. Identifying and understanding the complex socio-economic characteristics of farmers who may be willing to adopt or not adopt new conservation practices will be vital for developing community-focused approaches for the improved implementation of CA systems, addressing the needs of both the farmer and the environment.

The factors that determine adoption are highly contextual, based on local environmental and socio-cultural conditions at the community level (Halbrendt *et al.*, 2014). In terms of adoption decisions by farmers, a number of characteristics play a role at the individual level. Adoption of agricultural technologies, particularly for developing countries, depends on a wide range of factors, including socio-economic aspects of the individual farmer and household, physical characteristics of the farm, and access to resources (Kebede *et al.*, 1990). It can be assumed that farmers weigh the comparative advantage of a new technology against their capacity to implement the technology and their willingness to absorb the risk of decreased crop production (Tiwari *et al.*, 2008). A logistic regression model can identify the relevant farmer characteristics involved in the decision-making process and explore the interactions of those characteristics in terms of adopting CA practices.

Historically, the rate of agricultural technology transfer has been extremely slow in Nepal. Limited capability of the government extension system to transfer the new technologies is one of the main challenges for agricultural technology transfer in the country. Government extension has often been ineffective in disseminating information about new agricultural technologies, due to lack of resources (Shrestha, 2013). As a result, lack of information is a common constraint for adoption of new technologies (Floyd *et al.*, 2003). Globally, it has been recognized that informal social networks and connections play an important role for the transfer of information about sustainable agricultural technologies (Conley and Udry, 2001; Isaac *et al.*, 2007). Informal social networks also facilitate collaborative learning, evaluation, and adoption of new agricultural technologies, which is an essential element for successful adoption of CA by smallholder farmers (Erenstein *et al.*, 2012). Evidently, a few previous studies in Nepal have found that farmer-to-farmer technology transfer is successful in promoting conservation technologies (Shrestha, 2013). Evaluating the existing information-sharing networks, which are currently used to distribute information about agricultural technologies, and determining how often such networks provide information about sustainable agricultural technologies would aid in adoption of CA information. Additionally, analyzing stakeholders in the information networks is vital, because the presence (or absence) of different stakeholders affects the sustainability of information exchange in villages (Prell *et al.*, 2009). Hence, taking the case of the information network surrounding hedgerow management, establishing a sustainable information platform for exchanging information related to CA is important.

To understand better the issues of technology transfer and information flow, the research discussed in this chapter was designed to study the significant socio-economic factors contributing to an individual farmer's likelihood of adoption for introduced CA technologies under transitioning conditions or with weak government support. The results of this were assessed to identify and understand the major constraints leading to non-adoption of CA technologies. Furthermore, the structure of social networks for the transfer of CA and the major constraints due to network configuration were identified, with the goal of identifying implications for CA approaches and developing policy implications.

The central mid-hills of Nepal support 44% of the country's population and the region is dominated by smallholder farming. This area has been identified as a high priority for agricultural development to improve food security and mitigate the effects of climate change (Thapa and Paudel, 2002). In recent decades, increasing populations have led to a transition from shifting cultivation to intensified, continuous cultivation of marginal highly sloping land, resulting in a need for improved agricultural technologies to increase soil conservation and crop yields.

Two rural Chepang villages in the mid-hills region, Kholagaun (Tanahun District) and Thumka (Gorkha District) (Fig. 11.1), were selected based on past and present engagement with a local non-governmental organization (NGO) in conservation projects, as well as their high risk of food insecurity and limited opportunities for income generation. The two villages are comparable in terms of ethnic group, economic status, farm size, and farming practices. The Chepang are indigenous to Nepal and are one of the most marginalized ethnic groups in terms of geographic location and socio-economic status (Luni *et al.*, 2012). The villages predominantly consist of subsistence farming households and are located on the mountainside, lacking direct access to road networks and markets. The isolation of these villages limits opportunities for off-farm income generation as well as access to agricultural extension services. Farming in this area traditionally consists of a maize-based system on terraced land using sole cropping of local varieties, continuous cultivation, and draft plowing. However, the practices of conventional tillage, relatively low inputs of fertilizer, and leaving land fallow in the dry season have slowly degraded the land quality and increased soil vulnerability to erosion. It has been estimated that food scarcity exists for approximately 6 months of the year in Chepang households and few are

Fig. 11.1. The Chepang villages of Kholagaun and Thumka, Nepal, where the social networks and constraints to adoption of conservation agriculture were evaluated. (Coordinates for central market at Mugling: latitude 27° 50′ 56″ N, longitude 84° 33′ 03″ E.)

self-sufficient in agricultural production (Piya *et al.*, 2011).

11.2 Binomial Logit Model to Determine Constraints to Adoption

To determine the significant characteristics for identifying adopters and non-adopters of CA practices in a remote developing area, a binomial logistic regression model framework was developed. The model incorporates a dichotomous dependent variable, adoption or non-adoption of an introduced CA practice, and 12 explanatory variables derived from the literature to identify the significant contributing factors in the adoption decision. The explanatory variables are categorical or continuous. Past studies were used to develop a hypothesis for each of the explanatory variables to predict the direction of influence (+/–) on adoption. Each variable is *a priori* associated with hypotheses on its predicted influence on adoption (Table 11.1).

Theoretical Logit Model Equation

$$Y_i = \beta_0 + \beta_1 age_i + \beta_2 gender_i + \beta_3 education_i + \beta_4 trust_i + \beta_5 income_i + \beta_6 farmsize_i + \beta_7 labor_i + \beta_8 foodsecurity_i + \beta_9 information_i + \beta_{10} experience_i + \beta_{11} enviroconcern_i + \beta_{12} landtenure_i + \varepsilon_i$$

Face-to-face interviews were conducted with the primary male or female heads of household in the two Chepang villages in Nepal. A total of 56 surveys were conducted, representing 82% of total households. The surveys included questions to assess the dependent adoption variable and explanatory variables for individual farmer characteristics and farm/resource characteristics. All data were analyzed using SPSS statistical software (IBM Corp., 2013). For variables using an index (TRUST, FOODSECURITY, ENVIROCONCERN), a Cronbach's alpha internal consistency analysis was conducted to determine if it is justifiable to aggregate the responses for a suite of related questions (Cronbach, 1951). Questions found to reduce consistency (contributing to a Cronbach's alpha value < 0.7; George and Mallery,

Table 11.1. Summary of variables in the logit model.

Variable	Definition	Predicted direction of influence
Dependent variable		
Y_i	1 = ADOPTER, has implemented hedgerow technology on farm 0 = NON-ADOPTER, has never implemented hedgerow technology on farm	
Explanatory variables		
Farmer characteristics		
AGE	Age of farmer in years	–
GENDER	1 = Female, 0 = Male	–
EDUCATION	Years of formal education completed	+
TRUST	Index of trust in NGO staff, projects, and expertise	+
Resource/economic characteristics		
INCOME	Total on- and off-farm annual household income	+
FARMSIZE	Total hectares of farm	+
LABOR	Number of adult household members contributing to agricultural labor	+
FOODSECURITY	Household Food Insecurity Access Scale	–
INFORMATION	Level of interaction with agricultural extension or NGOs in past 2 years: 0 = none/low, 1 = moderate, 2 = high	+
EXPERIENCE	Number of years involved in farm decision making (shared or total control)	+
ENVIROCONCERN	Index of farmer perception of environmental degradation and need for conservation	+
LANDTENURE	1 = owns land title, 0 = does not own land title	+

2003; Tavakol and Dennick, 2011) were considered unacceptable and removed from the respective index.

Next, a stepwise regression using a bidirectional elimination approach was conducted to assess the fit and appropriateness of each explanatory variable in the model and to analyze the influence of the explanatory variables on the dependent adoption variable. While the appropriateness of using a stepwise approach is often debated, citing statistical bias and data mining (Whittingham *et al.*, 2006), in cases where the study population has been under-researched or the existing literature is not inclusive of the study demographic, there is justification for using this approach to test the validity and fit of the variables. Moreover, in areas where large data samples are not possible, as is often the case with remote rural villages, it is important to understand the interaction effects of the independent variables and ensure that data anomalies such as outliers and collinearity are not adversely impacting the precision of the model (NCSS). The adjusted logit model was then run to determine the significance and direction of influence for each of the remaining explanatory variables.

Implementation of hedgerow technology was used as a proxy for the adoption of CA practices. A local NGO (Local Initiatives for Biodiversity Research and Development, Pokhara, Nepal) worked with the study communities to introduce and provide technical assistance for hedgerow technologies from 2002 to 2005. At the time of the project, the NGO provided an equal opportunity for all households to participate in trainings, view demonstration plots, and receive seeds for implementation. The adoption of this technology over the subsequent years can be used to measure actual adoption outcomes (and delineate adopter and non-adopter groups) following multiple years of intervention and the completion of the NGO's active involvement. The use of this proxy variable is analogous for consideration of CA adoption due to the overlap in conservation goals, namely mitigating soil erosion and improving soil and water conservation on the farm (Sharma, 2013). Moreover, both hedgerow and CA technologies have similar requirements in terms of minimal required inputs, primarily initial seeds and additional labor. While it is recognized that hedgerow technologies address slightly different concerns in terms of soil and water conservation and the agricultural landscape as compared with CA, nevertheless this approach allows for the tangible characterization of adoption for an interrelated practice. In the model, this variable is defined as binary, with a value of 1 for respondents who have implemented hedgerows on their farm at any time since the training was provided (ADOPTERS), and a value of 0 for respondents who have never implemented hedgerows on their farm (NON-ADOPTERS). Since the survey did not explore the reasons why individuals may have stopped using hedgerows, any individual who took the initiative to implement hedgerows at any point was included in the adopter group.

Explanatory variables cover a broad range of factors that have been identified as contributing to the likelihood of adoption of CA practices. Characteristics of the individual farmer were evaluated using the demographic variables of age, gender, and education, as well as the latent variable of trust in NGOs. Access to resources and economic variables include income, farm size, agricultural labor, food security, access to information, farming experience, concern over the environment, and land tenure. The relevance of each of these variables to the model is described in further detail below.

11.2.1 Relevance of variables

Age

Measured as a continuous variable, increasing age is expected to affect adoption negatively. Older age can be associated with a decreased likelihood of benefitting from long-term conservation efforts, discouraging adoption (Bohlen *et al.*, 1961; D'Souza *et al.*, 1993; Neupane *et al.*, 2002). In contrast, younger household heads are expected to be more willing to innovate and have more time to accrue CA's long term benefits, thus would be more likely to adopt (Stark, 1996; Daberkow and McBride, 1998).

Gender

As a categorical variable, women are expected to be less likely to adopt conservation practices due to disproportionate burdens of household and agricultural labor (Lubwama, 1999). Additionally, due to persisting patriarchal traditions and often less engagement of women in trainings, women may have less access to or awareness of introduced technologies (Tiwari *et al.*, 2008).

Education

A continuous variable measured as years of formal education completed, more education is expected to increase adoption of CA. An assumed linkage of education with knowledge and less conservative behavior may result in the higher likelihood of adopting new technologies (Bohlen *et al.*, 1961; D'Souza *et al.*, 1993; Stark 1996; Knowler and Bradshaw 2007; Tiwari *et al.*, 2008).

Trust

Trust is an index consisting of 12 questions developed to measure the level of trust in NGO staff and projects. Questions were derived from the literature and assessed farmers' perceptions of NGO projects in terms of expertise, values, benefits, and reliability (Box 11.1). A low index score signifies less trust in NGO staff and is expected to correlate with decreased adoption (Bohlen *et al.*, 1961).

Income

Higher income can be associated with more financial stability and a greater capacity to absorb the risks of adopting a new agricultural practice. Therefore, higher income households are expected to not only be more likely to adopt CA, but will be among the early adopters of new technologies (Bohlen *et al.*, 1961; Kebede *et al.*, 1990; Daberkow and McBride, 1998; Knowler and Bradshaw, 2007; Tiwari *et al.*, 2008; AFT, 2013).

Farmsize

Similar to higher income, larger farms are hypothesized to better absorb the risks of adopting new practices and thus are associated with increased adoption (Bohlen *et al.*, 1961; Kebede *et al.*, 1990; Daberkow and McBride, 1998; Tiwari *et al.*, 2008; AFT, 2013).

Labor

A continuous variable, labor indicates the number of adults contributing to household agricultural labor. It is expected that increased labor availability will correlate with increased adoption and an increase in adoption is expected (D'Souza *et al.*, 1993; Stark, 1996).

Foodsecurity

This variable was measured using the Household Food Insecurity Access Scale (HFIAS), a widely accepted index comprising 9 questions related to household food availability, quantity, and diversity over the past month (Coates *et al.*, 2007). While scores may differ depending

Box 11.1. NGO Trust Index (questions used to determine TRUST in the logit model survey).

1. In general, do you think that NGO projects meet the stated expectations or goals?
2. Do you think that the goals of NGO projects are the same as your own goals for your farm?
3. When someone from an NGO makes a promise, does it usually happen? (e.g. arranging a meeting, organizing project activities).
4. Do NGO projects result in positive outcomes that benefit you or your family?
5. Do NGO workers have the same values as you for your farm? (e.g. income, food production, maintaining tradition).
6. Do NGO workers provide technical knowledge or show expertise in the project?
7. Are NGOs able to provide information that cannot be found in your community?
8. Do you think this information is beneficial to your household?
9. How satisfied have you been with NGO project activities in the past?
10. How confident are you that a technology introduced by an NGO will meet the stated goals?
11. In general, do you think that most people can be trusted?
12. In general, are you willing to try a new activity without seeing the outcome first?

on the season when the scale is accessed, the results will nevertheless indicate a measure of food security relative to others in the same study community. A higher score indicates greater food insecurity (maximum score = 27), and is expected to have a negative correlation with adoption. Low food security may limit adoption due to enhanced sensitivity to short-term production losses (Shively, 2001).

Information

The categorical variable of information was measured as the level of interaction with external sources of agricultural information and technology. A frequency distribution of the number of interactions with either NGO or agricultural extension officers over 2 years was used to define categories of none/low, moderate, or high levels of access to external sources of information. It is expected that greater access to information increases the likelihood of adoption (Bohlen *et al*., 1961; Kebede *et al*., 1990; Daberkow and McBride, 1998; Knowler and Bradshaw, 2007; AFT, 2013). Similarly, a lack of access to information and technical support systems may result in decreased adoption (USDA NRCS, 2005).

Experience

Measured as the number of years engaged in on-farm decision making, experience with farm management provides confidence and skills that will increase the likelihood of CA adoption (Kebede *et al*., 1990; USDA NRCS, 2005).

Enviroconcern

Environmental concern was measured as a 12-question index to assess farmers' levels of awareness and concern regarding their crop yield, soil erosion, and water availability. The questions were derived from concerns expressed in the literature regarding sloping land cultivation (Box 11.2). A high score in the index indicates increased awareness of local environmental conditions, which is expected to correlate with increased adoption of conservation practices (D'Souza *et al*., 1993; Knowler and Bradshaw, 2007). Similarly, a lack of awareness of erosion can result in non-adoption of mitigating practices (Stark, 1996; USDA NRCS, 2005; AFT, 2013).

Landtenure

Measured as a categorical variable, land tenure indicates whether the household holds a

Box 11.2. Environmental Concern Index (questions used to determine ENVIROCONCERN in the logit model survey).

1. How has the amount of your crop yields changed in the past 5 years?
2. How has the condition of your soil changed in the past 5 years (in terms of crop benefit)?
3. What is the condition of soil erosion on your farm in the past 5 years?
4. How has the amount of the water available to your household (from nearby streams and taps/hoses) changed over the past 5 years?
5. Do you think that the actions you take on your farm can impact crop yield? *(In either a positive or negative way)*
6. Do you think that the actions you take on your farm can impact soil quality? *(In either a positive or negative way)*
7. Do you think that the actions you take on your farm can impact soil erosion? *(In either a positive or negative way)*
8. Do you think that the actions you take on your farm can impact water availability for crops? *(In either a positive or negative way)*
9. Do you think there is a need to improve the water system (i.e. household and irrigation water) in your village?
10. Do you think there is a need to increase water availability for crops on your farm?
11. Do you think there is a need to improve the soil on your farm?
12. Do you think there is a need to improve the general farming practices on your farm?

government-recognized land title. Evidence has shown that lack of secure ownership of farmland acts as a disincentive for adoption and farmers are less likely to invest in long-term conservation efforts (Stark, 1996; Lee, 2005; Knowler and Bradshaw, 2007; Cox, 2011).

11.3 Social Network Analysis

In addition to individual farmer characteristics pertaining to integration of new practices, access to information and information networks is a critical component for the development of individual innovation and development. The data for analyzing the information network of conservation agriculture were collected through a household survey. Following standard social network analysis (SNA) protocol, a structured survey was administered to capture agricultural information sharing networks in the two rural Nepal villages during June–July 2013.

Broadly two types of measures are commonly used in network studies: (i) "whole network" measures; and (ii) "individual network" measures (Borgatti *et al.*, 2013). Whole network measures (e.g. fragmentation, density) provide information about the network as a unit of information flow among a select group (or at the graph level), while individual network measures (e.g. centralities) provide specifics on where an individual is located in the flow of information. This study selected two whole-network measures of information networks (i.e. fragmentation and density) and one individual-network measure (i.e. degree centrality).

- Network **density** is the simplest measure of network connectivity, which is simply the number of ties (connections) in the network, expressed as a proportion of the total number of possible ties (Borgatti *et al.*, 2013). Density provides a measure of network cohesion, which is important for understanding collective learning. Density is represented as a number between 0 and 1, showing the ratio of existing ties in the network to the total number of possible ties (Hanneman and Riddle, 2005). The density can also be reported by percentage, where a score of 100% indicates that every actor in the network is directly connected to every other actor whereas a score of 0% indicates all the actors are isolated and do not share any information.
- Network **fragmentation** is a measure of connectedness. Fragmentation refers to the number of pairs of nodes that cannot exchange information through any routes within the network (Borgatti *et al.*, 2002). If the network has many distinct (or unconnected) sub-groups or persons, such a network has higher fragmentation, thereby lower connectedness. Thus, higher network fragmentation can indicate greater constraints in information exchange among the community.
- **Degree centrality** can be broadly understood as the number of connections an individual has in the network. This can be divided into two categories based on the direction of information flow: (i) out-degree centrality of a farmer shows how many people a farmer gives information to in the community; and (ii) in-degree centrality refers to the number of incoming sources of information a farmer has. Generally, actors with a high degree centrality are in a better position to receive information and share among others (Freeman, 1979; Hanneman and Riddle, 2005). Higher degree centrality has also been associated with proactive technology adoption (Vilpponen *et al.*, 2006; Mikhail *et al.*, 2010). Therefore, degree centrality was selected as the individual network measure for this study.

To collect data for the information networks, both male and female decision makers in each farming household were asked to nominate at least three but up to ten individuals with whom they discussed useful information regarding any new agricultural technology. Next, respondents were asked to identify whether or not they sought advice about hedgerow technology from each nominated person. The specific questions were:

- Question 1: Can you name 3–10 people (including other farmers) whom you go to for information about new agricultural technology?
- Question 2: Do you seek advice about hedgerow practices with the person/s you mentioned in question 1?

The first step in the analysis was to create separate "adjacency" matrices for the information network. An "adjacency" matrix (A) is a square matrix, where number of rows (*i*) and columns (*j*) are equal to number of actors (vertex or nodes) and entry in each cell (a_{ij}) represents social relationship between i^{th} and j^{th} actors (Hanneman and Riddle, 2005). The adjacency matrices for hedgerow information networks were defined separately for each village, as follows:

- a_{ij} = 1 if i^{th} actor and j^{th} actor exchange information about hedgerow technology;
- a_{ij} = 2 if i^{th} actor and j^{th} actor exchange information about any other agricultural technology but not hedgerow technology; and
- a_{ij} = 0 otherwise.

The information network comprised by the condition a_{ij} =1 is referred to as the hedgerow information network (*hedgerow network*); and network comprised by conditions a_{ij} =1 and a_{ij} =2 together is referred to as the general agricultural information (*agricultural network*).

The next step in the analysis was visualization of networks and calculation of network measures. The network was visualized in NetDraw software (Borgatti *et al*., 2002). The calculation of network measures, i.e. density, fragmentation, and centrality, were conducted by standard methods in UCINET software version 6, taking data for one village at a time. Finally, to assess information sources in hedgerow information network, the actors were grouped in different stakeholder categories; and frequency and percentage values were calculated and compared with the benchmark (i.e. values for general agricultural network).

11.4 Discussion

In this discussion, we describe the key findings of the logit model to identify characteristics for the adopter and non-adopter farmer groups, outlining the individual farmer characteristics followed by the resource and economic characteristics for each group. We then explore the resulting social networks for technology information transfer, and discuss the structure of the networks, and the implications for information flow.

11.4.1 Logit model analysis

In the logit model, three of the explanatory variables used an index to measure latent constructs: TRUST; FOODSECURITY; and ENVIROCONCERN. FOODSECURITY showed high internal consistency with a Cronbach's alpha value of 0.904. This falls into the acceptable range of 0.70–0.95 (Tavakol and Dennick, 2011). Since the food security index has been vetted and is a widely accepted measure, this internal consistency is to be expected (Coates *et al*., 2007). The initial Cronbach's alpha coefficient for TRUST was 0.594, falling outside of the acceptable range; however, the analysis showed that three of the 12 questions were reducing consistency of the index. These questions, which were in regards to generalized trust in all people, as opposed to NGOs, were removed from the index. This resulted in an improved Cronbach's alpha coefficient of 0.694. While this is slightly out of Tavakol and Dennick's (2011) acceptable range, other researchers have indicated that values above 0.6 and approaching 0.7 may still be considered acceptable (George and Mallery, 2003), particularly in cases where data samples are small.

For ENVIROCONCERN, the Cronbach's alpha coefficient was 0.454, with no distinct questions reducing the consistency of the index. This inconsistency may be due to combining the topics of yield, erosion, and water availability in the same index, resulting in low overall correlation. A few independent questions showed significant correlation with the adoption variable (Table 11.2); however, as an index, these were not comprehensive enough to represent an individual's general level of concern over environmental degradation and the ENVIROCONCERN index was removed as a variable from the logit model.

The descriptive statistics of the explanatory variables (Table 11.3) give a general overview of the characteristics for adopters and non-adopters of hedgerow technologies. From the survey population (*n*=56), 73% of respondents had adopted hedgerow technology at some point, with the remaining 27% characterized as non-adopters. Due to this disparity in distribution of cases between adopter and non-adopter groups, some of the means may be falsely skewed due to the high/low number of cases in the group and may not accurately represent

Table 11.2. Correlation of environmental concern variables and adoption.

Environmental concern	Pearson correlation	Significance (1-tailed)	Adopters (*n*)	Non-adopters (*n*)
Observed change in soil erosion	–0.242	0.036[b]	9	7
Cultivation practices impact crop yield	0.363	0.003[a]	34	7
Need for improved water systems	0.235	0.041[b]	39	12
Need for increased water availability for crops	0.302	0.012[b]	40	12

[a]Significant at the 0.01 level.
[b]Significant at the 0.05 level.

Table 11.3. Descriptive statistics of variables used in the logit model.

Variable	Population (mean)	Adopters (mean)	Non-adopters (mean)
Dependent variable			
Y_i	*n*=56	*n*=41 (73%)	*n*=15 (27%)
Explanatory variables			
Farmer characteristics			
AGE (years)	37.8	39.2	34.3
GENDER[a]	0.41	0.39	0.47
EDUCATION (years)	1.8	1.4	3.0
TRUST[b]	6.5	6.7	5.7
Resource/economic characteristics			
INCOME (1000 NPR)	87.15	83.95	95.88
FARMSIZE (ha)	0.37	0.40	0.26
LABOR (no. people)	3.9	3.9	3.7
FOODSECURITY[c]	9.0	8.2	11.1
INFORMATION[d]	1	1.1	0.9
EXPERIENCE (years)	11.4	13.3	6.4
ENVIROCONCERN[e]	7.5	7.6	7.1
LANDTENURE[f]	0.8	0.8	0.8

[a]Values approaching 0 indicate a higher proportion of men, values approaching 1 indicate a higher proportion of women.
[b]Higher values indicate greater levels of trust, max value = 9.
[c]Higher values indicate greater food insecurity, max value = 27.
[d]Values assigned as 0 = none/low, 1 = moderate, 2 = high.
[e]Higher values indicate more concern over environment, max value = 12.
[f]Values approaching 0 indicate less land title ownership, values approaching 1 indicate more land title ownership.

Adjusted logit model equation:

$$Y_i = \beta_0 + \beta_1 \text{AGE}_i + \beta_2 \text{GENDER}_i + \beta_3 \text{EDUCATION}_i + \beta_4 \text{TRUST}_i + \beta_5 \text{INCOME}_i + \beta_6 \text{FARMSIZE}_i + \beta_7 \text{LABOR}_i + \beta_8 \text{FOODSECURITY}_i + \beta_9 \text{INFORMATION}_i + \varepsilon_i$$

adopter or non-adopter characteristics. Nevertheless, analysis of the logistic regression will determine the significant variables in the model.

The results from the stepwise regression showed that the explanatory variables of LANDTENURE and EXPERIENCE were not significant ($p<0.1$) and did not substantially contribute to the model. LANDTENURE had low correlation with the other variables and resulted in marginal changes when removed from the model, therefore it was excluded from the logit analysis. This may be attributed to the relatively short time period in which land ownership has been a common practice for the Chepang. Until a few generations ago, the Chepang were semi-nomadic, practicing some shifting cultivation, but subsisting primarily by foraging in the forest. Furthermore, the Chepang have historically been excluded from formal land ownership processes, relying on customary and oral traditions for land use and management, so they may not place the same weight on land titles as those in urban areas (Sharma, 2011). In the analysis of the EXPERIENCE variable, it was determined that this was highly correlated ($p<0.05$) with the AGE variable, resulting in a multi-collinearity effect. Since age and experience can be considered analogous to one another, and farming experience essentially begins at childhood in the study areas, the EXPERIENCE variable was omitted from the final model.

With the exclusion of the variables ENVIROCONCERN, LANDTENURE, and EXPERIENCE resulting from the prior analyses, the adjusted logit model equation includes nine explanatory variables (Table 11.3).

The logistic regression to determine farmer characteristics for adoption of CA practices was statistically significant, explaining 29.9% to 43.4% (Cox and Snell R^2 and Nagelkerke R^2, respectively) of the variance in adoption (Table 11.4). The model correctly classified 81.8% of cases. The results of the analysis showed EDUCATION, INCOME, and FOODSECURITY as significant at the 5% level, with the variables AGE, GENDER, and TRUST significant at the 10% level. FARMSIZE, LABOR, and INFORMATION were correlated with the model but were not significant.

Farmer characteristics leading to adoption of CA

Each of the farmer characteristics (AGE, GENDER, EDUCATION, TRUST) was found to be significant. While EDUCATION was highly significant ($p<0.05$), the direction of influence was opposite from the hypothesis. EDUCATION shows that more than 53% of respondents indicated having had no formal education, and only one respondent had more than 6 years in school. Due to the high number of cases with 0 years of education, the skewed distribution of data could have affected the directionality of the variable in the model. Nonetheless, much of the literature with positive correlations between education and higher rates of CA adoption considers areas, such as North America, where access to more than a primary level of education is standard (D'Souza *et al.*, 1993; Knowler and Bradshaw, 2007). However, there have been other cases in the developing world that have shown a negative correlation of education and adoption (Gould *et al.*, 1989; Okoye, 1998). In this case, more education showed a decreased willingness to adopt CA such that with every additional year of education, farmers were 0.57 times less likely to become adopters of CA. This result is a reflection of the relatively low levels of access to both primary and secondary education. Moreover, due to the subsistence nature of the study communities, education is often a tradeoff for meeting immediate agricultural and household labor demands (Tryndyuk, 2013). This effect could also be a function of less education leading to greater levels of reliance on external sources of information, such as NGOs, and a greater willingness to adopt introduced agricultural practices. This signifies

Table 11.4. Analysis of determinant variables in the logit model.

Variable	β	SE	Sig.	e^{β}
Farmer characteristics				
AGE	–0.096	0.055	0.082[b]	0.908
GENDER	–1.906	1.076	0.076[b]	0.149
EDUCATION	–0.559	0.260	0.031[a]	0.572
TRUST	0.383	0.226	0.090[b]	1.466
Resource/economic characteristics				
INCOME	–0.017	0.009	0.048[a]	0.983
FARMSIZE	2.459	2.200	0.264	11.689
LABOR	0.399	0.295	0.176	1.490
FOODSECURITY	–0.193	0.083	0.020[a]	0.825
INFORMATION	0.574	0.556	0.302	1.775
Constant	4.692	3.096	0.130	109.122

Notes: Cox and Snell R^2 = 0.299; Nagelkerke R^2 = 0.434; overall percentage correct = 81.8%.
[a]Significant at 5%.
[b]Significant at 10%.

that there is a distinction between formal and informal education in terms of adoption. Formal education is a departure from the applied knowledge gained from agricultural trainings and other community-targeted education programs and may not provide the practical skills needed for subsistence livelihoods. For the other farmer characteristic variables, the hypotheses were correct, with the model confirming that younger, male farmers and those with high levels of trust in NGOs were associated with an increased likelihood of adopting CA.

The significance of the TRUST index, including indicators of NGO values, expertise, and accountability, highlights the importance of the relationship between an NGO and the local community. The transfer of new knowledge is reliant on whether the information comes from a trustworthy social network (Carolan, 2006), resulting in a strong interconnection between trust, knowledge, and application of that knowledge (i.e. adoption of introduced practices). Furthermore, trust is created through a history of relational experiences and therefore must be built and maintained over time through a series of interactions (Hardin, 2001).

Resource/economic characteristics leading to CA adoption

Both FOODSECURITY and INCOME were strong indicators ($p<0.05$) of adoption in the logit model. The results showed that higher levels of food security (indicated as low food insecurity in the HFIAS index) were correlated with a willingness to adopt CA, which supports the initial hypothesis. However, the variable for INCOME showed another anomaly where the variable was highly significant, yet the direction of influence contrasted with the hypothesis. The data for income showed a wide range of values, from less than 500 NPR (Nepalese rupee) (US$5.09) to almost 300,000 NPR (US$3043), though the average household income was 87,150 NPR (US$887). The households with higher incomes were earning the bulk from sales of livestock and off-farm wage earning, with a small proportion of households (16%) receiving remittance from family members who have migrated for work. A study of Chepang communities showed that remittance and skilled non-farm jobs were the most remunerative sources of income; however, few households had the resources to take advantage of such opportunities (Piya *et al.*, 2011). This indicates that, as households gain the potential to earn more income through livestock or off-farm activities, they may become less invested in improving the cultivation of crops. Furthermore, among those with access to off-farm income-generating opportunities, the economic benefits of soil conservation efforts may not be large enough to justify investments in CA integration (Gould *et al.*, 1989).

FARMSIZE was shown to be insignificant in the model. The descriptive statistics show that the largest farm was 1.65 ha, with an average farm size of 0.37 ha. All the farms in the study fall within the designation of smallholder, thus even the largest farm in the villages does not have much of a comparative advantage in terms of excess land. LABOR was another factor that was insignificant. In this case, the availability of labor did not significantly impact the willingness to adopt CA. One aspect of labor that was not reflected in this model was the frequent use of exchange labor on farms in these communities. This unmeasured component of labor availability may be the reason that limited household labor is not indicated as a significant factor when considering new agricultural practices that would likely require additional labor. INFORMATION was not found to be significant in that greater access to sources of agricultural information did not correlate with higher rates of adoption. As described above, TRUST in NGOs was a positive indicator of adoption, yet greater access to NGO or extension staff did not result in significant levels of adoption. This signifies a possible disconnection between the intentions of NGO projects and their implementation. However, the inclusion of agricultural extension in the access to INFORMATION variable also contributes to the lack of influence on the model. Agricultural extension services in the Chepang villages are often infrequent and, due to government policies restricting cultivation in the mid-hills, do not always share the same values and goals as the farmers. In sum, with regards to resource and economic characteristics, adopters of CA practices have been identified as likely to have higher food security, though with moderate to low household incomes and thus still reliant on sustainable farming practices to maintain their livelihoods.

11.4.2 Numbers of ties and nature of stakeholders in hedgerow information networks

Important implications can be drawn from the visual observation of the hedgerow information networks in comparison with the general agricultural information network (Fig. 11.2). First, very few individuals in the agricultural information network will exchange information regarding hedgerow technologies. Out of the network ties providing general agricultural information, only 20% of ties in Thumka and 49% of ties in Kholagaun exchange information on hedgerow technologies. Second, the types of information source for hedgerow information networks are clearly different than the general agricultural information network. Even after 10 years of

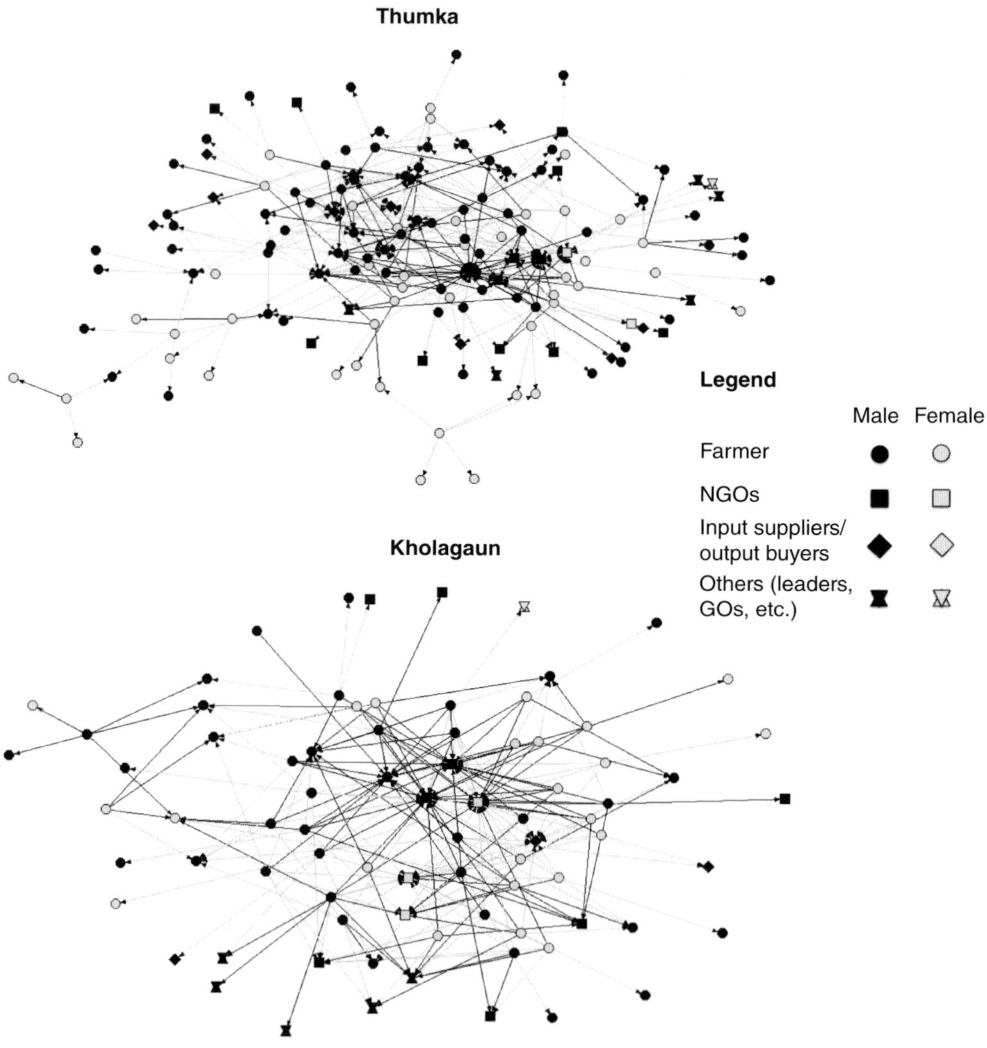

Fig. 11.2. Hedgerow information network in the backdrop of general agriculture information networks in the study villages of Thumka and Kholagaun. The dark lines in the network are ties that exchange the information about hedgerow technology, while the faded connecting lines indicate those ties that provide information about other agricultural technologies (such as new varieties, new crops, fertilization methods, use of pesticides, commercial vegetable farming, value addition, etc.) but not about hedgerow technologies.

participatory implementation and evaluation of the hedgerow technology, most farmers still depend on NGOs and other outside sources as dominant information sources for information related to hedgerow technology. NGOs constituted 42% and 53% of ties in the hedgerow information network of Thumka and Kholagaun, respectively, compared with 49% and 28% in the general agricultural information network. On the other hand, market-based information sources (input suppliers and output buyers) constituted only 5% and 0.8% of ties in the hedgerow information networks for Thumka and Kholagaun, respectively, compared with 20% and 13%, respectively, for the general agricultural information network. Government sources were almost absent in the hedgerow information network for Thumka but constituted about 8% of ties in Kholagaun. The higher presence of government extension in Kholagaun is due to the implementation of the leasehold forestry program by the Department of Livestock of Nepal. The leasehold forestry program also promotes the hedgerow technology in its program; hence farmers from Kholagaun get information about hedgerow from government sources, which is not the case in Thumka.

11.4.3 Comparing network measures of hedgerow information network with general agricultural information network

Since only small numbers of ties in the agricultural information networks provided information about hedgerow technologies, the hedgerow information networks are less dense and far more fragmented than the general agricultural information network. Table 11.5 shows that the densities of hedgerow information networks in Thumka (0.8%) and Kholagaun (3%) are significantly lower than the density of the general agricultural information network ($p<0.001$ in both cases, as indicated using the bootstrap density comparison of UCINET). Similarly, Table 11.5 also shows that hedgerow information networks have higher fragmentation compared with the general agricultural information network irrespective of the village (99% versus 88% in Thumka, and 92% versus 81% in Kholagaun).

The individual-level network measures for hedgerow information network are also lower than in the general agricultural information network. The average out-degree centrality of hedgerow information networks in Thumka (1.4) and Kholagaun (2.75) were significantly lower than the general agricultural information networks in both villages ($p<0.001$ in both cases, as indicated using the t-test for equality of means). This result suggests that farmers share hedgerow technology information with fewer individuals and at a lesser degree as compared with information on other agricultural technologies. Similarly, the in-degree centralities of hedgerow information networks are significantly lower than the general agricultural information network (0.7 versus 1.9 in Thumka, 1.1 versus 1.9 in Kholagaun), indicating that farmers have fewer sources from which to receive information on hedgerow technologies.

When the actors are categorized into two groups—core actors (who are central to the

Table 11.5. Graph-level network measures and individual-level network measures of hedgerow information network compared with general agricultural information network by study villages.

	Thumka		Kholagaun	
Network measures	Hedgerow information network	Agricultural information network	Hedgerow information network	Agricultural information network
Density (%)	0.8	2.3	2.9	6.1
Fragmentation (%)	98.5	87.7	91.7	81.2
Outdegree centrality (average)	1.4	4.26	2.75	5.75
Indegree centrality (average)	0.72	1.9	1.11	1.93

network), and periphery actors—the percentage of core actors are lower in the hedgerow information network than in the general agricultural information network (13% versus 24% in Thumka, and 23% versus 27% in Kholagaun) (Table 11.6). Additionally, the percentage of farmers who are core actors is also lower in the hedgerow information network compared with the general agricultural information network (60% versus 79% in Thumka, and 57% versus 65% in Kholagaun). The control of farmers in the hedgerow information network is less than in the general agricultural information network.

As the use of hedgerows is an imported technology in both villages, the development of a hedgerow information network provides an indication of how information networks of any new technologies, such as CA, develop over time in the villages. The integration of new technologies into government programs is very important for the sustainability of the information delivery for new technologies. As in the case of Kholagaun, when government owns and integrates the technology in their operational plan, there is a high probability that the information network of the new technology remains intact.

Another pertinent issue in the information network of new technology is the presence (or absence) of market-based agents. Input suppliers (e.g. agro-vets, shops, cooperatives) and output buyers (vegetable shops, cooperatives) play significant roles in the information exchange in general agricultural information network; however, they are largely neglected in the process of developing/disseminating new technologies. Market-based agents were not properly integrated in the process of hedgerow technology development and dissemination. Market is an integral part of sustainability for any new agricultural technology, hence capacity building and engagement of market-based agents are required for a sustainable flow of information related to new technology.

Finally, capacity building of the farmers and training of local resource persons are important for sustainability of complex process-based technologies such as CA. It takes time for farmers to enter the core group of an information network of new technologies. Therefore, careful and strategic efforts to identify the motivated and promising local resource persons are essential from the beginning of developing, piloting and promoting CA.

Our earlier analysis indicated that lack of access to information from formal agricultural extension was not a significant inhibitor for adoption of conservation technologies. However, the farmers who trusted the information sources (e.g. NGO) tended to be earlier adopters than other farmers. This relates directly to the results from information networks of the hedgerow. The information networks of hedgerow technology are dominated by NGOs; hence farmers who trust the NGOs are likely to have better access to information exchange than other farmers. The information delivery under formal extension systems is rather weak, specifically because it is one-way, one-time information delivery rather than two-way learning, sharing and continuous engagement. Such types of information sharing may not be effective for the transfer of complex, knowledge-based technologies like CA. While information delivery

Table 11.6. Numbers of actors in core and periphery of general agriculture information network and hedgerow information network by village.

	Thumka		Kholagaun	
Particular	Hedgerow information network	Agricultural information network	Hedgerow information network	Agricultural information network
Actors in core (number)	15	28	14	17
Actors in periphery (number)	104	91	48	45
Total actors	119	119	62	62
% of core actors	12.6	23.5	22.6	27.4
Number of farmers in core	9	22	8	11
% of farmers in core actors	60.0	78.6	57.1	64.7

focuses on providing readymade information to farmers, information networks engage farmers for potential for joint learning, evaluation and actions for social learning (Bodin and Crona, 2009). Therefore, facilitation of the information network is an improvement over conventional information delivery because information network facilitates two-way information exchange, where farmers are considered as both providers and receivers of information (Cramb, 2005).

11.5 Conclusion

As the logit model has shown, characteristics inherent to individual farmers can be valuable indicators of CA adoption particularly in a country transitioning out of poverty and instability. While age and gender cannot be changed, development practitioners can use this information to identify and understand the reasons why particular demographic groups may prefer not to adopt CA. On the other hand, factors such as education and trust can be fostered and investments can be made to enhance these characteristics and bridge the gaps in understanding between farmers and NGOs. Furthermore, education (formal or informal) can be targeted to better align with the skill needs and livelihood priorities of subsistence farmers. A better understanding of the dynamic relationship between factors such as income, food security, and interest in conservation agriculture can also provide insights into the drivers of farmer priorities and motivations. As we have shown, increased food security can allow farmers to invest in long-term sustainable farming practices, while increased off-farm income may hinder interest in farm management.

Due to the inherent characteristics (e.g. waiting for information delivery from external sources rather than initiating information exchange within the community), existing extension systems in developing and transitional countries show limited capacity to foster behavioral changes and motivate farmers to adopt complex, knowledge-based technologies such as CA. Unlike input-based technologies (such as improved seed, fertilizers, pesticides), which increase demand for market-based agents, promoting knowledge-based technologies does not provide direct incentives for market agents. Therefore, facilitating the sustainable exchange of information for knowledge-based technologies among farmers is a challenge and possible hurdle for adoption of such technologies. Promotion of technologies also requires significant efforts on facilitation and capacity building of local institutions (such as farmers' groups, cooperatives, women's groups, etc.), which could play a vital role in facilitating adaptation and adoption of new technologies. Similarly, identification of motivated and promising farmers and building their capacity as a local information resource is also essential for longer-term sustainability of the technology and its network until a strong extension system is developed. Generally, NGOs take care of the promotion of technologies and follow-up consultation.

This farmer-centric approach to understanding CA adoption highlights the barriers that may inhibit integration of new practices, while allowing for development practitioners to modify and adapt programs to better incorporate farmer and community needs until Nepal becomes a more stable and peaceful country. The study also provides recommendations for creating a congenial environment for information exchange among farmers and stakeholders for adaptation and adoption of knowledge-intensive technologies such as conservation agriculture. This type of understanding is applicable to identify progressive and adaptive individuals in a multitude of settings, including areas in transition where a change of practices is often imperative, as well as recognizing social network structures that can allow for more rapid and adaptive transfer of new information.

Acknowledgments

The authors thank LI-BIRD, Juni Maharjan, Bir Bahadur Tamang, and Catherine Chan for their invaluable contributions to this research. Additionally, we thank the survey participants and enumerators for their time and inputs.

References

AFT (2013) *The Adoption of Conservation Practices in Agriculture*. American Farmland Trust, Center for Agriculture in the Environment, DeKalb, Illinois. Available at: https://4aa2dc132bb150caf1aa-7bb737f4349b47aa42dce777a72d5264.ssl.cf5.rackcdn.com/Adoption-of-Conservation-Practices-in-Agriculture.pdf (accessed 8 December 2016).

Bodin, Ö. and Crona, B.I. (2009) The role of social networks in natural resource governance: what relational patterns make a difference? *Global Environmental Change* 19, 366–374.

Bohlen, J.M., Coughenour, C.M., Lionberger, H.F., Moe, E.O. and Rogers, E.M. (1961) *Adopters of new farm ideas: characteristics and communications behavior*. Michigan Co-operative Extension Service, Michigan State University, East Lansing, Michigan. *See also* North Central Regional Extension Publication No. 13. Farm Foundation and Federal Extension Service Cooperating. Available at: http://www.soc.iastate.edu/extension/pub/comm/NCR13.pdf (accessed 8 December 2016).

Borgatti, S.P., Everett, M.G. and Freeman, L.C. (2002) *Ucinet for Windows: Software for Social Network Analysis*. Analytic Technologies, Harvard, Massachusetts.

Borgatti, S.P., Everett, M.G. and Johnson, J.C. (2013) *Analyzing Social Networks*. SAGE Publications Ltd, London.

Bunch, R. (1999) Reasons for non-adoption of soil conservation technologies and how to overcome them. *Mountain Research & Development* 19(3), 213–220.

Carolan, M.S. (2006) Social change and the adoption and adaptation of knowledge claims: whose truth do you trust in regard to sustainable agriculture? *Agriculture and Human Values* 23, 325–339.

Coates, J., Swindale, A. and Bilinsky, P. (2007) Household Food Insecurity Access Scale (HFIAS) for measurement of food access: indicator guide. *Washington, DC: Food and Nutrition Technical Assistance Project, Academy for Educational Development*.

Cochran, J. (2003) Patterns of sustainable agriculture adoption/non-adoption in Panamá. Thesis. McGill University, Montreal, Canada.

Conley, T. and Udry, C. (2001) Social learning through networks: the adoption of new agricultural technologies in Ghana. *American Journal of Agricultural Economics*, 668–673.

Cox, E. (2011) *The Landowner's Guide to Sustainable Farm Leasing*. Drake University Agricultural Law Center, Des Moines, Iowa, 56 pp.

Cramb, R.A. (2005) Social capital and soil conservation: evidence from the Philippines. *Australian Journal of Agricultural and Resource Economics* 49, 211–226.

Cronbach, L.J. (1951) Coefficient alpha and the internal structure of tests. *Psychometrika* 16(3), 297–334.

Daberkow, S.G. and McBride, W.D. (1998) Socioeconomic profiles of early adopters of precision agriculture technologies. *Journal of Agribusiness* 16(2), 151–168.

D'Souza, G., Cyphers, D. and Phipps, T. (1993) Factors affecting the adoption of sustainable agricultural practices. *Agricultural and Resource Economics Review* 22(2), 159–165.

Erenstein, O., Sayre, K., Wall, P., Hellin, J. and Dixon, J. (2012) Conservation agriculture in maize- and wheat-based systems in the (sub) tropics: lessons from adaptation initiatives in South Asia, Mexico, and Southern Africa. *Journal of Sustainable Agriculture* 36, 180–206.

FAO (2014) *What Is Conservation Agriculture?* Agriculture and Consumer Protection Department, Conservation Agriculture, Food and Agriculture Organization, Rome. Available at: http://www.fao.org/ag/ca/1a.html (accessed 8 December 2016).

Floyd, C., Harding, A.H., Paudel, K.C., Rasali, D.P., Subedi, K. and Subedi, P.P. (2003) Household adoption and the associated impact of multiple agricultural technologies in the western hills of Nepal. *Agricultural Systems* 76, 715–738.

Freeman, L.C. (1979) Centrality in social networks conceptual clarification. *Social Networks* 1, 215–239.

George, D. and Mallery, P. (2003) *SPSS for Windows Step by Step: a Simple Guide and Reference*. 11.0 update (4th edn). Allyn & Bacon, Boston, Massachusetts.

Giller, K.E., Witter, E., Corbeels, M. and Tittonell, P. (2009) Conservation agriculture and smallholder farming in Africa: the heretic's view. *Field Crops Research* 114, 23–34.

Gould, B.W., Saupe, W.E. and Klemme, R.M. (1989) Conservation tillage: the role of farm and operator characteristics and the perception of soil erosion. *Land Economics* 65(2), 167–182.

Halbrendt, J., Gray, S.A., Crow, S., Radovich, T., Kimura, A.H. and Tamang, B.B. (2014) Differences in farmer and expert beliefs and the perceived impacts of conservation agriculture. *Global Environmental Change* 28, 50–62.

Hanneman, R.A. and Riddle, M. (2005) *Introduction to Social Network Methods*. University of California, Riverside, California.

Hardin, R. (2001) Conceptions and explanations of trust. In: Cook, K. (ed.) *Trust and Society*. Russell Sage, New York, pp. 3–39.

Hobbs, P.R., Sayre, K. and Gupta, R. (2008) The role of conservation agriculture in sustainable agriculture. *Philosophical Transactions of the Royal Society* 363, 543–555.

IBM Corp. (2013) *IBM SPSS Statistics for Macintosh, Version 22.0*. IBM Corp., Armonk, New York.

IFAD (2011) *The world's population is about to hit 7 billion*. International Fund for Agricultural Development. Available at: http://www.ifad.org/media/events/2011/7billion.htm (accessed 8 December 2016).

IFAD (2014) *Rural poverty in Nepal*. Rural Poverty Portal, International Fund for Agricultural Development, Rome. Available at: http://www.ruralpovertyportal.org/country/home/tags/nepal (accessed 8 December 2016).

Isaac, M.E., Erickson, B.H., Quashie-Sam, S.J. and Timmer, V.R. (2007) Transfer of knowledge on agroforestry management practices: the structure of farmer advice networks. *Ecology and Society* 12, 32.

Kebede, Y., Gunjal, K. and Coffin, G. (1990) Adoption of new technologies in Ethiopian agriculture: the case of Tegulet-Bulga district Shoa province. *Agricultural Economics* 4(1), 27–43.

Knowler, D. and Bradshaw, B. (2007) Farmers' adoption of conservation agriculture: a review and synthesis of recent research. *Food Policy* 32(1), 25–48.

Lee, D.R. (2005) Agricultural sustainability and technology adoption: issues and policies for developing countries. *American Journal of Agricultural Economics* 87(5), 1325–1334.

Lubwama, F.B. (1999) Socio-economic and gender issues affecting the adoption of conservation tillage practices. In: Kaumbutho, P.G. and Simalenga, T.E. (eds) *Conservation Tillage with Animal Traction. A Resource Book of the Animal Traction Network for Eastern and Southern Africa*. ATNESA, Harare, Zimbabwe.

Mikhail, M., Fencl, A., Kemp-Benedict, E. and Naruchaikusol, S. (2010) *Innovation and diffusion of sustainable agricultural water resource management in a changing climate: a case study in Northeast Thailand*. Project Report, Stockholm Environment Institute, Stockholm, Sweden. Available at: http://www.icarus.info/wp-content/uploads/2011/610-Fencl_ICARUS_paper.pdf (accessed 8 December 2016).

NCSS. *Stepwise regression*. NCSS Statistical Software. Available at: http://www.ncss.com/wp-content/themes/ncss/pdf/Procedures/NCSS/Stepwise_Regression.pdf (accessed 8 December 2016).

Neupane, R.P., Sharma, K.R. and Thapa, G.B. (2002) Adoption of agroforestry in the hills of Nepal: A logistic regression analysis. *Agricultural Systems* 72(3), 177–196.

Okoye, C. (1998) Comparative analysis of factors in the adoption of traditional and recommended soil erosion control practices in Nigeria. *Soil and Tillage Research* 45, 251–263.

Piya, L., Maharjan, K.L. and Joshi, N.P. (2011) Forest and food security of indigenous people: a case of Chepangs in Nepal. *Journal of International Development and Cooperation* 17(1), 113–135.

Prell, C., Hubacek, K. and Reed, M. (2009) Stakeholder analysis and social network analysis in natural resource management. *Society and Natural Resources* 22, 501–518.

Shang, X. (2014) *International year of family farming*. International Food Policy Research Institute, Washington, DC. Available at: http://www.ifpri.org/blog/international-year-family-farming (accessed 8 December 2016).

Sharma, D.P. (2011) Understanding the Chepangs and shifting cultivation: a case study from rural village of central Nepal. *Dhaulagiri Journal of Sociology and Anthropology* 5, 247–262.

Sharma, G. (2013) *Natural resource management approaches and technologies in Nepal: Technology – Hedgerow technology*. International Centre for Integrated Mountain Development, Kathmandu, Nepal. Available at: http://lib.icimod.org/record/28276/files/Technology%20(4).pdf (accessed 9 December 2016).

Shively, G.E. (2001) Poverty, consumption risk, and soil conservation. *Journal of Development Economics* 65, 267–290.

Shrestha, D.P., Zinck, J.A. and Van Ranst, E. (2004) Modelling land degradation in the Nepalese Himalaya. *Catena* 57(2), 135–156.

Shrestha, S.K. (2013) *Decentralizing the farmer-to-farmer extension approach to the local level*. Paper submitted to CAPSA for the 'Technology Transfer for Smallholder Farmers' Policy Dialogue. UN Centre for Alleviation of Poverty through Sustainable Agriculture, Bogor, Indonesia. Available at: http://www.satnetasia.org/sites/default/files/8-Decentralizing_farmer-to-farmer_extension-Helvetas_Nepal_paper.pdf (accessed 8 December 2016).

Stark, M. (1996) Adoption and adaptation of contour hedgerow farming in the Philippine uplands: results of an early case study. *Der Tropenlandwirt, Beittrage zur tropischen Landwirtschaft und Veterinarmedizin* 97, 3–16.

Tavakol, M. and Dennick, R. (2011) Making sense of Cronbach's alpha. *International Journal of Medical Education* 2, 53–55.

Thapa, G.B. and Paudel, G.S. (2002) Farmland degradation in the mountains of Nepal: a study of watersheds 'with' and 'without' external intervention. *Land Degradation and Development* 13, 479–493.

Tiwari, K.R., Sitaula, B.K., Nyborg, I.L.P. and Paudel, G.S. (2008) Determinants of farmers' adoption of improved soil conservation technology in a middle mountain watershed of central Nepal. *Environmental Management* 42, 210–222.

Tryndyuk, I. (2013) Isn't it too early to drop out of school? A study of girl's education in the Chepang community of Nepal. Thesis. University of Tromsø, Tromsø, Norway.

USDA NRCS (2005) *The Adoption and Diffusion of Conservation Technologies*. People, Partnerships, and Communities series, issue 7. US Department of Agriculture, Natural Resources Conservation Service, Washington, DC.

Vilpponen, A., Winter, S. and Sundqvist, S. (2006) Electronic word-of-mouth in online environments: exploring referral networks structure and adoption behavior. *Journal of Interactive Advertising* 6, 8–77.

Whittingham, M.J., Stephens, P.A., Bradbury, R.B. and Freckleton, R.P. (2006) Why do we still use stepwise modelling in ecology and behaviour? *Journal of Animal Ecology* 75(5), 1182–1189.

Yadav, R.P. (1987) *Agricultural research in Nepal: resource allocation, structure, and incentives*. Research Report 62, International Food Policy Research Institute, Washington, DC.

12 Understanding Conservation Agriculture Adopter's Information Network to Promote Innovation and Agricultural Entrepreneurship: The Case of Tribal Farmers in the Hill Region of Nepal

Bikash Paudel,[1]* Katherine A. Wilson,[2] Catherine Chan[2] and Bir Bahadur Tamang[1]
[1]*Local Initiatives for Biodiversity, Research and Development (LI-BIRD), Pokhara, Nepal;* [2]*University of Hawai'i at Mānoa, Honolulu, Hawaii*

12.1 Introduction

Small-scale rural entrepreneurships are crucial for improving livelihood and reducing poverty in the developing world (Barrett, 2008; Tieguhong *et al.*, 2012). Agricultural or forest-based small enterprises help reduce poverty by building local wealth and creating local job opportunities while also promoting the utilization of local stewardship for local natural resources (Kaaria *et al.*, 2008; Koirala *et al.*, 2013). Small and medium-sized enterprises are important for economic growth worldwide. About 92.1% of firms in European countries are small to medium-sized enterprises which collectively contribute to 29% of jobs in the industrial sector and share about 21.1% of value added business (Gagliardi *et al.*, 2013). There are great differences in the types and scales of small enterprises in developing countries as different nations define them differently. For the majority of the developing world, small-scale rural enterprises include very simple changes in farming systems such as: growing fresh vegetables and linking them to markets; marketing and value addition of local products; creating small cottage industries based on local resources; creating small businesses such as grocery shops, tea shops, restaurants, and skill-based shops; and collecting, processing, and marketing non-timber forest products. Thus, most of these small enterprises require the adoption of innovative methods, practices, and technologies in farming systems in order for their enterprise to be successful.

Globally, adoption of agricultural innovation is a successful modus operandi for smallholder farmers to alleviate poverty (Minten and Barrett, 2008; Kassie *et al.*, 2011; Asfaw *et al.*, 2012; Yoshida *et al.*, 2014). Although there is a reduction in the population living under extreme poverty (14.5% in 2014), there is still a striking regional imbalance in the distribution of global poverty as Africa and Asia combined contribute 80% (in 2014) of the globally classified poor people (FAO *et al.*, 2013; World Bank Group, 2015). Since Africa and Asia also have an increasing population growth rate, reducing poverty in these continents will be a daunting challenge for decades to come (Grafton *et al.*, 2015).

*bpaudel@libird.org

Being a small country in Asia (147,181 km^2 in area and about 31 million population), Nepal faces high levels of national poverty as well as high levels of poverty imbalances among regions and different societal groups within the country. While about 26% of the population live under poverty conditions at a national level, about 55% of the farmers in the rural areas remain in poverty (UNDP, 2009). Heavy reliance on traditional agriculture due to lack of alternative employment opportunities caused by a scant number of rural enterprises is the primary reason for high rural poverty rates in Nepal. Traditional agriculture in Nepal employs about 70% of the working population (CBS, 2008b) despite per capita land holding being half of the global average (0.011 ha versus 0.02 ha) (CBS, 2008a; Worldstat, 2014). Due to the seasonal nature of farming, subsistence farmers are heavily unemployed or underemployed during winter and fall months. These extended periods of unemployment and lack of income sources increase poverty and deprivation among rural farmers. Engaging rural farmers in agricultural and forest-based enterprises and small-scale business can be a huge boost for economic development in Nepal.

There are several reasons for the low level of entrepreneurship development in rural areas of Nepal. The existence of traditional management practices, underdeveloped entrepreneurial culture, small capital base, obsolete production technologies, and poor business and marketing methods all present challenges to starting an enterprise in rural areas (Rimal, 2008). In addition, lack of innovation and non-adoption of improved technologies caused by a deficiency of information are often attributed to the shortage of sufficient rural enterprises in Nepal and other developing nations. Evidently, lack of information has been widely documented as a cause for slow adoption of improved agricultural technologies in Nepal (Floyd *et al.*, 2003; Paudel and Thapa, 2004).

During the early days of agricultural extension, an innovation-diffusion model stressed that receiving information is the first step in the adoption process of agricultural innovations (Rogers, 1962). Although the model is highly criticized for being technology biased, lack of access to knowledge and information is still regarded as a primary hurdle to the adoption of agricultural innovations by smallholder farmers in the developing world (Zhao, 2005; Wall, 2007; Aker, 2011). Policy makers and development planners in Nepal, in general, believe that farmers in rural areas of Nepal have limited sources of information about innovations and entrepreneurial opportunities. This is partly due to weak formal and institutional extension services. However, formal means of communications are only a part of an information network. Farmers in rural areas may well get information about new agricultural innovations and entrepreneurial opportunities from informal sources. Furthermore, informal social ties and connections are possibly more important than formal sources due to the weak presence of formal systems in Nepal. Even at the global scale, the role of informal social networks in exchange of information regarding agricultural innovations has been widely recognized (Warriner and Moul, 1992; Isaac *et al.*, 2007). Social networks are understood as a collection of individual members and the social ties among them (Maertens and Barrett, 2013). Social networks are often considered a form of capital for smallholder farmers (Hoang *et al.*, 2006) as they expose farmers to new ideas and provide them with access to information when other formal information channels are not present (Bandiera and Rasul, 2006; Bodin and Crona, 2009). Similarly, with greater network connections, farmers will be able to access accurate information about entrepreneurial opportunities, which engenders sound decisions minimizing the risk of being an entrepreneur. The connection to the "outside world" is essential to increase the chance of the isolated communities collectively breaking from the poverty cycle; therefore, establishing a comprehensive network with outside information sources can be vital in poverty reduction schemes (Smith and Johnson, 2001).

Nepalese society, particularly the people in rural areas, has a high level of social capital due to a strong sense of membership to a common and interdependent community where there is shared knowledge and a high level of trust (Pariyar, 2008). However, there is lack of research on evaluating the role social networks have in making decisions about adoption of agricultural innovations and small-scale enterprises in rural areas of Nepal. The study in this chapter, therefore, was conducted to understand how social networks lead to access to information, thus helping farmers make decisions on adoption of agricultural innovations and entrepreneurial activities in isolated rural areas of Nepal. The findings of this study fulfill a knowledge gap regarding how social networks can be used to transfer agricultural innovations and promote entrepreneurship activities among poor, disadvantaged, and marginalized farmers, thereby improving their livelihood status and reducing poverty in rural areas.

This chapter illustrates the importance of social networks on entrepreneurial decisions in remote areas of developing countries. Specifically, the arguments are based on empirical analyses of: status of adoption of agricultural innovations and entrepreneurial activities by farmers; comparison of information networks of adopter farmers with non-adopter farmers, using social network analysis methodology; and evaluating influence of social networks on farmers' decision to adopt innovations and entrepreneurial activities, using general linear models.

12.2 Methodology

12.2.1 Study sites

The sites for this study were three villages in the central mid-hills region of Nepal: Hyakrang village (in Dhading district), Kholagaun village (in Tanahun district), and Thumka village (in Gorkha district) (Fig. 12.1). The study villages were selected based on dominance of poor and marginalized Chepang tribal farmers who do not interact with other communities other than Chepang. The Chepang are very marginalized and disadvantaged tribal communities in Nepal who are predominantly found in the hill districts of central Nepal, particularly in Dhading, Makwanpur, Chitwan, Gorkha, and Tanahun. The population of the Chepang tribal community is slightly less than 100,000 people (CBS, 2012).

The altitude of these villages ranges from 1000 m to1600 m above mean sea level. All the villages have a sub-tropical climate, where the average temperature decreases with increase in altitude. The rainfall pattern is skewed to a few rainy-season months (July–September) when about 75–80% of total annual rainfall occurs. The average total rainfall in these villages is in the range of 1000–1500 mm. The vegetation in the region is dominated by deciduous forest, hedges and deciduous fodder trees, which shed their leaves during the winter season.

All three villages are relatively accessible in terms of road access and travel time. Although the villages do not have year-round roads, they are within 1–2 hours walking distance of national highways. This distance does not hinder the potential conversion of traditional agriculture to market-based commercial agriculture. Additionally, the government of Nepal and other

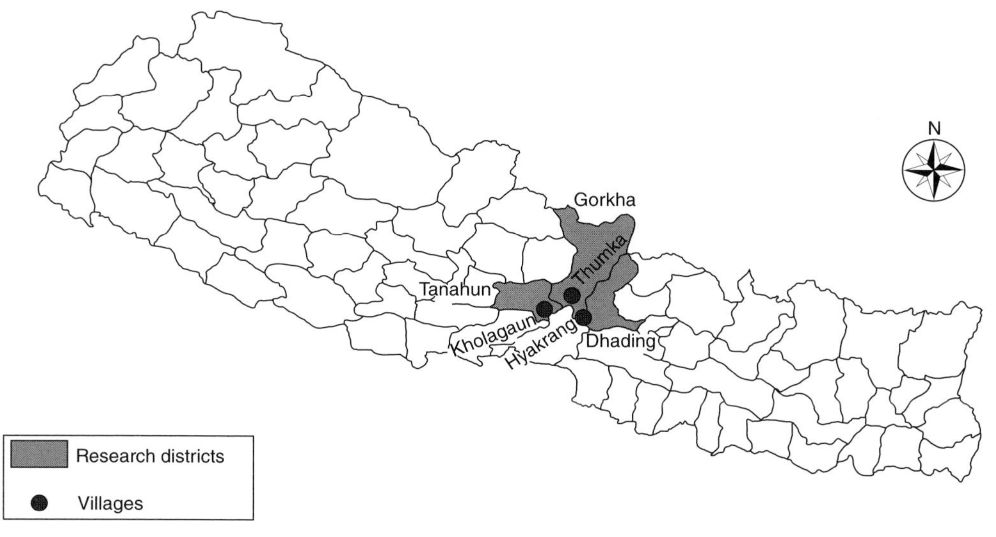

Fig. 12.1. Map of Nepal showing districts and villages where the information networks were studied.

agencies have targeted programs for such road corridors where they support farmers to start off-season vegetable cultivation, large-scale plantations, improved and/or commercialized livestock rearing and local enterprises. These government programs have already started to show increased adoption of entrepreneurial activities in the villages, albeit in small scale.

Demography

The Chepang community has larger family sizes than average in Nepal, possibly due to lack of knowledge, awareness, and access to family planning services. Additionally, agriculture is the primary livelihood in their community and larger family sizes means a higher and more reliable labor supply. The average household sizes of villages are 10, 7 and 6, respectively for Thumka, Hyakrang and Kholagaun (Chan-Halbrendt *et al.*, 2011), which is significantly higher than the national average (4.88) of Nepal (CBS, 2012). In contrast, the average land holding in the villages is in the range of 0.46–0.62 ha/household, which is lower than the national average of 0.78 ha/household (CBS, 2008a). The level of education in the study sites is also below the national average: illiteracy is in the range of 74.4–80.4%, which is significantly higher than the national level of 34.1%.

Chepang: people in transition

Chepang people inhabit the steep-sloping terrains of the hill region, causing them to rely on sloping, marginalized, unfertile, and degradation-prone lands for agriculture. Due to these inherent characteristics of available agricultural land and slow adoption of improved agricultural practices, only a few Chepang households are self-sustainable. To supplement food shortages, Chepang people collect foods (e.g. yams, tubers, asparagus, banana) from the forest and hunt in the forest (Kafle *et al.*, 2009; Piya *et al.*, 2011b). Additionally, they practice "shifting cultivation" on non-agricultural sloping lands. Shifting cultivation practiced by Chepang communities is a unique practice where available lands are divided into three to four patches and crop production activities are rotated among the patches. When one crop is growing in one land patch, all other patches remain fallow. After 1 to 2 years of cultivation, the patch is left fallow and crop production is shifted to an adjacent patch after clearing the vegetation by burning. This rotation allows for each patch to have a few years of fallow where vegetation will grow and increase soil nutrients. However, due to an increase in the demand for food caused by rapid population growth and a decrease in forest-based sources of food, Chepang people are forced to shorten the fallow period in shifting cultivation land. These changes in the land-use system and agricultural practices have led to problems of soil erosion and land degradation. Chepang communities are in acute need of new agricultural technologies and innovations to increase agricultural productivity in a sustainable manner. At the same time, it is impossible to fulfill the increased demand for food and income only by changing current agricultural practices. In this context, promotion of agro-based rural enterprises is an ultimate goal to reduce the people's dependency on agriculture and diversification of the livelihood strategies in the villages.

Household economy and income

Most of the Chepang households practice subsistence agriculture, where their primary objective is to fulfill the household food and feed demand, rather than selling into the market. Yet, some agriculture-based activities provide much-needed cash income for the households. A baseline survey (December, 2011) showed that the contribution of income from agriculture to total household cash income was about 60.6–66.9%. The majority of the agriculture-based cash income was derived from selling grain legumes such as black gram and cowpea (46–57%), livestock and livestock products (33–49%) and fresh vegetables (3–15%) (Chan-Halbrendt *et al.*, 2011). In addition to agriculture, a few farmers had other income sources such as wage laboring, hunting or collection of food from forest, handicrafts, skilled non-farm jobs, and remittance (Piya *et al.*, 2011a). Although many of these activities may not qualify as innovation or entrepreneurial activities to some, considering the high level of subsistence agriculture in Chepang communities, adoption of new crops or new income sources qualifies as being innovative in the villages.

12.2.2 Social network analysis

The data required for this study were collected from two sources. Firstly, a social network survey was conducted in the three villages in June and July 2013 to gather information about the agricultural information network (through which farmers get information about new agricultural technologies) in the villages. Secondly, three key informant surveys were conducted in all villages in June 2015 to identify the adopters or non-adopters of innovations and agro-entrepreneurial activities.

Following standard social network analysis (SNA) protocol, structured household surveys were administered to capture agricultural information-sharing networks. Both male and female decision makers in the household were asked to nominate at least three, and up to ten, individuals with whom they discussed useful information regarding any new agricultural technology. To avoid bias and maintain the anonymity of the nominee as well as nominator, all the farmers in the villages were coded and the nominations were recorded in code. For the survey, commercial vegetable production, off-season production, use of agro-chemicals, use of modern varieties, methods for grading, standardization and value-added agro-based enterprises were all regarded as new technologies. Altogether, 136 farmers (from 92 out of 102 households) including 69 females (50%) were interviewed during the SNA survey.

Key informant interview

A key informant interview was conducted during June 2015 for adoption (or lack of adoption) of entrepreneurial activities by farming households. This interview was conducted as an open-ended question, where the informants were asked to mention the names of the farmers who had adopted certain entrepreneurial activities. For simplicity, the entrepreneurial activities were placed into two groups. Commercial vegetable cultivation/off-season vegetable cultivation was taken as a separate entrepreneurial activity due to the relatively higher abundance of farmers adopting this activity. Other entrepreneurial activities such as collection of non-timber forest products, value addition, starting of small business (tea shops, grocery shops, etc.) were grouped together since there were very few farmers in each group. This information was collected from various key informants such as output buyers, input sellers, local resource persons, local non-governmental organization (NGO) workers, and progressive farmers. All households were listed (n=102) and information regarding their adoption (or lack of adoption) of agricultural innovation and agricultural enterprise was collected. The process was continued until information about adoption (or lack of adoption) was collected for every household in the villages. For ensuring the reliability of the information, at least two key informants were asked about the adoption (or lack of adoption) of a technology by a household. If their response matched, that was taken as final. However, if their responses contradicted each other, the status of adoption (or lack of adoption) was finalized after asking a third key informant from the same village. Since there are not many households (total 102) in the villages, key informants used in the interview were mostly spot-on about the economic activities of the households. For more than 90% of the households, the third key informant was not needed. The information collected in both surveys was pooled for analysis.

12.2.3 Social network analysis

Social network analysis uses socio-grams and graphs to describe complex social relationships such as information sharing. In SNA, relationships among people are visualized by graphs made of nodes and ties, where actors are shown as nodes and the relationships between the actors are represented by connecting lines between the nodes. For this study, each farmer and sources of information (all individuals) were represented by nodes/actors and information exchange between actors (presence) were represented as ties.

Broadly there are two types of network measures: "whole-network" measures and "individual-network" measures (Borgatti *et al.*, 2013). The whole-network measures (e.g. density) provide information about the network as a whole (or at graph level), while individual-network measures (e.g. centralities) provide information about how an individual is placed in the network.

Similarly, whole-network measures have higher relevance in terms of generating collective learning and governing collective behavior, while individual-network measures are directly connected with individual farmers' behavioral changes such as adoption (or non-adoption) of new technologies. This study determined the agricultural information networks in all three villages and visualized those networks by using Netdraw software (Borgatti *et al.*, 2002). The analysis compared degree centralities and betweenness centrality of individual networks to compare between adopter and non-adopters of agricultural innovations and agricultural entrepreneurial activities.

The network was analyzed to categorize the farmers in the core and periphery of the network to compare how the position of adopter farmers and non-adopter farmers differed in the network. A core–periphery analysis of actors was conducted using UCINET software. Eventually, the core–periphery analysis categorized the farmers and other actors into two groups: core actors and periphery actors. After identification of the core and periphery actors, simple descriptive tools were used to compare percentage of adopter farmers in core groups versus percentage of adopter farmers in periphery actors. This way, the role of farmers' positions in the core versus periphery of the information network was related to farmers' adoption behavior.

12.2.4 General linear model

A general linear model was used to analyze the influence of outdegree, indegree and betweenness degree centralities (see Table 12.1 for definitions) of the adoption (or lack of adoption) of agricultural innovations and agricultural entrepreneurship.

Specification of dependent variables

Two dependent variables were used in the general linear model. The adoption of commercial vegetable farming was selected as the proxy measure of agricultural innovation adoption in the villages. The formal extension systems as well as non-governmental organizations have promoted commercial (including off-season)

Table 12.1. List of dependent variables, explanatory variables, description, and predicted direction of influence on adoption of agricultural innovation.

Variables	Definition	Hypothesized direction of influence
Dependent variables		
Adoption (or lack of adoption) of commercial vegetable farming	Y_i =1 for adopter farmer Y_i = 0 otherwise	
Adoption (or lack of adoption) of value-added enterprises	Y_i =1 for adopter farmer Y_i = 0 otherwise	
Explanatory variables		
Farmer characteristics		
AGE	Age of household head (years)	–
GENDER	Gender (categorical): 1 = Female, 0 = Male	–
EDUCATION	Level of education (categorical): 0 = illiterate; 1 = primary; 2 = secondary +	+
GROUP	Prior group membership status (categorical) 0 = non-member, 1 = member	+
TRAINING	Prior agricultural training status (categorical): 0 = non-trained, 1 = trained	+
Network variables		
OUTDEGREE	Value of outdegree centrality (continuous)	+
INDEGREE	Value of indegree centrality (continuous)	+
BETWEENNESS	Value of betweenness centrality (continuous)	+

vegetable production in these villages. Hence, adoption (or non-adoption) of commercial vegetable farming provides an indication of farmers' willingness to adopt new agricultural innovations; therefore it is taken as a dependent variable in the analysis.

As mentioned above, large-scale enterprises and business activities do not exist in the studied villages. Therefore, the study selected small-scale value-addition works and small-level rural enterprises to study the factors affecting adoption of agricultural enterprises by farmers. Adoption (or non-adoption) of agricultural and forest-based enterprises (including non-timber forest product collection and value addition) as well as small business ventures (e.g. small grocery shop, tea shop, agricultural marketing) are considered as the proxy for adoption of rural entrepreneurship. Several development agencies have tried to motivate farmers to start small-scale agriculture-based rural enterprises in recent years, making this a suitable proxy for farmers' entrepreneurship in villages. Therefore, analysis of farmers' adoption behavior for these two innovations and entrepreneurial activities is analogous for consideration of adoption of major agricultural interventions and rural entrepreneurial activities in the region.

Explanatory variables

Both dependent variables were regressed with eight explanatory variables at the beginning of modeling. Out of the eight explanatory variables, four are categorical, which were treated as factors, while the remaining four were continuous variables and treated as covariates in the model. Past studies and general understandings were used to develop a hypothesis for predicting the direction of influence (+/–) of the variable to adoption decision (Table 12.1).

Theoretical Logit Model Equation

$$Y_i = \beta_0 + \beta_1 \text{AGE}_i + \beta_2 \text{GENDER}_i + \beta_3 \text{EDUCATION}_i + \beta_4 \text{GROUP}_i + \beta_5 \text{TRAINING}_i + \beta_6 \text{OUTDEGREE}_i + \beta_7 \text{INDEGREE}_i + \beta_8 \text{BETWEENNESS}_i + \varepsilon_i$$

Although all eight explanatory variables were included in the base model, the best-fit model was selected using the Akaike Information Criteria (AIC) value. While selecting the model with lowest AIC value for improving overall model fit, some of the explanatory variables were dropped.

Distribution of respondents by personal characteristics

Gender was balanced in the distribution of the respondents based on the significant farmers' characteristics in the best-fit logit model (Table 12.2). The majority of the farmers in Hyakrang and Kholagaun were illiterate (69–70%), while the majority of respondents in Thumka were under the primary education category (47%). The majority of farmers in all three villages were members of at least one farmers' group (58–64%) and the majority had taken at least one agricultural training in the past (55–81%). The average age of respondents in Hyakrang and Kholagaun was a little higher (41 and 43 years, respectively) than in Thumka (34 years). As this survey covered more than 90% of households in the three villages, this distribution is close to the actual reality in the villages.

12.3 Findings

12.3.1 Status of adoption of agricultural innovations and entrepreneurial activities in study villages

Commercial vegetable production is a useful proxy measure for adoption of innovative agricultural technology by farmers. The status of adoption of commercial vegetable farming differed among the villages (Fig. 12.2). Almost two-thirds of the households had adopted commercial vegetable farming. Specifically, 63%, 59% and 69% of the households in Hyakrang, Kholagaun and Thumka, respectively, were growing fresh vegetables for commercial marketing. This level of adoption of commercial vegetable farming is overwhelming, considering the geographic remoteness, transportation challenges and technical know-how of the farmers. A 2011 survey showed that only 23%, 16%, and 34% of households in Hyakrang, Kholagaun and Thumka, respectively, grew commercial vegetables (Chan-Halbrendt *et al.*, 2011). Thus, a comparison of the level of adoption between 2011

Table 12.2. Characteristics of respondents in the three Nepalese villages evaluated for agricultural innovation and information networks.

Characteristics	Village					
	Hyakrang		Kholagaun		Thumka	
Gender		%		%		%
Male	18	(45.0)	18	(50.0)	31	(51.7)
Female	22	(55.0)	18	(50.0)	29	(48.3)
Level of education						
Illiterate	28	(70.0)	25	(69.4)	23	(38.3)
Primary	8	(20.0)	8	(22.2)	28	(46.7)
Secondary and higher	4	(10.0)	3	(8.3)	9	(15.0)
Prior group membership status						
Non-member	17	(42.5)	13	(36.1)	22	(36.7)
Member	23	(57.5)	23	(63.9)	38	(63.3)
Prior agricultural training status						
Non-trained	14	(35.0)	7	(19.4)	27	(45.0)
Trained	26	(65.0)	29	(80.6)	33	(55.0)
Average age (years)	41.2	±1.9	43.1	±2.5	33.6	±1.7

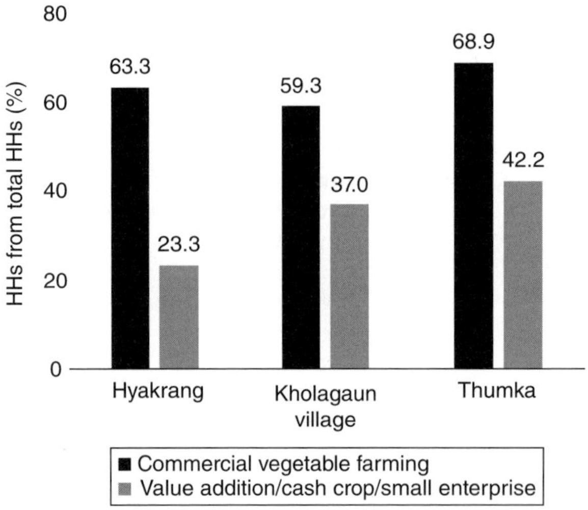

Fig. 12.2. Percentage of farming households in three Nepalese villages engaged in commercial vegetable farming and value-added/cash-crop/small enterprises.

and 2015 shows that the majority of farmers have started growing commercial vegetables recently. In general, commercial vegetable production has higher profitability compared with cereals, but farmers in the villages were either unaware of the profitability, uninformed about the technology or had no access to market before. Lately, there has been widespread adoption of growing commercial vegetables in nearby villages. The support of commercial vegetable schemes in road corridors by the government and increased information about technology and profitability as well as better access to markets (due to proximity to Nepal's capital, the Kathmandu market) may have increased the adoption of commercial vegetable farming in these villages. These villages are within the 1–2 hours walking distance from the Prithivi Highway, which joins Kathmandu, Pokhara and Narayangadh, the three major cities in central Nepal. With improved communication facilities and easier mobility, farmers are aware of the marketing opportunity and increasingly being attracted to commercial vegetable farming.

The adoption of value addition and small rural enterprises is lower than the adoption of commercial vegetable production. About 23%, 37% and 42% households in Hyakrang, Kholagaun, and Thumka, respectively, have adopted at least one activity related to value addition of agricultural and non-timber forest products or agricultural marketing or establishing any small-scale rural business enterprises (Fig. 12.2). No baseline is available to compare these percentages; however, considering the economic history of the Chepang communities, it can be assumed that these are also very recent changes.

In addition to the above-mentioned agro-based entrepreneurial activities, farmers (particularly youth) are increasingly looking for livelihood strategies outside agriculture. Most young people have migrated to cities and foreign nations for education and work. Some of them have also migrated to the terai and other agricultural pocket areas in Nepal to work as agricultural labor. These types of livelihood diversification activities have not been accounted for this study because they were out of scope for village boundaries.

12.3.2 Agricultural information networks of adopter versus non-adopter farmers

The increasing adoption of agricultural innovations and entrepreneurial activities could also be the result of stronger informal information networks in the villages compared with the past. Since the Chepang community formerly lived a semi-nomadic tribal life until recently, their informal networks related to agricultural innovation and entrepreneurial activities were extremely limited. However, with permanent settlement, slow but gradual empowerment of Chepang communities, and entry of various government agencies as well as NGOs promoting modern development in the villages, the Chepang communities are now much better connected with other communities than in the past. Therefore, it is very likely that, with the improved social network of the Chepang communities, they have become more aware of new agricultural technologies and entrepreneurial opportunities.

These results for agricultural information network in three Chepang villages provide evidence that the entry of external agencies provides information to Chepang farmers. The information network is highest in Kholagaun compared with the other villages (Fig. 12.3), and this is mainly due to most of the farming households being connected in the network due to presence of NGOs. Careful observation of the network figures also indicates that adopter farmers have multiple incoming and outgoing ties as compared with non-adopter farmers. In addition, the figures show that female farmers are less connected to the network compared with male farmers. In all villages, neighboring farmers are the people with whom farmers commonly exchange information about agricultural technology. These results support previous claims that progressive farmers and neighbors are the most important information sources for rural farmers in Nepal (Ghimire *et al.*, 2012; Shrestha, 2013).

The information network of adopters also shows that they are more often connected with other adopter farmers rather than non-adopter farmers. This observation supports the idea that having an early adopter in the information network increases the likelihood of connecting farmers adopting new technology (Bandiera and Rasul, 2006; Magnan *et al.*, 2013).

NGO workers and market-based agents such as input sellers and output buyers share a significant number of ties in each network; hence, they represent important sources of information for farmers. Furthermore, careful observation of out-of-village sources (NGOs, market based) versus within-village sources (i.e. farmers) reveals that farmers having more direct connections with out-of-village information sources are more likely to be adopters. These results also support the previous findings suggesting that farmers with direct ties to outside sources of information are more likely to be adopters of agricultural innovations (van Rijn *et al.*, 2012). This is because lack of connections to outside information sources hinders access by smallholder farmers to information about new technologies (Wall, 2007).

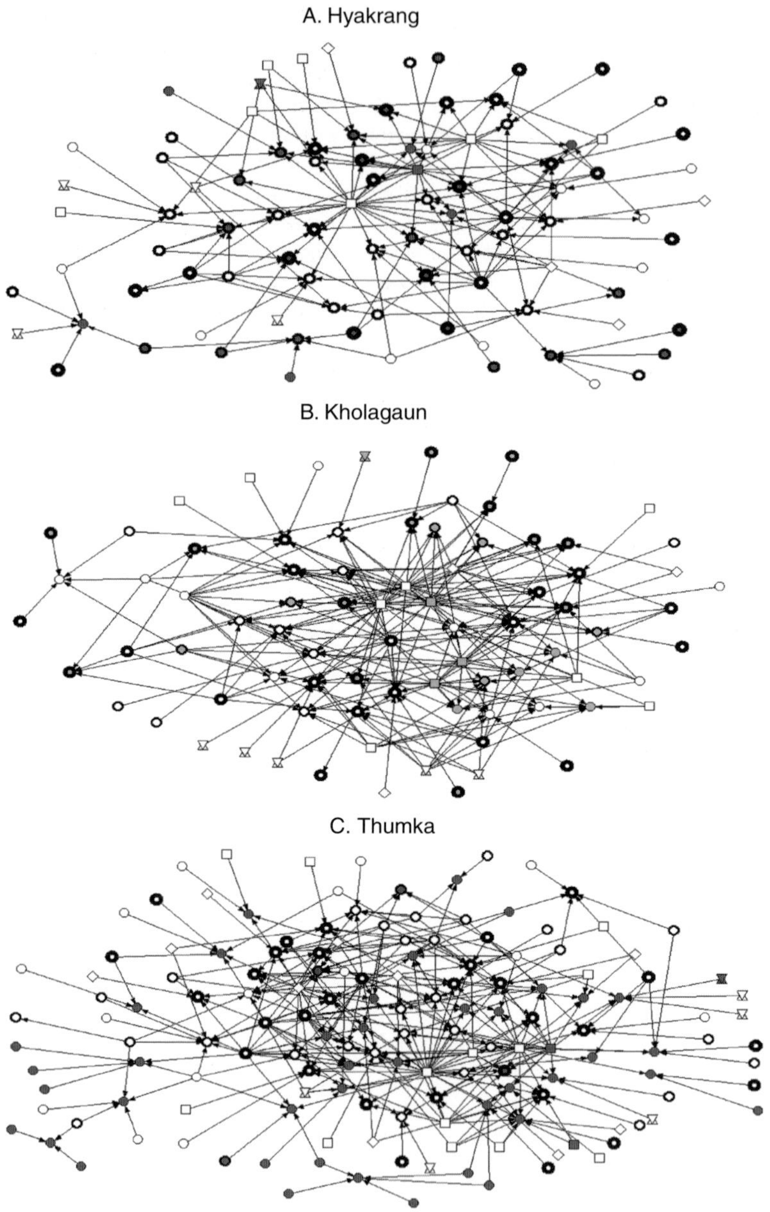

Legend:

Example symbols	Meaning	Description
	Adopter farmers	nodes with bold margin
	Non-adopter farmers	nodes with thin margin
	Female	nodes with dark fill
	Male	nodes with transparent fill
	Farmers	circular nodes
	NGOs	square nodes
	Market-based agents	diamond nodes

Fig. 12.3. Farmers' agricultural information networks in three Nepalese villages.

Centralities of information networks of adopter versus non-adopter farmers

The general observations of information networks in previous sections provide meaningful insights and relations between social networks of farmers and the likelihood of the adoption (or lack of adoption) of new agricultural technology or entrepreneurial activities. The conclusions deducted from visual observations, however, warrant empirical verification. Therefore, the next set of analysis is a comparison of networks of adopter farmers with those of non-adopter farmers.

To compare the networks of adopter and non-adopter, three types of network measures are used. *Degree centrality* is the most appropriate network measure to compare between networks of adopter and non-adopter networks. Degree centrality can be broadly understood as the number of connections an actor has in the network (Freeman, 1979). Degree centrality is further divided into two categories based on the direction of ties: (i) *outdegree*, which is number of outgoing ties from a node; and (ii) *indegree*, which is number of incoming ties to a node. For farmers' information networks, a farmer who receives information from many sources has high outdegree centrality and a farmer who provides information to many other farmers will have high indegree centrality. Therefore, higher outdegree centrality means better access to information while higher indegree centrality indicates that the individual is a source of information for many other individuals (Borgatti, 2005; Borgatti *et al.*, 2013). Generally, actors with a high degree of centrality are considered in a better position to receive information and share among others (Freeman, 1979). Higher degree centrality has also been associated with proactive technology adoption (Vilpponen *et al.*, 2006; Mikhail *et al.*, 2010).

Along with indegree and outdegree centrality, *betweenness* centrality is also very closely related to adoption or non-adoption of technologies by farmers. Betweenness centrality is defined as the number of times an actor falls on the shortest path between two other actors (Freeman, 1979). Betweenness centrality is associated not only with diverse information sources but also with ability to control information flow (Bodin *et al.*, 2006; Borgatti *et al.*, 2013). An actor with high betweenness centrality essentially acts as a bridge for information exchange between others (Bodin and Crona, 2009; Barnes-Mauthe *et al.*, 2014). To compare the mean values of centralities between adopters and non-adopters, a bootstrap probability test was applied as recommended by Hanneman and Riddle (2005).

The comparison of outdegree, indegree, and betweenness degree centralities between adopters and non-adopters of commercial vegetable farming and adoption of rural enterprises is shown in Table 12.3. The results verified that all the centralities (i.e. outdegree, indegree

Table 12.3. Information network centralities between adopters and non-adopters of entrepreneurial activities in three Nepalese villages.

	Centralities								
	Outdegree			Indegree			Betweenness		
Entrepreneurial activities	Non-adopter	Adopters	*p*	Non-adopter	Adopters	*p*	Non-adopter	Adopters	*p*
Commercial vegetable farming									
Hyakrang	4±0.5	4.3±0.3	0.59	0.7±0.4	1.8±0.5	0.11	9.8±5.6	13.3±4.1	0.62
Kholagaun	5.9±0.4	6.3±0.4	0.48	1.8±1.2	1.8±0.7	0.99	25.5±16.5	19±8.3	0.73
Thumka	4.8±0.2	5.7±0.3	0.02	0.8±0.3	2.8±0.7	0.01	9.3±4.1	95±30	0.01
Value-added and small rural enterprises									
Hyakrang	4.3±0.3	4.2±0.4	0.81	1.0±0.2	2.8±1.1	0.03	8.0±2.1	23.5±10	0.03
Kholagaun	6.1±0.3	6.4±0.5	0.60	1.6±0.6	2.1±1.0	0.67	19.5±8.3	21.9±12	0.87
Thumka	5.1±0.3	5.8±0.4	0.12	0.9±0.2	4.1±1.1	0.01	27.7±12.3	113±45	0.05

Notes: *p* values based on bootstrap probability test in UCINET; numbers show mean ± SE of mean.

and betweenness) of the adopter farmers were generally equal or higher than the non-adopter farmers, irrespective of the village.

The comparisons of mean outdegree, indegree and betweenness degree between adopters and non-adopters of commercial vegetable farming showed that mean values of all measures were significantly higher for adopters compared with non-adopters in Thumka (*p* values 0.02, 0.01 and 0.01, for outdegree, indegree and betweenness degree, respectively). The average number of centrality values of adopter farmers was higher than the average number of non-adopter farmers in the remaining two villages; however, the differences were not significant.

The comparisons of average centralities of adopters and non-adopters of value-added and small rural enterprises showed that the adopters' outdegree centralities were not different to those of non-adopters, irrespective of village (Table 12.3). However, the indegree and betweenness centralities of adopters were significantly higher than those of non-adopters in Hyakrang (*p* values 0.03 for both indegree and betweenness centrality) and Thumka (*p* values 0.01 and 0.05 for indegree and betweenness centrality, respectively). In Kholagaun, none of the centralities of the adopter group were significantly higher than those of the non-adopter group. As indicated above, the network in Kholagaun has higher connectedness than other villages; hence this difference of network between adopter farmers and non-adopter farmers makes it logically not significant.

Overall, it was apparent that adopters were likely to have higher or similar numbers of network centralities than non-adopter farmers, albeit the differences were not always significant. However, the non-adopter groups never had a greater average number of centralities than adopters. Thus, the results of the empirical analysis further verified that farmers with higher network centrality are more likely to be innovative and entrepreneurial.

Farmers' position in the information network and adoption of agricultural innovations and entrepreneurial activities

The analysis of core and periphery actors in the information network further provided valuable insights on the nature of adopter farmers. The analysis of the percentage of adopter farmers among core actors (who bridge the network) and periphery actors (advice seekers) in the network showed that farmers in the core category were more likely to be adopters of the innovations and entrepreneurial activities compared with farmers in the periphery category. The percentages of adopter farmers were always higher among core network actors compared with periphery actors, irrespective of village (Table 12.4). The percentage of adopter farmers among core actors was unanimously higher than the percentage of adopters among periphery actors (89.8% versus 82.8% in Hyakrang, 81.8% versus 73.1% in Kholagaun, and 87.5% versus 33.3% in Thumka). Similarly, in the case of value-added and small enterprise, the percentage of adopter farmers was higher among core actors than periphery actors in Hyakrang (45.5% versus 24.1%) and Thumka (56.3% versus 22.2%).

Thus, this result suggested that it was not only the centrality, but how these centralities

Table 12.4. Farmers' position in core or periphery of information network and adoption of new commercial vegetable farming and value-added and small enterprises in three Nepalese villages.

	Commercial vegetable farming (A)		Value-added and small enterprise (B)		Both (A) and (B)		Total numbers of persons	
Village	Core	Periphery	Core	Periphery	Core	Periphery	Core	Periphery
Hyakrang	10 (89.8%)	24 (82.8%)	5 (45.5%)	7 (24.1%)	5 (45.5%)	7 (24.1%)	11 (27.5%)	29 (72.5%)
Kholagaun	9 (81.8%)	19 (73.1%)	5 (45.5%)	12 (46.2%)	3 (27.3%)	6 (23.1%)	11 (29.7%)	26 (70.3%)
Thumka	14 (87.5%)	15 (33.3%)	9 (56.3%)	10 (22.2%)	4 (25.0%)	3 (6.7%)	16 (26.2%)	45 (73.8%)

were positioned that also affected the adoption (or lack of adoption) behavior of farmers. Farmers who were core network actors were more likely to adopt more than one innovative practice, compared with farmers who were periphery network actors.

12.3.3 Influence of network centrality measures to farmers' adoption decisions

This section describes how network centrality influences a farmer's decision to adopt commercial vegetable farming and the adoption of value-added and rural enterprises.

Influence of network centralities on farmers' adoption of commercial vegetable farming

The best-fit binomial logit models developed for identifying significant variables affecting the adoption of commercial vegetable farming have three to five explanatory variables (Table 12.5). Variables that had highest p values were dropped one by one from the model to develop the best-fit binomial logit model.

Out of the retained explanatory variables, respondent's age was negatively related to adoption of commercial vegetable farming in Kholagaun ($\beta = -0.11$, $p=0.01$) and Thumka ($\beta = -0.06$, $p=0.05$) but it was not related in Hyakrang (Table 12.5). Respondent's gender was a significant predictor in Thumka ($\beta = -5.83$, $p<0.01$), indicating that female farmers were less likely to be adopters than male farmers were. However, gender was not a significant predictor in the other two villages. Level of education, group membership, and prior training were not significant explanatory variables for any village, probably due to less variation among the farmers according to these variables.

Out of the three network-related explanatory variables, outdegree centrality was retained in the final model (the one that is the best-fit model according to AIC criteria) for Hyakrang and Thumka but it was not a significant predictor

Table 12.5. Binomial logit model for factors influencing the adoption of commercial vegetable farming practice in three Nepalese villages.

	Hyakrang		Kholagaun		Thumka	
Predictors	β	*p* value	β	*p* value	β	*p* value
Intercept	–3.813	0.053	2.761	0.090	2.269	0.206
Age	0.072	0.110	–0.115	0.011	–0.061	0.055
Centralities						
Outdegree	–0.183	0.590	d		0.291	0.540
Indegree	0.446	0.062	0.136	0.627	0.495	0.019
Betweenness	d		0.020	0.433	0.011	0.294
Gender						
Female	–0.467	0.607	–0.028	0.866	–5.835	0.001
Male	0[a]		0[(a)]		0[a]	
Level of education	d		d		d	
Illiterate	–		–		–	
Primary	–		–		–	
Secondary +	–		–		–	
Group membership	d		d		d	
Member	–		–		–	
Non-member	–		–		–	
Prior training status	d		d		d	
Trained	–		–		–	
Non-trained	–		–		–	

Notes: *p* value based on Wald's test; d = variables dropped during model selection to increase the model fit.
[a]β value was 0 because this was reference category.

for adoption in either of them. Indegree centrality was retained in the best-fit models of all three villages but it was significantly and positively associated with the adoption of commercial vegetable farming in Hyakrang (β = 0.44, p=0.06) and Thumka (β = 0.49, p=0.01) but not in Kholagaun. The low level of influence of indegree centrality to adoption of commercial vegetable farming in Kholagaun was possibly due to other constraints for adoption of technology rather than information constraints. Kholagaun is further from Kathmandu than the other two villages, which is likely discouraging the adoption of technology even though farmers have the information about it. Finally, betweenness centrality was retained in Kholagaun and Thumka but it was not a significant predictor in both cases.

Among the three network-related variables, indegree centrality was positively related to farmers' adoption decision for commercial vegetable farming in some villages, but other centralities were not associated with proactive adoption. The farmers who had more connections and served as information providers in villages were more likely to be the adopters of the agricultural innovations.

Influence of network centralities on farmers' adoption of value-added and rural enterprises

The best-fit binomial logit models (the model with least AIC) identified significant variables affecting the adoption of value-added and small rural enterprises. The model retained three, seven, and five explanatory variables in best-fit model for Hyakrang, Kholagaun and Thumka, respectively (Table 12.6). Among five variables related to respondents' personal traits, age was negatively and significantly associated with higher adoption of entrepreneurial activities in Kholagaun (β = –0.147, p<0.01) but it was positively and significantly associated in Thumka (β = 0.037, p=0.01) and not a significant predictor in Hyakrang. Thus, results were inconsistent and contradictory amongst villages.

Respondent's gender was retained in the final model of all villages. It was negatively and significantly associated with adoption of

Table 12.6. Binomial logit models for factors influencing the adoption of value addition and small rural enterprises in three villages in Nepal.

	Hyakrang		Kholagaun		Thumka	
Predictors	β	*p* value	β	*p* value	β	*p* value
Intercept	–4.866	0.006	18.418	0.547	–1.773	0.179
Age	0.076	0.170	–0.147	0.002	0.037	0.012
Centralities						
Outdegree	d		0.997	0.005	0.017	0.893
Indegree	0.279	0.082	1.883	0.024	0.337	0.011
Betweenness	d		0.169	0.023	–0.005	0.185
Gender						
Female	0.518	0.564	–2.440	0.015	–1.468	0.004
Male	0[a]		0[a]		0[a]	
Level of education	d		d		d	
Illiterate	–		–		–	
Primary	–		–		–	
Secondary +	–		–		–	
Group membership	d				d	
Non-member	–		–2.584	0.014	–	
Member	–		0[a]		–	
Prior training status	d				d	
Trained	–		0.179	0.451	–	
Non-trained	–		0[a]		–	

Notes: *p* value based on Wald's test; d = variables dropped during model selection to increase the model fit.
[a]β value is 0 because this was reference category.

agricultural entrepreneurial activities in Kholagaun (β = −2.44, p=0.01) and Thumka (β = −1.46, p<0.01) indicating that females were less likely to start entrepreneurial activities compared with males. Gender, although not a significant predictor for adoption of entrepreneurial activities in Hyakrang, still had a negative relationship with adoption of entrepreneurial activities, suggesting that there was not a difference in the direction by villages, although there was some difference in the intensity.

Group membership was significantly negatively related in Kholaguan, indicating that people who were non-members were less likely to start entrepreneurial activities compared with members. Group membership was not significant in Hyakrang or Thumka, and was not retained in the model. Prior training status was significant in Kholagaun, with trained individuals more likely to start entrepreneurial activities than non-trained, but it was not significant in Hyakrang or Thumka. Positive influence of group membership and training in Kholagaun, in contrast to non-significant influence in Thumka and Hyakrang, was possibly due to the fact that Kholagaun was the project site for a leasehold forestry project while Thumka and Hyakrang were not. The forestry project is being implemented by government to promote forest-related entrepreneurial activities for the poor, which could explain the positive influence.

Indegree centrality was a significant predictor of adoption of entrepreneurial activities in all villages (β = 0.279, p=0.08 in Hyakrang; β = 1.883, p=0.02 in Kholagaun; and β = 0.337, p=0.01 in Thumka). The outdegree centrality was retained in the final model of Kholagaun and Thumka even though it was only a significant predictor of adoption in Kholagaun (β = 0.997, p<0.01). Outdegree centrality was not significant in Hyakrang nor was it retained in Hyakrang's final model. Betweenness centrality was retained in Kholagaun and Thumka; however, it was only a significant predictor of adoption in Kholagaun (β = 0.169, p=0.02). These results demonstrated that different measures of network centrality were significant predictors for adoption behavior; yet higher network centralities were associated with greater likelihood for adoption of entrepreneurial activities.

12.4 Discussion and Implications

12.4.1 Adoption of agricultural innovations and entrepreneurial activities by farmers

Although conventional livelihood strategies of the Chepang people are changing, the rate of adoption of new technologies is rather slow. Defining entrepreneurship or agricultural innovation is challenging and the context is not the same as that of a developed country. Considering the initial status of entrepreneurship in the villages, any change in agricultural practice leading to marketing of the products and generating cash income can be regarded as entrepreneurial. Indeed, the entrepreneurial activities of farmers in the study sites are limited to the adoption of market-oriented agricultural production systems; such as, commercial fresh vegetable production, off-season vegetable production, cultivation of new cash crops (such as cowpea, banana), collection of non-timber forest products and starting small-scale value-addition enterprises. Some farmers have started small businesses such as a grocery store or teashop in villages or nearby highways, although the number is low.

The rate of adoption of entrepreneurial activities is even slower among Chepang communities compared with other communities. There are several possible reasons for slower adoption of enterprises in these villages. Firstly, the level of education of Chepang people is very low. The lack of education constrains their access to production resources, information, and opportunities for capacity building. Secondly, it is less likely for Chepang people to take risks due to a weak economic foundation, as most of them are resource poor and cannot afford to take risks. Thirdly, it is hard for them to invest in long-term enterprises, because available resources are insufficient to meet their immediate demands. Fourthly, Chepang communities were formerly hunter-gatherers or laborers for other communities. Since they were culturally directed to work for others, the motivation for starting their own enterprise is often weak. Lastly, lack of access to institutional credit is a universally discussed constraint for enterprise development. Since most Chepang households hold small and often unregistered marginal land, they lack land ownership certificates as collateral for loans, which severely jeopardizes their chance for receiving institutional credit. As historically marginalized from development interventions, only a few of the Chepang have the technical skills that are needed to start a local-scale enterprise. The access to opportunities to build capacity is also limited due to lack of education and access to those opportunities. Altogether, these constraints are too many and too strong for Chepang people to overcome and start even a small-scale enterprise. Therefore, the entrepreneurial activities of the Chepang households are limited to small-scale gradual market orientations of their traditional farming.

Recently, government and other development agencies have nationally targeted economic and social empowerment of marginalized and disadvantaged communities, such as Chepang, by implementing multifarious, multi-sectorial efforts. Conventional development agencies, including the government extension system, are also forced to target economically poor and socially disadvantaged groups systematically. The Chepang community has benefited to some extent from these policy changes. Such targeted efforts have increased the level of education, skill, access to resources, and resource endowment among Chepang communities. Chepang people have also started market-oriented agricultural production and other agro-based value-addition and cottage enterprises. These changes have been observed in the study sites as well. These insights can guide policies on how to be more efficient in their efforts and to benefit as many Chepang or other disadvantaged community members as possible.

12.4.2 Nature of adopter and non-adopter farmers

Different characteristics are associated with farmers' decisions to adopt or not adopt agricultural innovations and entrepreneurial activities. Gender, age, education, and information networks all are constraints to adopting agricultural innovations and entrepreneurial activities.

Women are less likely to adopt agricultural innovations and entrepreneurial activities, due to inherent characteristics of a patriarchal society where women are often less active in searching for new innovations and improved practices and receive little or no information on technological innovations in agriculture (Neupane *et al.*, 2002; Upadhyay, 2004; Tiwari *et al.*, 2008). Similarly, women farmers also lack access to working capital. Although women are theoretically equal to men in accessing institutional credit, several social and cultural barriers limit their access to credit. Collateral is needed to access institutional credit; land ownership certificates are commonly used as collateral; and women generally do not have these certificates, consequently their access to credit is constrained (Bushell, 2008). In addition to lack of capital, other reasons limiting entrepreneurship among women include limited access to formal education and small network and access to market (Bushell, 2008).

Older farmers are less likely to adopt commercial vegetable farming than younger farmers, consistent with the initial hypothesis that older farmers are less likely adopters than younger farmers (Stark, 1996; Daberkow and McBride, 1998). Since commercial vegetable farming requires farmers to learn new farming techniques and develop good communication skills with input suppliers as well as output buyers, this causes reluctance in older people to start an entrepreneurial activity (Raut *et al.*, 2013). In addition, commercial vegetable farming is a labor-intensive activity that may be more suitable for younger people rather than older people. Therefore, young people rather than old easily adopt commercial vegetable farming. However, the effect of a farmer's age on adoption of rural enterprise is not straightforward. Behaviors are inconclusive and contradictory across different villages. Age is positively associated with adoption of agricultural enterprises in Thumka but negatively associated in Kholagaun and non-significant in Hyakrang. Looking at the composition of respondents, a certain pattern emerges on the relational direction between a farmer's age and adoption of agricultural enterprises. It seems that young people are less likely adopters than adults due to constraints of investment capitals; however, after a certain number of years, age is negatively associated with adoption due to lack of motivation (Neupane *et al.*, 2002;

Diederen *et al.*, 2003). In Thumka, the average age of respondents is 33 years. Farmers below that age lack the capital for investment, which restricts starting new enterprises; hence lower age is associated with lower likelihood of adoption. In Kholagaun, the average age of respondents is 43. Above that age, farmers may have capital to invest but lack motivation for starting new ventures, hence age is negatively associated with their adoption of innovations.

It is expected that higher levels of education will be associated with earlier adoption of agricultural innovations due to a greater tendency to seek new information (Olajide, 2011; Babu *et al.*, 2012; Uddin *et al.*, 2014) and greater access to information (Deressa *et al.*, 2009; Zhou *et al.*, 2010; Mittal and Mehar, 2013). Contrary to the hypothesis, the influence of education was not evident from this study. The most probable reason for the deviation is the extremely low level of variation in level of education among the farmers. All Chepang people have low education, causing this variable not to have any significant impact on farmers' adoption decisions.

A positive influence of group membership to adoption of agricultural innovations is expected, based on previous studies. However, the influence of group membership to farmers' adoption decisions is not evident from this study. Previous reports suggested that membership in farmers' groups increased the adoption of agricultural innovations (ABTRCO, 2007; Abebaw and Haile, 2013) but in this study, membership did not affect the adoption of agricultural innovations. The level of influence of group membership on personal adoption decisions is dependent on the strength and trust between group members. Long-term mutual communication is required for building strength and trust in any group (Crona and Bodin, 2006). Since the study villages do not have a long development history, farmers' groups may not be strong and sufficiently influential to engender adoption decisions of individual farmers.

The influence of agricultural trainings is expected to be positive on farmers' adoption of agricultural innovations. Farmers with some level of training are more likely to be early adopters of agricultural innovations. Unlike the hypothesis, this study suggests that prior agricultural training is not associated with adoption or non-adoption of agricultural innovations and entrepreneurial activities. This may be due to lack of targeted efforts and practicality of agricultural trainings in Nepal. Generally, such trainings focus on the technical aspects of innovation rather than economic and business aspects. Therefore, traditional trainings often fail to motivate farmers to adopt new practices and enterprises. These results suggest that there is a need to reconstruct how trainings are conducted in Nepal.

All information network variables included in the model are expected to increase farmers' access to and control of information about new technologies. Therefore, it is hypothesized that farmers with higher outdegree, indegree and betweenness centralities are more likely to be adopters of agricultural innovations (Warriner and Moul, 1992; van Rijn *et al.*, 2012). This study supports that higher centralities are generally positively associated with the adoption of agricultural innovations and rural enterprises, although other factors such as distance from major market centers, and technical and financial constraints for adoption may cause differences between villages. Previous researchers have documented that lack of information is a cause for slower adoption of agricultural innovations in Nepal (Neupane *et al.*, 2002; Floyd *et al.*, 2003). Since informal information networks and connections play an important role in the transfer of information about agricultural innovations to farmers (Hoang *et al.*, 2006), the positive influence of network variables on farmers' adoption of new technologies is hypothesized and duly accepted by the findings. Similarly, the positive influence of social networks on rural enterprise development has also been recognized. Social networks facilitate entrepreneurship by either reducing the transaction costs through cooperation and collective marketing or increasing access to newer markets (Zhang and Fung, 2006).

The indegree centrality is the most commonly related network measure. Indegree centrality refers to the number of incoming ties to a farmer, indicating the measure of popularity of a person as the source of information for other farmers (Borgatti *et al.*, 2013). Farmers in villages generally listen to the person who has already adopted the technology and has first-hand experience. Therefore, it is understandable that indegree centrality emerged as the most consistent predictor for adoption of entrepreneurial activities by farmers.

12.5 Conclusion

A social network perspective can and should be applied to understand the different agricultural information networks of adopter and non-adopter farmers. Case studies like this one demonstrate that even small shifts in small rural enterprises, such as adopting commercial vegetable production, can have a positive impact on poor communities like Chepang, and it is possible to introduce a new technology successfully. Stronger information networks can lead to further adoption of small enterprises in these villages, thereby increasing their economic status and alleviating localized poverty in their villages. Critical insights on the structure of information networks of adopter farmers provides opportunities to restructure and strengthen the information networks in rural areas so that the farmers who are excluded from technology transfer can be integrated into the network and start receiving information. Preemptively embedding those people further into the information network can help increase the percentage of adopters. Information networks can be used to introduce new, beneficial technology into these villages and have a successful adaptation within the villages. Using information networks to increase technology, entrepreneurship, and adoption of agricultural innovation in rural villages will help alleviate poverty in these areas. The information network of adopters is not always greater in terms of connections; but at the same time, the adopter farmers are more likely to be at the core of the information network. Similarly, the adopters are more likely to have greater connections with outside sources of information such as NGO staff, government extension agents, and market agents. Therefore, it is key to bring people into the network and tie them to people with connections to outside sources. This will strengthen the network and increase the number of adopters.

Strengthening social network capitals among farmers could increase the rate of adoption of agricultural innovations and starting of rural enterprises. Social networks not only increase farmers' access to information but also encourage farmers to work collectively on the process of technology evaluation and adoption as well as small-scale enterprise development. The extension systems of Nepal already promote a group approach to extension where, in principle, there is enough scope for strengthening social networks and information networks. Similarly, the government of Nepal has adopted a public–private partnership (PPP) model (an extension model where public and private sectors collaborate to provide services to farmers) as the foundation of economic development in Nepal. Making PPP a priority provides opportunities for promoting rural enterprises by building multi-stakeholder networks in rural areas. However, care has to be taken to identify the key members in the information network and facilitate network building by providing them with technical and facilitation support. Women should be positively targeted in the process of their representation, as they are often hindered by social and cultural norms.

Information networks may not be enough to increase adoption of agricultural innovations or increase entrepreneurship in rural areas, but they are the first step where farmers can learn about the opportunities that are available. Other structures such as institutions providing credit, marketing facilities, and technical support should also be present to support innovative and interested farmers. In this way, the various specialized information networks may emerge and grow in the villages. After the growth of the network, it can be specialized to a certain kind of agricultural innovation (innovation network) or certain types of value addition and marketing activities (market network), leading to an establishment of self-sustained social network capital. Such social capital would increase access to information, support the extension system and markets and thus encourage farmers to adopt agricultural innovations and entrepreneurial activities, aiding in the alleviation of poverty in rural areas of Nepal.

References

Abebaw, D. and Haile, M.G. (2013) The impact of cooperatives on agricultural technology adoption: Empirical evidence from Ethiopia. *Food Policy* 38, 82–91.

ABTRCO (2007) Impact study on farmer's group approach adopted in agricultural extension system of Nepal. Agribusiness and Trade Promotion Multi-Purpose Co-operative, Department of Agriculture, Ministry of Agriculture Development, Lalitpur, Nepal.

Aker, J.C. (2011) Dial 'A' for agriculture: A review of information and communication technologies for agricultural extension in developing countries. *Agricultural Economics* 42, 631–647.

Asfaw, S., Kassie, M., Simtowe, F. and Lipper, L. (2012) Poverty reduction effects of agricultural technology adoption: a micro-evidence from rural Tanzania. *Journal of Development Studies* 48, 1288–1305.

Babu, S.C., Glendenning, C.J., Asenso-Okyere, K. and Govindarajan, S.K. (2012) *Farmers' Information Needs and Search Behaviors: Case Study in Tamil Nadu, India*. IFPRI discussion paper 1165. International Food Policy Research Institute, Washington, DC.

Bandiera, O. and Rasul, I. (2006) Social networks and technology adoption in northern Mozambique. *The Economic Journal* 116, 869–902.

Barnes-Mauthe, M., Gray, S.A., Arita, S., Lynham, J. and Leung, P. (2014) What determines social capital in a social–ecological system? Insights from a network perspective. *Environmental Management* 55, 392–410.

Barrett, C.B. (2008) Smallholder market participation: Concepts and evidence from eastern and southern Africa. *Food Policy* 33, 299–317.

Bodin, Ö. and Crona, B.I. (2009) The role of social networks in natural resource governance: What relational patterns make a difference? *Global Environmental Change* 19, 366–374.

Bodin, Ö., Crona, B. and Ernstson, H. (2006) Social networks in natural resource management: what is there to learn from a structural perspective? *Ecology and Society* 11, r2.

Borgatti, S.P. (2005) Centrality and network flow. *Social Networks* 27, 55–71.

Borgatti, S.P., Everett, M.G. and Freeman, L.C. (2002) *Ucinet for Windows: Software for Social Network Analysis*. Analytic Technologies, Harvard, Massachusetts.

Borgatti, S.P., Everett, M.G. and Johnson, J.C. (2013) *Analyzing social networks*. SAGE Publications.

Bushell, B. (2008) Women entrepreneurs in Nepal. *Gender and Development* 13, 459–464.

CBS (2008a) *National Agriculture Census Monograph*. Central Bureau of Statistics, National Planning Commission, Government of Nepal, Kathmandu.

CBS (2008b) *Nepal Labour Force Survey 2008 (NLFS II)*. Central Bureau of Statistics, National Planning Commission, Government of Nepal, Kathmandu.

CBS (2012) *National Population Census*. Central Bureau of Statistics, Nepal Planning Commission, Government of Nepal, Kathmandu., pp. 2–3.

Chan-Halbrendt, C., Halbrendt, J., Lai, C., Reed, B., Paudel, B. and Limbu, P. (2011) Unpublished baseline data of 'Sustainable Management of Agro-ecological Resources on Tribal Societies of India and Nepal' project. Department of Natural Resources and Environmental Management, University of Hawaii at Manoa.

Crona, B. and Bodin, Ö. (2006) What you know is who you know? Communication patterns among resource users as a prerequisite for co-management. *Ecology and Society* 11, 7.

Daberkow, S.G. and McBride, W.D. (1998) Socioeconomic profiles of early adopters of precision agriculture technologies. *Journal of Agribusiness* 16, 151–168.

Deressa, T.T., Hassan, R.M., Ringler, C., Alemu, T. and Yesuf, M. (2009) Determinants of farmers' choice of adaptation methods to climate change in the Nile Basin of Ethiopia. *Global Environmental Change* 19, 248–255.

Diederen, P., Van Meijl, H., Wolters, A. and Bijak, K. (2003) Innovation adoption in agriculture: innovators, early adopters and laggards. *Cahiers d'économie et sociologie rurales* 67, 30–50.

FAO, IFAD and WFP (2013) *The State of Food Insecurity in the World 2013. The multiple dimensions of food security*. FAO, Rome. Available at: http://www.fao.org/publications/sofi/2013/en/ (accessed 8 February 2015).

Floyd, C., Harding, A.H., Paudel, K.C., Rasali, D.P., Subedi, K. and Subedi, P.P. (2003) Household adoption and the associated impact of multiple agricultural technologies in the western hills of Nepal. *Agricultural Systems* 76, 715–738.

Freeman, L.C. (1979) Centrality in social networks conceptual clarification. *Social Networks* 1, 215–239.

Gagliardi, D., Muller, P., Glossop, E., Caliandro, C., Fritsch, M., Brtkova, G., Bohn, N.U., Klitou, D., Avigdor, G., Marzocchi, C. and Ramlogan, R. (2013) *A Recovery on the Horizon? Annual Report on European SMEs 2012/2013*. Publication office, European Commission, Brussels.

Ghimire, R., Adhikari, K., Chen, Z.-S., Shah, S. and Dahal, K. (2012) Soil organic carbon sequestration as affected by tillage, crop residue, and nitrogen application in rice–wheat rotation system. *Paddy Water Environ* 10, 95–102.

Grafton, R.Q., Williams, J. and Jiang, Q. (2015) Food and water gaps to 2050: preliminary results from the global food and water system (GFWS) platform. *Food Security* 7, 209–220.

Hanneman, R.A. and Riddle, M. (2005) Introduction to social network methods. University of California Riverside. Available at: http://faculty.ucr.edu/~hanneman/nettext/ (accessed 25 February 2015).

Hoang, L.A., Castella, J.-C. and Novosad, P. (2006) Social networks and information access: Implications for agricultural extension in a rice farming community in northern Vietnam. *Agriculture and Human Values* 23, 513–527.

Isaac, M.E., Erickson, B.H., Quashie-Sam, S.J. and Timmer, V.R. (2007) Transfer of knowledge on agroforestry management practices: the structure of farmer advice networks. *Ecology and Society* 12, 32.

Kaaria, S., Njuki, J., Abenakyo, A., Delve, R. and Sanginga, P. (2008) Assessment of the enabling rural innovation (ERI) approach: case studies from Malawi and Uganda. *Natural Resources Forum* 32(1), 53–63.

Kafle, G., Limbu, P., Pradhan, B. and Jing, F. (2009) Piloting ecohealth approach for addressing land use transition, climate change and human health issues. *Climate Change Bulletin* 3, 6–10.

Kassie, M., Shiferaw, B. and Muricho, G. (2011) Agricultural technology, crop income, and poverty alleviation in Uganda. *World Development* 39, 1784–1795.

Koirala, G., Acharya, R.P., Dhakal, S. and Karki, G. (2013) *A Rapid Assessment of Forest Based Enterprises in Nepal*. Multi Stakeholder Forestry Programme (MSFP), Babarmahal, Nepal.

Maertens, A. and Barrett, C.B. (2013) Measuring social networks' effects on agricultural technology adoption. *American Journal of Agricultural Economics* 95, 353–359.

Magnan, N., Spielman, D.J., Lybbert, T.J. and Gulati, K. (2013) *Leveling with friends: social networks and Indian farmers' demand for agricultural custom hire services*. International Food Policy Research Institute (IFPRI). Available at: http://www.ifpri.org/publication/leveling-friends (accessed 8 February 2015).

Mikhail, M., Fencl, A., Kemp-Benedict, E. and Naruchaikusol, S. (2010) *Innovation and diffusion of sustainable agricultural water resource management in a changing climate: Case Study in Northeast Thailand*. Available at: http://www.icarus.info/wp-content/uploads/2011/610-Fencl_ICARUS_paper.pdf (accessed 23 February 2015).

Minten, B. and Barrett, C.B. (2008) Agricultural technology, productivity, and poverty in Madagascar. *World Development* 36, 797–822.

Mittal, S. and Mehar, M. (2013) *Agricultural information networks, information needs and risk management strategies: a survey of farmers in Indo-Gangetic Plains of India. Mexico, DF (Mexico)*. Socioeconomics Working Paper no. 10. International Center for Maize and Wheat Improvement (CIMMYT), Mexico City, Mexico. Available at: http://repository.cimmyt.org/xmlui/handle/10883/3225 (accessed 23 February 2015).

Neupane, R.P., Sharma, K.R. and Thapa, G.B. (2002) Adoption of agroforestry in the hills of Nepal: a logistic regression analysis. *Agricultural Systems* 72, 177–196.

Olajide, B.R. (2011) Assessment of farmers' access to agricultural information on selected food crops in Iddo District of Oyo State, Nigeria. *Journal of Agricultural & Food Information* 12, 354–363.

Pariyar, D. (2008) *Country pasture/forage resource profiles*. Pasture & Fodder Research Division, Khumaltar, Kathmandu, Nepal. Available at: http://www.fao.org/ag/agp/AGPC/doc/Counprof/PDF files/Nepal.pdf (accessed 8 February 2015).

Paudel, G.S. and Thapa, G.B. (2004) Impact of social, institutional and ecological factors on land management practices in mountain watersheds of Nepal. *Applied Geography* 24, 35–55.

Piya, L., Maharjan, K.L. and Joshi, N.P. (2011a) Livelihood strategies of indigenous nationalities in Nepal: a case of Chapangs. *Journal of International Development and Cooperation* 17, 99–113.

Piya, L., Maharjan, K.L., Joshi, N.P. and Dangol, D.R. (2011b) Collection and marketing of non-timber forest products by Chapang community in Nepal. *Journal of Agriculture and Environment* 12, 10–21.

Raut, N., Raya, B., Sitaula, B.K., Bajracharya, R.M. and Kristjanson, P. (2013) *Gender roles and greenhouse gas emissions in intensified agricultural systems in the mid-hills of Nepal*. Working Paper No. 47, CGIAR Research Program on Climate Change, Agriculture and Food Security (CCAFS), Copenhagen.

Rimal, J. (2008) *Study on export potential: an analysis of selected sub-sectors of Nepal*. Enhancing the Capacity of Intermediary Business Organizations in Nepal (ECIBON) project, Kathmandu, Nepal. Available at: http://www.eec-nepal.org.np/eec/ecibon/report/211/Publication/chapter%201-2.pdf (accessed 8 December 2016).

Rogers, E.M. (1962) *The Diffusion of Innovations*. The Free Press, New York.

Shrestha, S.K. (2013) *Decentralizing the farmer-to-farmer extension approach to the local level*. Paper submitted to CAPSA for the "Technology Transfer for Smallholder Farmers" Policy Dialogue. Available at: http://www.satnetasia.org/sites/default/files/8-Decentralizing_farmer-to-farmer_extension-Helvetas_Nepal_paper.pdf (accessed 8 December 2016).

Smith, L.G. and Johnson, C. (2001) *Development of rural villages and social capital*. Report prepared for Social Capital and Poverty Reduction in Latin America and the Caribbean: Toward a New Paradigm. UN Economic Commission for Latin America and the Caribbean (ECLAC/CEPAL), Santiago, Chile, and Michigan State University.

Stark, M. (1996) Adoption and adaption of contour hedgerow farming in the Philippine uplands: results of an early case study. *Der Tropenlandwirt – Journal of Agriculture in the Tropics and Subtropics* 97, 3–16.

Tieguhong, J., Ndoye, O., Grouwels, S., Mala, W. and Betti, J. (2012) Rural enterprise development for poverty alleviation based on non-wood forest products in Central Africa. *International Forestry Review* 14, 363–379.

Tiwari, K.R., Sitaula, B.K., Nyborg, I.L. and Paudel, G.S. (2008) Determinants of farmers' adoption of improved soil conservation technology in a Middle Mountain watershed of Central Nepal. *Environmental Management* 42, 210–222.

Uddin, M.N., Bokelmann, W. and Entsminger, J.S. (2014) Factors affecting farmers' adaptation strategies to environmental degradation and climate change effects: a farm level study in Bangladesh. *Climate* 2, 223–241.

UNDP (2009) *Nepal human development reports 2009: state transformation and human development*. United Nations Development Programme, Kathmandu, Nepal. Available at: http://hdr.undp.org/en/content/state-transformation-and-human-development (accessed 8 February 2015).

Upadhyay, B. (2004) Gender aspects of smallholder irrigation technology: insights from Nepal. *Journal of Applied Irrigation Science* 39, 315–327.

van Rijn, F., Bulte, E. and Adekunle, A. (2012) Social capital and agricultural innovation in Sub-Saharan Africa. *Agricultural Systems* 108, 112–122.

Vilpponen, A., Winter, S. and Sundqvist, S. (2006) Electronic word-of-mouth in online environments: exploring referral networks structure and adoption behavior. *Journal of Interactive Advertising* 6, 8–77.

Wall, P.C. (2007) Tailoring conservation agriculture to the needs of small farmers in developing countries: an analysis of issues. *Journal of Crop Improvement* 19, 137–155.

Warriner, G.K. and Moul, T.M. (1992) Kinship and personal communication network influences on the adoption of agriculture conservation technology. *Journal of Rural Studies* 8, 279–291.

World Bank Group (2015) *Global Monitoring Report 2014/2015: Ending Poverty and Sharing Prosperity*. World Bank, Washington, DC. doi:10.1596/978-1-4648-0336-9.

Worldstat (2014) *Nepal: Agriculture Land*. Available at: http://en.worldstat.info/Asia/Nepal/Land (accessed 20 December 2012).

Yoshida, N., Uematsu, H. and Sobrado, C.E. (2014) *Is extreme poverty going to end? An analytical framework to evaluate progress in ending extreme poverty*. Policy Research Working Paper 6740. World Bank, Washington, DC.

Zhang, Q. and Fung, H.G. (2006) China's social capital and financial performance of private enterprises. *Journal of Small Business and Enterprise Development* 13, 198–207.

Zhao, J. (2005) *The role of information in technology adoption under poverty*. UNU-WIDER Research Paper No. 2005/41, World Institute for Development Economics Research/United Nations University, Helsinki. Available at: http://hdl.handle.net/10419/63317 (accessed 23 February 2015).

Zhou, Y., Yang, H., Mosler, H.J. and Abbaspour, K.C. (2010) Factors affecting farmers' decisions on fertilizer use: a case study for the Chaobai watershed in Northern China. *Journal of Sustainable Development* 4, 80–102.

Index

Page numbers in **bold** refer to illustrations and tables.